Springer Geology

The book series Springer Geology comprises a broad portfolio of scientific books, aiming at researchers, students, and everyone interested in geology. The series includes peer-reviewed monographs, edited volumes, textbooks, and conference proceedings. It covers the entire research area of geology including, but not limited to, economic geology, mineral resources, historical geology, quantitative geology, structural geology, geomorphology, paleontology, and sedimentology.

More information about this series at http://www.springer.com/series/10172

Chaplina Tatiana Olegovna
Editor

Processes in GeoMedia—Volume I

Springer

Editor
Chaplina Tatiana Olegovna
Institute of Problems in Mechanics
Russian Academy of Sciences
Moscow, Russia

ISSN 2197-9545 ISSN 2197-9553 (electronic)
Springer Geology
ISBN 978-3-030-38176-9 ISBN 978-3-030-38177-6 (eBook)
https://doi.org/10.1007/978-3-030-38177-6

This Springer imprint is published by the registered company Springer Nature Switzerland AG
The registered company address is: Gewerbestrasse 11, 6330 Cham, Switzerland

Contents

The Formation of the "Baer Hillocks" by the Peak Water Flow at a Sharple Reduction of the Caspian Sea Level 1
O. N. Melnikova and K. V. Pokazeev

Features of the Formation of Gold Manifestations in the Black Shale Deposits of the Kumakskoye Ore Field 11
P. V. Pankratiev, A. V. Kolomoets, S. V. Bagmanova, and V. S. Panteleev

Analytical Test Problem of Wind Currents 17
V. S. Kochergin, S. V. Kochergin, and S. N. Sklyar

Determination of the Propagation Speed of Wave Perturbations in a Two-Component Field 27
V. V. Ocherednik and A. S. Zapevalov

A Model of a Coastal Barrier and Its Application to the Anapa Bay-Bar Coasts 35
I. O. Leont'yev and T. M. Akivis

Testing a Mathematical Model of Air Movement in a Street Canyon Using the OpenFoam Package 43
M. V. Volik and N. S. Orlova

The Manifestation of Caldera-Forming Volcanism in the Formation of the Coast (On Example of Iturup Island of the Great Kuril Ridge) 51
V. V. Afanasiev, N. N. Dunaev, A. O. Gorbunov, and A. V. Uba

Comparative Evaluation of Accuracy of Numerical Wave Models Based on Satellite Altimetry Data 63
V. G. Polnikov, N. S. Zilitinkevich, F. A. Pogarskii, and A. A. Kubryakov

Study of Falling Rocks Using Discrete Element Method 75
N. S. Orlova and M. V. Volik

The Application of Mathematical Modelling to Assess Ecological Safety of the Coastal Area of the International Resort of Varadero (Cuba) 83
N. N. Dunaev, I. O. Leont'yev, T. Y. Repkina, and J. L. Juanes Marti

Anisotropy Study of Statistical Characteristics of Wind Waves Under the Influence of Hydrodynamic Perturbations in Laser-Reflective Method 93
V. N. Nosov, S. G. Ivanov, V. I. Timonin, and S. B. Kaledin

Comparative Analysis of Eutrophication Level of the Sevastopol Bay Areas Based on the Results of E-Trix Index Numerical Modeling 101
K. A. Slepchuk

Geyser Eruption Mechanism as a Way of Extracting and Utilizing Underground Heat 109
A. M. Nechaev, A. A. Solovyev, and D. A. Solovyev

Analytical Solutions of Internal Gravity Waves Equation for Model Buoyancy Frequency Distribution 119
V. V. Bulatov and Yu. V. Vladimirov

Dynamics of Riverine Water in the Black Sea Shelf Zone 127
M. V. Tsyganova and E. M. Lemeshko

Optimization of Pollutant Emissions for Air Quality Modeling in Moscow 137
N. A. Ponomarev, N. F. Elansky, V. I. Zakharov, and Y. M. Verevkin

On the Ratio of Biological and Aquatic Migration of Chemical Elements in the Continental Biosphere Sector 149
V. S. Savenko

Calculation and Analysis of the Average for Seasons Fields of the Currents Velocities of the Marmara Sea Waters 159
S. V. Dovgaya

Strain Properties of Materials with Gas-Filled Cracks 169
V. I. Karev

The Calculation of the Earth's Insolation for the Period 3000 BC–AD 2999 181
V. M. Fedorov and A. A. Kostin

Timan-Ural Stone Province—Mineral Resource Base of the Gemstones of the Russian Federation 193
Vitaly Gadiyatov, Peter Kalugin, and Alexander Demidenko

Interannual Variability of the Heat Exchange of the Ocean and the Atmosphere in the Arctic 205
A. E. Bukatov and A. A. Bukatov

Structural Features of the Dip Process of Thermics with Negative Buoyancy 215
A. A. Volkova and V. A. Gritsenko

Alternative Use of Satellite Maps of Yandex for Visualization of Geophysical Objects 223
N. Voronina and N. Inyushina

Mechanism for Sediment Transport at Tidal Estuaries Under Bore Formation 231
E. N. Dolgopolova

Intra-annual Variability of the Diurnal Water Temperature Variations on Sambian Plateau (South-Eastern Baltic Sea) in 2016 243
V. F. Dubravin, M. V. Kapustina, and S. A. Myslenkov

On the Analogy of Pore Pressure and Temperature Effects in the Elastic and Elastoplastic Problems Solutions 255
K. B. Ustinov and E. V. Stepanova

Total Duration of Arctic Air Outbreaks Over the Azov-Black Sea Region in 2000–2018 263
A. V. Kholoptsev, S. A. Podporin, and L. E. Kurochkin

Variations of Seismic Scaling Before Strong Earthquakes 271
I. R. Stakhovsky

Detecting and Analysis of Bubble Gas Emissions in Shallow Water by Method of Passive Acoustics 279
A. A. Budnikov, I. N. Ivanova, T. V. Malakhova, and V. V. Pryadun

Paleogeodynamics of the Drake Passage in the Scotia Sea 287
A. A. Schreider, A. E. Sazhneva, M. S. Klyuev, A. L. Brekhovskikh, I. Ya. Rakitin, E. I. Evsenko, O. V. Grinberg, F. Bohoyo, H. Galindo-Zaldivar, P. Ruano, J. Martos, and F. Lobo

Features Study of the Marks Movement on the Surface and in the Depth of Vortex Flow 299
T. O. Chaplina, E. V. Stepanova, and V. P. Pakhnenko

Influence of the Stresses on Flow Rate of Wells and Their Stability . . . 311
V. I. Karev and T. O. Chaplina

The Formation of the "Baer Hillocks" by the Peak Water Flow at a Sharple Reduction of the Caspian Sea Level

O. N. Melnikova and K. V. Pokazeev

Abstract On the basis of a physical model linking the formation of ridges at the bottom of rivers with the action of stationary waves with fixed crests on the water surface, it is proposed the hypothesis of the formation of Baer hillocks in the Volga delta with peak water flow during a sudden decrease in the level of the Caspian Sea. The proposed hypothesis makes it possible to describe all the features of the Baer ridges: in areas 10 km long, the ridge tops that exceed the middle line are located at the same distance from each other, the surface oscillations below this line do not have a regular component. An estimate of the flow Froude numbers was obtained for which the Baer ridges were formed: $0.68 < Fr < 0.73$.

Keywords Stationary waves on the surface of water flow · Bottom forms · Stability of waves on the surface of water · Three-dimensional stationary waves · Baer hillocks

1 Introduction

On the free surface of flows, whose velocity varies along the flow ($\frac{\partial u}{\partial x} \neq 0$), stationary waves with fixed crests appear. Wave crests do not move, since the phase velocity of the waves is equal in magnitude to the flow velocity and directed towards it (Lighthill 1978). Dispersing stationary waves exist if the flow velocity $u > 23.1$ cm c^{-1}, Froude number $Fr = u/\sqrt{gh} < 1$ (g—the acceleration of gravity, h is the depth of the water layer), with the values of the longitudinal velocity gradient $|u_x| > 0.001$ c^{-1}, which corresponds to the slope of the water surface $i \approx u_x h/u = 0.0001$. The steepness of the waves $ak = a\frac{2\pi}{\lambda}$ (λ—a wave length, a—an amplitude) grows with increasing of the velocity gradient u_x. We studied experimentally nonlinear stationary waves ($ak > 0.2$) on the surface of the flow in a channel of a finite width in (Melnikhova and Rykounov 1998). It was found that under the front slope of a stationary wave where the flow velocity decreases there is an intense separation of cylindrical vortices at the bottom that capture and carry away soil particles, which

O. N. Melnikova (✉) · K. V. Pokazeev
Faculty of Physics, M.V. Lomonosov Moscow State University, Moscow, Russia
e-mail: olamel@yandex.ru

C. T. Olegovna (ed.), *Processes in GeoMedia—Volume I*,
Springer Geology, https://doi.org/10.1007/978-3-030-38177-6_1

leads to the formation of a scour (Butov et al. 2000; Mel'nikova 2005). In the region of flow acceleration (backward slope of the wave), eddies do not form at the bottom (Mel'nikova 2005). The alternation of scours forms the bottom ridges. The distance between the ridges is equal to the length of the stationary wave. As the height of the ridges increases, the initial wave is amplified—its steepness is growing rapidly. When the critical wave slope is reached $0.31 < (ak)_{cr} < 0.37$ (it grows with an increase in the Froude number), a specific non-linear effect appears—a plane wave decays into two three-dimensional waves of greater length, with crests arranged in a staggered manner (Melnikhova and Rykounov 1998). The decay of a plane wave into three-dimensional waves is not observed if $Fr > 0.73$. New waves have the same phase velocity as the original wave, as a result, the crests of new waves do not move either. Three-dimensional stationary waves form three-dimensional ridges elongated along the transverse coordinate z and arranged in a checkerboard pattern. The ratio of the lengths of three-dimensional waves increases with the Froude number, and the longest wave has a maximum amplitude. This wave forms a ridge of maximum height at a peak water flow. After the fall of the peak water flow, the stream flows around the ridges (located in a checkerboard pattern) forming the sinusoidal river bends. In Mountain Rivers the number of Froude exceeds the value $Fr > 0.73$, dispersing stationary waves do not exist and sinusoidal bends are not formed.

In Melnikhova and Rykounov (1998), dependences of the lengths of three-dimensional waves (normalized to the depth of the stream) on the Froude number were experimentally obtained. Comparison of these dependencies with field data (5 rivers studied) showed good agreement. A good agreement of the field and experimental data allows us to predict the erosion of the river bottom at peak water consumption for a given Froude number. On the other hand, non deformed ridges of a maximum length and height formed at the peak of water flow allow us to estimate the characteristics of the flow that formed these ridges. If for some reason a peak water discharge occurs once in a long period of time (maybe hundreds or thousands of years), the ridges formed by this stream may persist during this time interval.

The famous "Baer hills" serve as an example of the existence of such ridges on the northern shore of the Caspian Sea. Mounds have the form of ridges, elongated along the geographic parallel and arranged in a checkerboard pattern. Academician Baer, who first described these "bumps" in the middle of the nineteenth century, put forward such a hypothesis for the origin of the ridges—a catastrophic discharge of sea water (Baer 1856). K. M. Baer suggested that they were formed as a result of rapid and "forced" flow of Caspian waters in the direction from the valley of Kuma and the Manych to the sea. "The level of the Caspian Sea did not decrease gradually, but suddenly. The monuments testifying to this are hillocks", he wrote. The history of the Caspian contains a number of powerful rises and falls in sea level, reaching 100 m from the zero mark of absolute altitudes (Svitoch 2016). Large-scale natural events—the draining of huge areas of the seabed, a change of landscapes and reliefs—were observed with changes in sea level. There are also other hypotheses of formation of the Baer ridges—wind, sea (waves running onto the shore), erosion (in river deltas). However, these hypotheses contradict one or other peculiarities of the Baer hillocks related to the composition of the ridges and their shape (Golovachev

2017). The Baer hypothesis has no such flaws, but it does not allow to explain the regular structure of the ridges—the distance between the ridges is approximately 1000 m, they are staggered. It can be assumed that the ridges could have been formed by three-dimensional stationary waves on the surface of a powerful flow of water that arose during a sudden drop in sea level. In this case, the length of three-dimensional waves will be close to the distance between the ridges. A good argument in favor of the hypothesis will be a description of all the features of the Baer ridges—the distance between the ridges and their height. The solution to this problem is the goal of this work.

2 Estimation of the Depth of the Water Layer Before a Catastrophic Discharge

To test the hypothesis, we will use the relief of the left bank of the Volga in the area of ridges (Fig. 1), obtained on the basis of satellite data (http://gis-lab.info/qa/srtm.html). The ridges are visible in the photo as light parallel stripes, and the Volga riverbed is a dark area. The location of the sections selected for the relief analysis is shown in the photograph by segments with the indication of the profile number. The vertical coordinates on the profiles are measured from the zero mark of the absolute height scale. The current water level in the Caspian Sea is lowered by 27 m from zero.

Section 7 is approximately 30 km from the bank of the Volga, 8–20 km, and 6 passes near the bank of the Volga. Profile 7 has the highest marks and lies on the

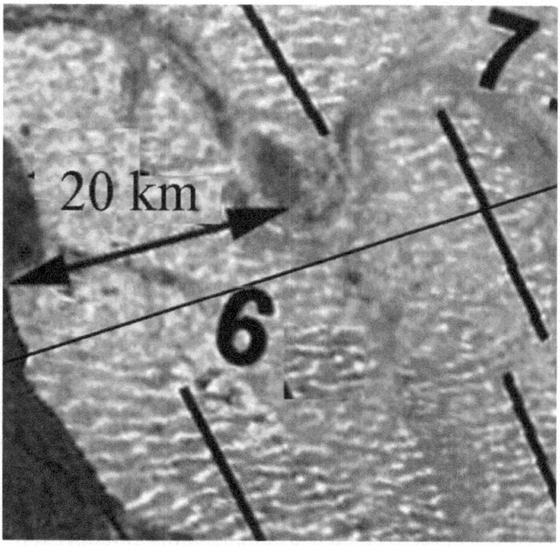

Fig. 1 Left bank of the Volga. Light stripes—Baer bumps. The dark zone is the channel of the Volga. The segments show the position of the cuts perpendicular to the axis of the ridges, the thin line is parallel to the axis of the ridges

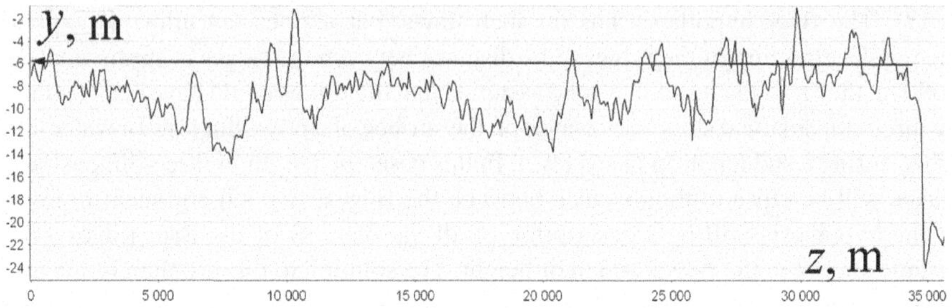

Fig. 2 Cross-section (parallel to the axis of the ridges) passes through the center of the profile 7

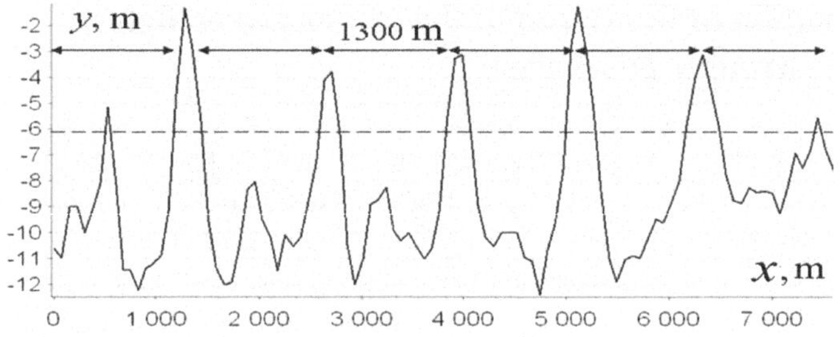

Fig. 3 Section profile 7, directed along the normal to the axis of the ridges

outer border of the hillock area. A cross-section parallel to the line of mounds (a thin line in Fig. 1) passes through the center of the profile 7 (Fig. 2).

Arrow in Fig. 2 shows the average level of the shore surface that existed before the catastrophic flow occurred. The position of the average level at minus 6 m was determined as follows. Figure 3 shows the profile of the section 7, directed along the normal to the axis of the ridges.

There are regular peaks on the profile, the distance between the peaks is 1300 m and does not change along the longitudinal coordinate, as noted in field studies (Baer 1856; Golovachev 2017). The dashed line cuts off the bottom of the ridges containing additional irregular modes. The upper regular part of the ridges is formed by the maximum water flow, which is no longer repeated. The lower part was deformed by flows, the discharge of which was much smaller and varied in a wide range of values over many years. We will consider the line of demarcation as the average level of the flat bottom that existed before the catastrophe. On the cuts presented in Figs. 2 and 3, the middle line is minus 6 m and does not change along the longitudinal axis x and the transverse axis z.

The minimum mark of the surface of the water before a catastrophe can be roughly estimated from the marks of the highest peaks (the level of the surface of the water should not be lower than the highest peak)—it coincides with the zero mark of the absolute scale. Then the minimum depth of the water layer above the flat bottom at section 7 before the catastrophe is approximately 6 m.

The section parallel to the axis of the ridges and passing through the centers of profiles 8 and 6 is shown in Fig. 4. The centers of profiles 8 and 6 have transverse coordinates $z = 2000$ m and $z = 17{,}000$ m respectively. On this section, the average level of the earth's surface decreases in the direction to the Volga channel—from −7 m at section 8 to −9 m at section 6.

Sections 8 and 6 directed along the normal to the axis of the ridges are shown in Fig. 5.

The data shows that the position of the midline, cutting off the regular ridges from irregular in section 8, is −7 m, and in section 6 it is −9 m. The position of the centerline does not change in selected areas along the longitudinal axis. The maximum height of the peaks in Figs. 4 and 5 again does not exceed the zero

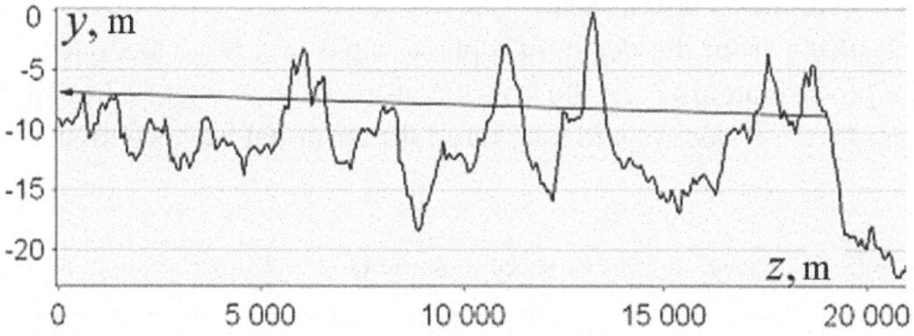

Fig. 4 Section, parallel to the axis of the ridges, passing through the center of the profiles 8 ($z =$ 2000 m) and 6 ($z = 17{,}000$ m)

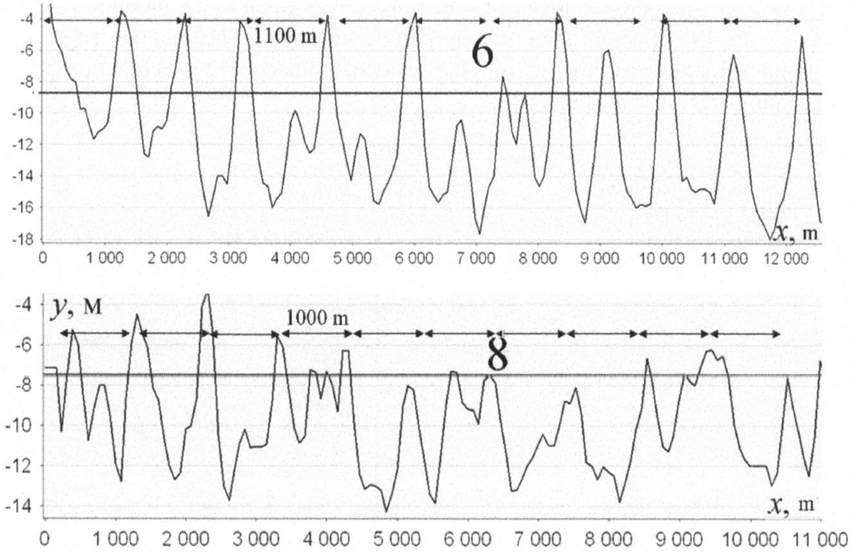

Fig. 5 Section 6 and 8, directed along the normal to the axis of the ridges

mark of absolute heights. We get the minimum depth for profiles 8 and 6: $h = 7$ m and $h = 9$ m respectively.

3 Determination of Froude Number by the Distance Between the Ridges

To test the proposed hypothesis of formation of Baer ridges, let us estimate the Froude number of the catastrophic flow. This can be done by determining both the length of the three-dimensional wave and the critical height of the ridges, at which the wave of a critical steepness decays. If the results are close, then an argument in favor of the proposed hypothesis will appear.

Waves resulting from the decay of a plane wave of a large steepness satisfy the condition of phase matching and the law of conservation of energy. Planar wave with wave number $k_1 = \frac{2\pi}{\lambda_1}$ decays into long three-dimensional waves with wave numbers k_2 and k_3:

$$\vec{k}_1 \pm \vec{k}_2 \pm \vec{k}_3 = 0$$
$$\omega_1 \pm \omega_2 \pm \omega_3 = 0 \quad (1)$$

$$a_1^2 = a_2^2 + a_3^2 \quad (2)$$

where $\omega = kc$, c—phase velocity of stationary waves equal to the modulus of flow velocity with the opposite sign, a_i—amplitude of waves. The relative lengths of three-dimensional waves arising from the decay of a plane wave, as a function of the Froude number, is shown in Fig. 6.

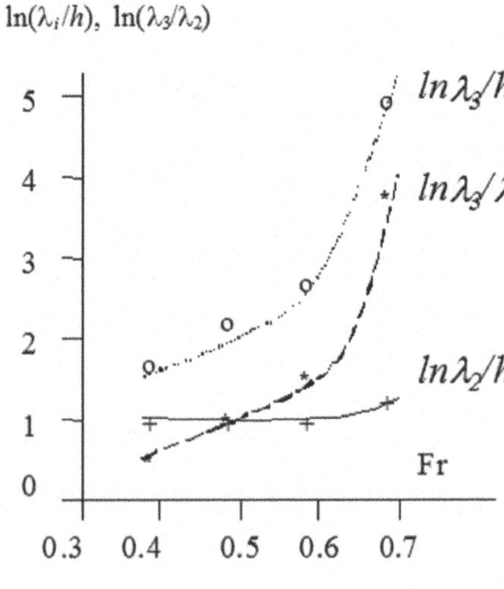

Fig. 6 The relative length of three-dimensional waves arising from the decay of a plane wave, as a function of the Froude number

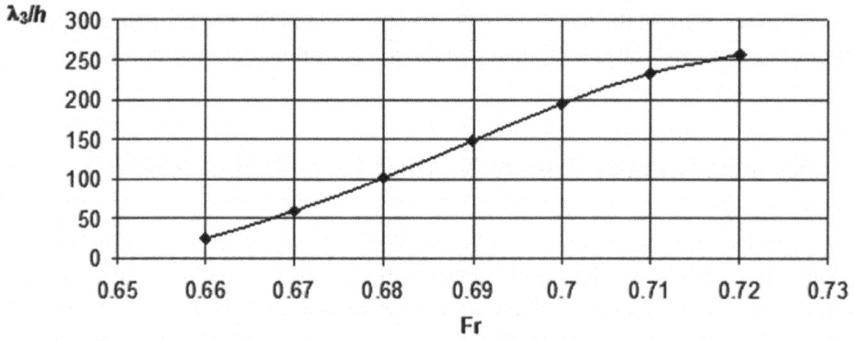

Fig. 7 Experimental dependence of the length of the longest three-dimensional wave on the Froude number

The length of the longest three-dimensional wave exceeds the depth of the water layer by 2 orders of magnitude for Froude numbers Fr > 0.68 (Fig. 7). The amplitude of a long three-dimensional wave is several times larger than the amplitude of a short wave. The longest wave λ_3 forms the ridge of maximum height. At smaller Froude numbers, Fr < 0.6, the wavelengths after the decay differ only several times from each other (Fig. 6) (Svitoch 2016): in this case, the spectrum of distances between the ridges contains several close modes. Field data with two modes in the spectrum of ridges for such Froude numbers are given in Svitoch (2016).

To estimate the Froude number of the catastrophic flux at the time of the decay of the plane wave of the limiting steepness, we used the profiles of the earth's surface, obtained from a model based on satellite data [10]. The profile of the section 7, directed along the normal to the axis of the ridges, is shown in Fig. 3. The average height of regular ridges rising above the average level is 2.5 m, the distance between the peaks of the ridge is 1300 m. The presence of one mode in the spectrum of the peak ridges allows us to limit the range of possible Froude numbers: Fr > 0.68. We assume that the distance between the regular ridges is equal to the length of the three-dimensional wave that occurs after the collapse $\lambda_3 = 1300$ m. In this case we have $\frac{\lambda_3}{h} = 217$ for section 7. We can determine the Froude number of the flow, based on the experimental dependence $\frac{\lambda_3}{h}$ on Froude number shown in Fig. 7. We have Fr = 0.703.

The presence of only one mode in the spectrum of ridges in the region of long waves allows us to determine the region of possible values of the Froude numbers of the catastrophic flux Fr > 0.68. The value of Fr = 0.703 lies in the allowed domain, which confirms the proposed hypothesis.

Let's define the parameters of ridges and waves on sections 8 and 6. The distance between the peaks of the regular ridges on section 8 is 1000 m, and on section 6 is 1100 m, depth is 7 and 9 m, respectively (Fig. 5). In the spectrum of the ridges, there is only one long-wave mode, which makes it possible to limit the range of possible Froude numbers: Fr > 0.68. For these sections we determine the relative lengths of the longest waves: $\left.\frac{\lambda_3}{h}\right|_8 = 143$ and $\left.\frac{\lambda_3}{h}\right|_6 = 122$. The data shown in Fig. 7 allow us to determine: Fr = 0.688 and Fr = 0.685 (sections 8 and 6). The estimates obtained fall into the allowed range of Froude numbers, confirming our assumptions.

4 Determination of Froude Number by Height of Ridges

The critical value of the steepness of a plane wave is reached due to amplification on a ridge. As a result of the decay, a long three-dimensional wave is formed. This wave forms a new relief, and as the height of the ridges increases, the wave amplifies. There is a critical value of the relative height of the ridge $\left(\frac{h_r}{h}\right)_{cr}$, at which a new wave decay occurs. The critical value can be approximately determined from the experimental dependence of the critical value of the relative height of the ridge on the Froude number constructed for the decay of a plane wave and shown in Fig. 8 (Svitoch 2016).

We assume that the critical relative height of the ridge for the first and second decay coincide at the same depth of the water layer. The condition for the absence of a new decay is

$$\frac{h_r}{h} < \left(\frac{h_r}{h}\right)_{cr} \tag{3}$$

It is possible to determine the maximum Froude number for the second decay Fr_2 based on the data shown in Figs. 3 and 5. The results are shown in Table 1.

It follows from the presented data that the results of Froude obtained on the basis of an estimate of the length of the three-dimensional wave and on the basis of the height of the ridges were close. This confirms the proposed hypothesis.

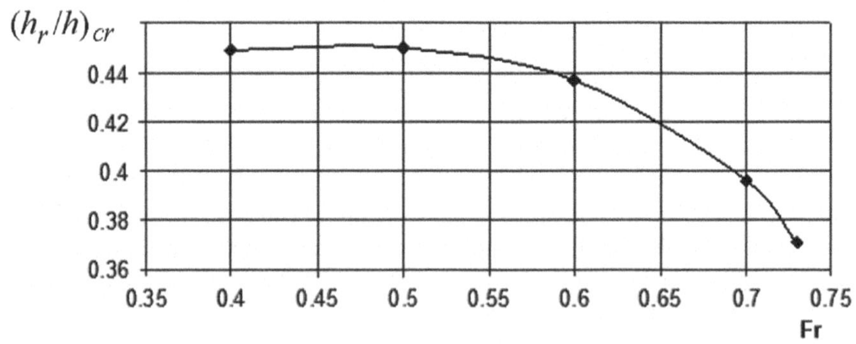

Fig. 8 Experimental dependence of the critical value of the relative height of the ridge on the Froude number

Table 1 Flow parameters and characteristics of ridge formed by a three-dimensional wave

Section	h (m)	$\frac{\lambda_3}{h}$	Fr	h_r (m)	$\left(\frac{h_r}{h}\right)_{cr}$	Fr2
6	9	122	0.685	3.7	0.411	0.68
7	6	217	0.703	2.4	0.4	0.69
8	7	143	0.688	2.88	0.411	0.68

5 Conclusion

Estimates of the flow parameters of water arising from a sudden decrease in the level of the Caspian Sea have been obtained. The Froude number of the flow is close to the value 0.7. For this Froude number, the length of a stationary three-dimensional wave on the water surface is calculated. The wavelength coincides with the distance between the ridges, determined from the field data. The height of the ridges is close to the calculated value. This allows us to conclude that ridges are formed by three-dimensional waves, and the proposed hypothesis is confirmed.

References

Baer KM (1856) Scientific notes on the Caspian Sea and its environs. Notes imperat. Russian geographer of the society. Book, vol XI. St. Petersburg, p 181

Butov SA, Melnikhova ON, Zhmur VV (2000) Cylindrical vortexes moving in flow with steady waves. Phys Vib 8(1):42

Golovachev IV (2017) Baer hillocks and their origins. Geol Geogr Glob Energy 67(4):140

Lighthill J (1978) Waves in fluids. Cambridge University Press, Cambridge

Mel'nikova ON (2005) Ridge formation on the bottom of a straight channel by a stationary flow with a flat free surface. Izv Atmos Ocean Phys 41(5):620

Melnikhova ON, Rykounov LN (1998) Sand wave formation by stationary waves. Adv Water Resour 21(3):193

Svitoch AA (2016) Regressive periods of the Great Caspian. Water Resour 43(2):270

http://gis-lab.info/qa/srtm.html

Features of the Formation of Gold Manifestations in the Black Shale Deposits of the Kumakskoye Ore Field

P. V. Pankratiev, A. V. Kolomoets, S. V. Bagmanova, and V. S. Panteleev

Abstract In this paper the perspective gold ore district of the Southern Urals is considered—Kumakskoe ore field. In structural terms, it represents the articulation of the East Ural uplift and Tobolsk anticlinorium. Within the Kumakskoe ore field, a number of gold deposits and occurrences are extended (Kumak, Kommercheskoye, Vasin, Caesar, Baikal, Lun, Amur). The main structure elements of the field are two large tectonic faults, limiting it from the East and West—Anikhovsky and Staro-Karabutak grabens, which are confined to one large tectonic seam, having commonality in the mechanism and time of formation. The paper shows their geological structure. Both grabens, which are made of Lower and Middle Paleozoic volcanogenic-sedimentary formations, which are represented by three structural floors separated by regional stratigraphic disagreements. A feature of the internal structure of grabens is noted—presence of faults in several directions and of different nature. Gold-bearing black slate deposits are studied. They show the presence of packs of contrasting interlayering of clay and terrigenous rocks with significant development carbonaceous mica-carbonate-quartz shales, the presence of sulfides and native gold in them, the wide development of crush zones, the proximity of Upper Paleozoic granitoid massifs.

Keywords Gold · Black shales · East-Urals uplift · Kumakskoe ore field · Anikhovsky graben · Old-Karabutakskoe graben · Shebekinskaya formation · Balajadia formation · New Orenburg formation

The East-Urals uplift is broken up by a system of faults corresponding to structural formative zones, subzones and blocks. From North to South, Suunduk, Adamov, Akkuduk and Ushkatinskaya structural formative zones of anticlinical type and Aydyrlinskaya, Elenovsko-Kumakskaya of sinclinor type are distinguished. The original rocks are mainly sialic blocks with characteristic granite-gneiss domes, large intrusions of granodiorite and granite formations. The second are small deflections, laid on the sialic base of the median massif, controlled by a system of limiting faults

P. V. Pankratiev · A. V. Kolomoets (✉) · S. V. Bagmanova · V. S. Panteleev
Orenburg State University, Orenburg, Russia
e-mail: kolomoyets56@mail.ru

C. T. Olegovna (ed.), *Processes in GeoMedia—Volume I*,
Springer Geology, https://doi.org/10.1007/978-3-030-38177-6_2

and made of volcanogenic basaltoid formations with a capacity of 2–3 km. Tectonic faults are fixed by linear bodies of ultramafic, less often of basic and acidic composition. Along the faults separating the sinclinoric and anticlinoric zones, deep grabensinclinic structures were developed, made by terrigenous-carbonate, less often of volcanogenic-sedimentary rocks D3-C1. These structures form superimposed structural formational zones (Kvarkenskaya, Djerlinsko-Dombarovskaya), smaller linear blocks (Shotinskaya) and separate local structures, blocks, of high order.

The Kumak ore field is structurally—a junction of the East Ural Rise and the Tobolsk Anticlinorium, separated by the Anikhovsky Graben-Sinclinorium. The latter is complicated by the Kumak-Kotansun crush zone, which is one of the fragments of the—Chelyabinsk deep fault. A number of gold deposits and manifestations of the Kumak ore field stretch along its strike for tens of kilometers in the central part of the buckling zone (Fig. 1).

The main structure elements of the Kumak ore field are two large tectonic disturbances limiting it from the East and West. These are the Anikhovsky and Staro-Karabutak grabens, which have a North-North-Eastern and submeridional direction, respectively.

One of the main features of the internal structure of both grabens is the presence of bursting disorders of several directions and different nature. The most clearly in the structures of grabens, the disturbances of the submeridional and North-Western directions are manifested. The sublatitudinal, North-Eastern and meridional values are subjected. Their intersections are most favorable for the localization of gold mineralization.

Both grabens are made by Lower and Middle Paleozoic volcanogenic-sedimentary formations, which are represented by three structural floors separated by regional stratigraphic disagreements (Arifulov 2005; Loshinin 2006, 2008).

The lower structure floor is represented by sediments of the Shebekta and Baldyk strata of the middle Ordovician rocks, which are age analogues of the New Orenburg strata, widespread in the North of the East-Ural uplift in the area of the Orenburg Urals (Loshinin 2008). The Shebekta strata is present in the extreme Southern part of this structure and in the Eastern-Mugojar zone in the territory of the Staro-Karabutak graben. It's made up of two undercutters: lower-volcanogenic-sedimentary (quartz sandstones, quartz-mica, quartz-chlorite and main composition effusives) 400–600 m and higher carbonaceous-terrigenous-shale (intermittent quartz sandstones, carbonaceous clayey, sericite quartz and quartz-sericite shales) 300–600 m.

Formations of the Balaldyk strata are developed in the East of the Staro-Karabutak graben. They are noticed in the form of tectonic blocks represented by conglomerates, quartz sandstones, carbonaceous phyllite shales. Total sediment formation is up to 600 m (Loshinin 2006, 2008).

The middle structure floor is marked only within the limits of Anikhovsky graben. It is composed of Middle-Upper Devonian volcanogenic-sedimentary rocks. Its lower boundary is not opened, but overlapped with the erosion of carbonaceous-carbonate-terrigenous formation of the lower carbon. Its base (200–300 m) is mainly shale, and in the middle and upper part there are tuffs of acidic composition, tuffaceous stones,

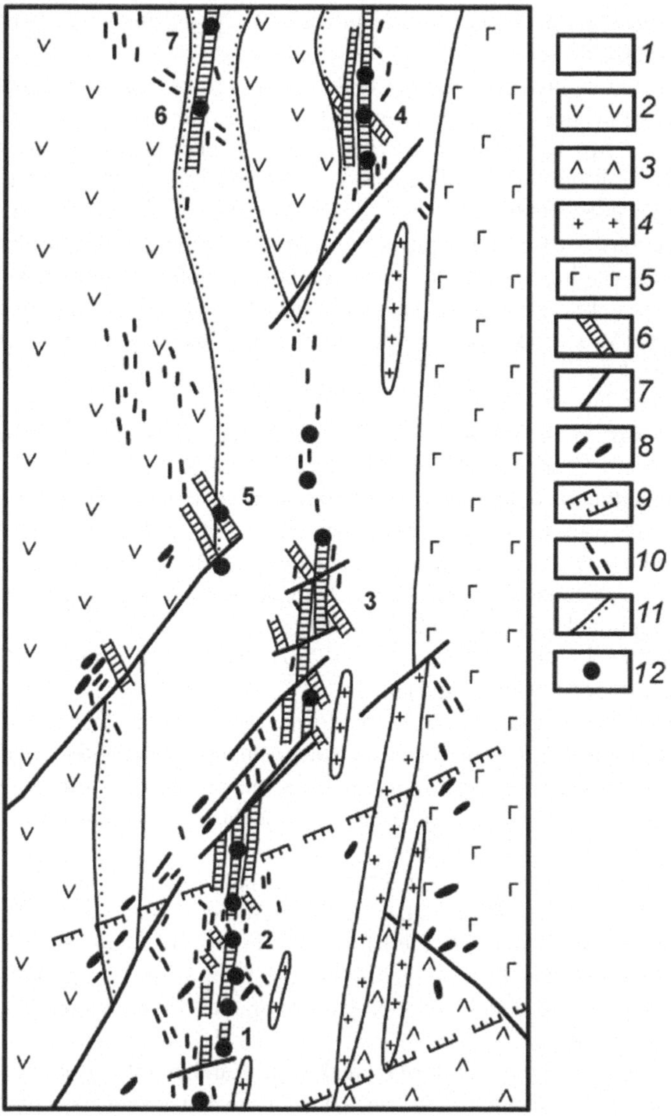

Fig. 1 Layout of gold deposits and occurrences in the Kumak ore field [Warrior, 1966]: 1—Kumak "black shale" strata; 2—volcanogenic—sedimentary deposits (C1); 3—main effusives (D1); 4—granite—porphyries (P); 5—gabbroids (PZ3); 6—mineralized zones of buckling; 7—separate disturbances; 8—small intrusions and large dikes of the Kumak complex; 9—contours of small NW intrusions belt; 10—large quartz veins; 11—geological boundaries; 12—gold deposits and ore occurrences. Include: 1—Kumak South, 2—Kumak North, 3—Baikal, 4—Commercial, 5—Amur, 6—Mile, 7—Moon

tuffa-alavrolites, tuffoconglomerates with interlayers of siltstones and carbonaceous clay shales (500–600 m).

The upper structural floor is fixed both in Anikhovsky and Staro-Karabutaksky grabens. It's thickness represented by carbonaceous carbonate-terrigenous rocks. It consists of two subtopics: the lower one carbonate-carbonate-shale (middle-Tourney-Visaean age) of 350–400 m and the upper one—volcanogenic-sedimentary (upper-Visavian-Serpukhov time) 80–150 m. The composition of the lower subthrust is carbonaceous-mica-quartz shales (prevailing), conglomerates, gravelites, sandstones, siltstones, limestones; the upper subthrust is composed of conglomerates, tuffoconglomerates, tuffa-alavrolites, sandstones, limestones (Pankratiev 2005).

Within the Anikhovsky Graben-Sinclinorium, the lower underbody is also divided into three packs.

The lower bundle—carbon shale (200–300 m) is a gold-hosted and composed mainly of carbon micaceous quartz, quartz-micaceous and carbon micaceous carbonate-quartz siltstones with interlayers of gravelites, sandstones and siltstones. Small and medium pebbles conglomerates are present at the base.

The middle pack terrigenous-carbonate (50–150 m) is distributed in the North of the Anikhovsky graben and is represented by limestone with rare interlayers of carbonaceous siltstones and sandstones. Carbonate bedding with nesting colonies of corals, sponges and foraminifera is observed.

The upper pack—clay-terrigenous (100–200 m) is composed of fine and medium-grained sandstones, siltstones and clays.

In the Staro-Karabutak graben, only the lower carbon shale sequence, the power of which increases to 400–600 m, can be traced. It contains layered tuffy rocks of acidic and basic composition. The upper volcanic-sedimentary substage with conglomerates at the base lies directly on the washed out surface of the lower carbon-carbonate-shale substage of the Middle-Turne-Visacean tier.

The above formations are broken through by numerous intrusions of Upper Paleozoic age. In the Anikhovsky graben, these are microcline granites of the Adamovsky complex and granitoid dykes of the Kumaksky complex of small intrusions. In the Staro-Karabutak graben it is a group of intrusions forming the East Mugodzhar belt, composed of leucocratic and biotite-amphibole granites of the Shotka, Aktastinsky and Kokpasai massifs.

On the squares of the Anikhovsky and Staro-Karabutaksky grabens, the main mass of gold mineralization is connected with the lower sequence of the lower carbonaceous-carbonate-terrigenous formation of the lower carbon. Thicknesses are grayish-black, sometimes black, fine-grained, shale, and easily split in shale planes, consisting of fine grains of quartz, sericite, and biotite in a mass of carbonaceous matter and graphite dust. Carbon content according to chemical analyses reaches 8.7%. This makes it possible to classify them as normal carbonaceous rocks (Kolomoets 2018).

Tracking the ore-bearing deposits of the South Kumak ore field along the Staro-Karabutak graben, there is a slight decrease in concentrations of noble metal in them, and then, as approaching the development of the Balkanbaisk graben in the Eastern Mughodzhars, increase again. The reason of the wide spread of gold-bearing deposits of the lower subfloor of the lower carbon-carbonate-terrigenous layer of C1 (Loshinin 2006; Pankratiev 2005) are: (1) presence of contrasting interlayers of clayey and terrigenous rocks with a significant development of carbonaceous mica-carbonate-quartz shales (mineralization concentrators) among them; (2) presence of sulfides and native gold; (3) wide development of deposit crumpling zones at the object, the most favorable for ore-localization, coordinated with the areas of articulation of differently directed tectonic ruptures; (4) proximity of Upper Paleozoic granitoid massifs, the main sources of hydrothermal gold.

Thus, the study of black shale deposits of the Kumak ore field of the Orenburg region is now becoming important in the prediction and evaluation of industrial gold ores. The meridional strip along of the Anikhovsky and Staro-Karabutaksky grabens in the development area including ore-bearing packs of Lower Carboniferous age should be searched up to Mugodzhar of Kazakhstan, where new manifestations with significant concentrations of gold may be revealed at depth.

Acknowledgements The work is done with the financial support of the regional grant in the field of scientific and scientific-technical activities in 2018 (Decree of July 19, 2018, No. 444-p).

References

Arifulov ChH (2005) Black shale gold deposits of various geological settings. Ores Met 2:9–19 (in Russian)

Kolomoets AV (2018) Conditions of formation of the Kumak deposit of the black shale formation (Orenburg region). Vestnik Zabaikalsk State Univ 24(6):28–35 (in Russian)

Loshinin VP (2006) Gold-bearing capacity of the Lower Middle Paleozoic Black shale formations of the Eastern Orenburg Region. In: Pankratiev PV, Loshinin VP (eds) Strategy and processes of georesources development. Perm, pp 79–82 (in Russian)

Loshinin VP (2008) Golden manifestations of stratiform type in the paleozoic of the Vostochno-Uralsky uplift (in Russian). In: Loshinin VP, Pankratiev PV (eds) Water management problems and rational nature use. Materials of the all-Russian Auch-practical conference. Perm, p 119 (in Russian)

Pankratiev PV (2005) Golden mineralization of riftogenic basins of the Orenburg Region. In: Pankratiev PV, Loshinin VP (eds) Strategy and processes of georesources development. Perm, pp 13–15 (in Russian)

Analytical Test Problem of Wind Currents

V. S. Kochergin, S. V. Kochergin, and S. N. Sklyar

Abstract The paper presents an analytical solution for a three-dimensional model of wind currents. The obtained expressions for barotropic velocity components, additional velocities and vertical components can be used in testing and analysis of difference schemes and algorithms in the construction of hydrodynamic models of water dynamics.

Keywords Dimensionless problem · Wind currents · Test problem · Analytical solution

1 Introduction

It is difficult to overestimate the importance of numerical modeling of ocean dynamics for solving urgent problems of environmental monitoring and tasks related to improving the accuracy of weather prediction. In recent years, the development of computer technology has made significant progress in this direction, especially in terms of increasing the discretization of the problems to be solved for a more correct description of the dynamic processes taking place in the ocean. However, the direction associated with the correction of the models themselves, the methods of their numerical implementation is far from exhausted, because with the combined use of modern computer power and new computational schemes, and algorithms can lead to a significant effect in solving such problems. When choosing a model that describes the dynamics of the ocean, the results obtained from one model are most often compared with others, which in General is often subjective. Therefore, the presence of an accurate (analytical) solution of a problem allows you to make the correct choice of the used schemes and algorithms for its numerical implementation. Models of ocean dynamics are quite complex. There are analytical solutions for the simplest productions, for example, the Stommel model (Stommel 1948, 1965; Kochergin 1978). In

V. S. Kochergin (✉) · S. V. Kochergin
Marine Hydrophysical Institute of the RAS, Sevastopol, Russia
e-mail: vskocher@gmail.com

S. N. Sklyar
American University of Central Asia, Bishkek, Kyrgyzstan

C. T. Olegovna (ed.), *Processes in GeoMedia—Volume I*,
Springer Geology, https://doi.org/10.1007/978-3-030-38177-6_3

Eremeev et al. (2008) such a problem is realized by the method of dynamic operator inversion (Kochergin 1978) for the study of applied computational schemes of a special kind for the calculation of velocity fields. In this paper we consider a more complex problem the solution of which allows us to obtain an analytical solution for the barotropic component of the velocity, its three-dimensional additional part and the vertical component. The importance of the latter is perhaps clear to everyone who has faced the problem of calculating the vertical velocity from the continuity equation in one way or another.

2 The Problem in a Dimensionless Form

Let the surface of the water basin in the $x0y$ plane have the shape of a rectangle:

$$\Omega_0 \times [0, r][0, q].$$

its depth $H > 0$. The axes of the Cartesian coordinate system are directed as follows: $0x$—to the East, $0y$—to the North, $0z$—vertically down. In the three-dimensional domain $\Omega = \{(x, y, z)|(x, y) \in \Omega_0, 0 \le z \le H\}$, we consider a system of equations of motion in a dimensionless form:

$$\begin{cases} -lv = -\frac{\partial P^s}{\partial x} + \frac{\partial}{\partial z}\left(k\frac{\partial u}{\partial z}\right) \\ lu = -\frac{\partial P^s}{\partial y} + \frac{\partial}{\partial z}\left(k\frac{\partial v}{\partial z}\right), t > 0, (x, y, z) \in \Omega^0 \\ \frac{\partial u}{\partial x} + \frac{\partial v}{\partial y} + \frac{\partial w}{\partial z} = 0 \end{cases}, \tag{1}$$

with boundary conditions:

$$\left\{t > 0, z = 0, (x, y) \in \Omega_0^0\right\}: k\frac{\partial u}{\partial z} = -\tau_x, \quad k\frac{\partial v}{\partial z} = -\tau_y, \quad w = 0, \tag{2}$$

$$\left\{t > 0, z = H, (x, y) \in \Omega_0^0\right\}: k\frac{\partial u}{\partial z} = -\tau_x^b, \quad k\frac{\partial v}{\partial z} = -\tau_y^b, \quad w = 0, \tag{3}$$

$$\left\{t > 0, 0 \le z \le H, (x, y) \in \partial\Omega^0\right\}: U \cdot n_x + V \cdot n_y = 0, \tag{4}$$

and initial data:

$$\{t = 0, (x, y, z) \in \Omega\}: u = u^0, v = v^0, w = w^0, \tag{5}$$

where Ω_0^0 and Ω^0 are interior points of Ω and Ω_0.

In (4) the dimensionless integral velocity are defined as follows:

$$U(t,x,y)=\int_0^H u(t,x,y,z)dz,\quad V(t,x,y)=\int_0^H v(t,x,y,z)dz,$$

and in (3) the following variant of parametrization of bottom friction is used:

$$\tau_x^b=\mu\cdot U,\quad \tau_y^b=\mu\cdot V,\quad \mu\equiv const>0. \tag{6}$$

In accordance with the model Stommel, suppose:

$$l=l_0+\beta\cdot y,\quad k\equiv const; \tag{7}$$

$$\tau_x=\frac{F\cdot q}{\pi}\cos\left(\frac{\pi y}{q}\right),\quad \tau_y=0.$$

We will look for the horizontal components of the velocity vector in the form:

$$u=U\cdot H^{-1}+\hat{u},v=V\cdot H^{-1}+\hat{v}, \tag{8}$$

where the first terms are called barotropic, and the second are called additional components of speed.

3 Analytical Solution. Barotropic Components

In the system of Eqs. (1)–(5), we integrate each equation by "z" in the range from 0 to H. Taking into account the boundary conditions, we obtain the problem for integral velocities:

$$\begin{cases}\mu U-lV=-H\frac{\partial p^s}{\partial x}+\tau_x\\ lU+\mu V=-H\frac{\partial p^s}{\partial y}+\tau_y\\ \frac{\partial U}{\partial x}+\frac{\partial V}{\partial y}=0,(x,y)\in\Omega_0^0\\ U\cdot n_x+V\cdot n_y=0,(x,y)\in\Omega_0\end{cases} \tag{9}$$

For the stream function, we obtain the following problem:

$$\begin{cases}\mu\left(\frac{\partial^2\psi}{\partial x^2}+\frac{\partial^2\psi}{\partial y^2}\right)+\beta\frac{\partial\psi}{\partial x}=-F\cdot\sin\left(\frac{\pi y}{q}\right),(x,y)\in\Omega_0^0\\ \Psi=0,(x,y)\in\partial\Omega_0\end{cases}. \tag{10}$$

Finally, the integral of the velocity found by the formula:

$$\begin{cases} U(x,y) = -\frac{F}{\mu(\pi/q)} \cdot \left(C_1 \cdot e^{Ax} + C_2 \cdot e^{Bx} - 1\right) \cdot \cos\left(\frac{\pi y}{q}\right) \\ V(x,y) = \frac{F}{\mu(\pi/q)^2} \cdot \left(C_1 \cdot Ae^{Ax} + C_2 \cdot Be^{Bx}\right) \cdot \sin\left(\frac{\pi y}{q}\right) \\ C_1 = \frac{1-e^{Br}}{e^{Ar}-e^{Br}}, C_1 + C_2 = 1 \end{cases}, \tag{11}$$

where $A = -\frac{\beta}{2\mu} + \sqrt{\left(\frac{\beta}{2\mu}\right)^2 + \left(\frac{\pi}{q}\right)^2}$, $B = -\frac{\beta}{2\mu} - \sqrt{\left(\frac{\beta}{2\mu}\right)^2 + \left(\frac{\pi}{q}\right)^2}$.

4 Analytical Solution. Additional Components

Substitute in the first two equations of the system (9) the horizontal components of the velocity vector written in the form (8). Taking into account the equations for integral velocities:

$$\begin{cases} -lVH^{-1} = -\frac{\partial P^s}{\partial x} + \frac{\tau_x - \tau_x^b}{H} \\ lUH^{-1} = -\frac{\partial P^s}{\partial y} + \frac{\tau_y - \tau_y^b}{H} \end{cases}$$

we obtain the problem for additional components of horizontal velocities:

$$\begin{cases} -l\hat{v} = \frac{\partial}{\partial z}\left(k\frac{\partial \hat{u}}{\partial z}\right) + \frac{\tau_x^b - \tau_x}{H} \\ l\hat{u} = \frac{\partial}{\partial z}\left(k\frac{\partial \hat{v}}{\partial z}\right) + \frac{\tau_y^b - \tau_y}{H} \end{cases} \quad (x,y) \in \Omega_0, \quad 0 < z < H, \tag{12}$$

$$\{z = 0, (x,y) \in \Omega_0\}: k\frac{\partial \hat{u}}{\partial z} = -\tau_x, \quad k\frac{\partial \hat{v}}{\partial z} = -\tau_y, \tag{13}$$

$$\{z = H, (x,y) \in \Omega_0\}: k\frac{\partial \hat{u}}{\partial z} = -\tau_x^b, \quad k\frac{\partial \hat{v}}{\partial z} = -\tau_y^b. \tag{14}$$

We write the problem (12)–(14) in a complex form, introducing the following functions:

$$\theta = \hat{u} + i \cdot \hat{v}, \quad \tau^b = \tau_x^b + i \cdot \tau_y^b = \mu(U + i \cdot V), \quad \tau = \tau_x + i \cdot \tau_y = F\frac{q}{\pi}\cos\left(\frac{\pi y}{q}\right),$$

as a result we have:

$$\begin{cases} il\theta - k\frac{\partial^2 \theta}{\partial z^2} = \frac{\tau^b - \tau}{H}, 0 < z < H \\ z = 0: k\frac{\partial \theta}{\partial z} = -\tau \\ z = H: k\frac{\partial \theta}{\partial z} = -\tau^b \end{cases}. \tag{15}$$

The General solution of the equation from (15) is found as the sum of the particular solution of this equation and the General solution of the corresponding homogeneous equation. We introduce the following functions:

$$C(z) = \cos(H - z)\eta \cdot \cosh(H + z)\eta - \cos(H + z)\eta \cdot \cosh(H - z)\eta,$$
$$S(z) = \sin(H - z)\eta \cdot \sinh(H + z)\eta - \sin(H + z)\eta \cdot \sinh(H - z)\eta,$$

where $\eta = \sqrt{\frac{l}{2k}}$; $L(z) = \frac{C(z)}{C(H)} - \frac{z}{H}$, $M(z) = \frac{S(z)}{C(H)}$.

$$Cs(z) = \cos(H - z)\eta \cdot \sinh(H + z)\eta + \cos(H + z)\eta \cdot \sinh(H - z)\eta,$$
$$Sc(z) = \sin(H - z)\eta \cdot \cosh(H + z)\eta + \sin(H + z)\eta \cdot \cosh(H - z)\eta,$$

Taking into account the boundary conditions on the surface and bottom of (15) can be obtained:

$$\hat{u} = \frac{\mu V}{lH} + \frac{\eta}{lC(H)} \cdot \left\{ -\frac{Fq}{\pi} \cdot \cos\left(\frac{\pi y}{q}\right) \cdot [Sc(H - z) - Cs(H - z)] \right.$$
$$\left. - \mu(U + V) \cdot Cs(z) + \mu(U - V) \cdot Sc(z) \right\} \quad (16)$$

$$\hat{v} = -\frac{\mu U - \frac{Fq}{\pi} \cdot \cos\left(\frac{\pi y}{q}\right)}{lH} + \frac{\eta}{lC(H)} \left\{ -\frac{Fq}{\pi} \cdot \cos\left(\frac{\pi y}{q}\right) \cdot [Sc(H - z) \right.$$
$$\left. + Cs(H - z)] + \mu(U - V) \cdot Cs(z) + \mu(U + V) \cdot Sc(z) \right\} \quad (17)$$

5 Analytical Solution. Vertical Component of the Velocity Vector

The vertical component w of the velocity vector determined from the continuity equation, this will integrate the continuity equation by z from 0 to z, taking into account boundary condition (2) and the continuity equation for the integral velocities:

$$\frac{\partial U}{\partial x} + \frac{\partial V}{\partial x} = 0.$$

Due to this the following formula can be obtain:

$$w(z) = D + C, \quad (18)$$

where

$$D = \left[\frac{F}{l} \cdot \sin\left(\frac{\pi y}{q}\right) + \frac{\beta q F}{\pi l^2} \cdot \cos\left(\frac{\pi y}{q}\right)\right] \cdot L(H - z)$$
$$+ \left[\frac{B\mu}{l^2} \cdot U - \frac{\beta V - F \cdot \sin(\pi y/q)}{l}\right] \cdot L(z) + \frac{\beta\mu}{l^2} \cdot V \cdot M(z) \quad (19)$$

For the convenience of calculating the value of C, we introduce the following functions:

$$P(z) = (H + z) \cdot [\cos(H - z)\eta \cdot \sinh(H + z)\eta + \sin(H + z)\eta \cdot \cosh(H - z)\eta],$$
$$Q(z) = (H + z) \cdot [\sin(H - z)\eta \cdot \cosh(H + z)\eta - \cos(H + z)\eta \cdot \sinh(H - z)\eta],$$

then finally we have:

$$C = -\frac{Fq}{\pi l} \cdot \cos\left(\frac{\pi y}{q}\right) \cdot \left\{CY(H - z) - CY(H) \cdot \left[L(H - z) + \frac{H - z}{H}\right]\right\}$$
$$- \frac{\mu U}{l} \cdot \left\{CY(z) - CY(H) \cdot \left[L(z) + \frac{z}{H}\right]\right\} - \frac{\mu V}{l} \cdot [SY(z) - CY(H) \cdot M(z)]. \quad (20)$$

6 Numerical Example (Black Sea)

To illustrate, consider as a model object a rectangular basin of water with a flat bottom with characteristic dimensions of the Black sea:

$$a = 11 \times 10^7 \text{ (sm)} = 1100 \text{ (km)},$$
$$b = 5 \times 10^7 \text{ (sm)} \approx 500 \text{ (km)}$$
$$D = 2 \times 10^5 \text{ (sm)} = 2000 \text{ (m)},$$
$$R = 0.02 \text{ (sm/s)}, \quad G = 1 \text{ (dyn/sm}^2\text{)},$$
$$E = 1 \text{ (sm}^2\text{/s)}, \quad \rho_0 = 1 \text{ (g/sm}^3\text{)},$$
$$f_0 = 10^{-4} \text{ (1/s)}, \quad f_1 = 1 \times 10^{-13} \text{ (1/sm s)}.$$

We choose characteristic scales:

$$L = 10^7 \text{ (sm)}, \quad h = 2 \times 10^5 \text{ (sm)}, \quad u_0 = 10 \text{ (sm/s)}.$$

Then we have:

$$r = 11, \quad q = 5, \quad H = 1, \quad k = 0.05,$$
$$l_0 = 1, \quad \beta = 0.002,$$
$$\beta = 0 \text{ without } \beta\text{-effect},$$

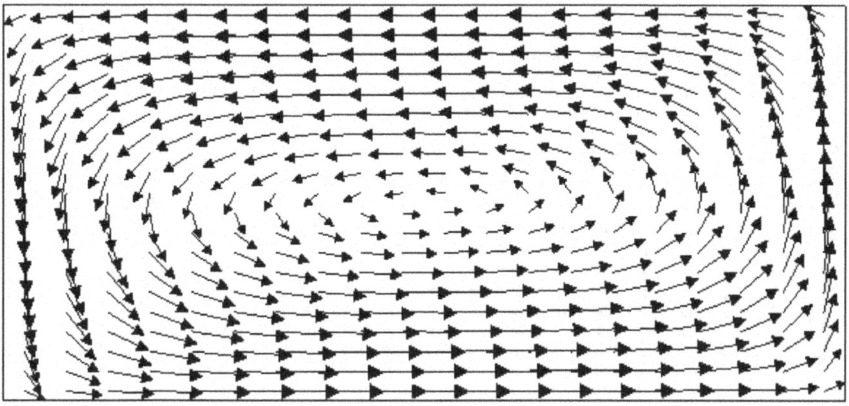

Fig. 1 Barotropic field of currents ($\beta = 0$)

$$F = \frac{\pi}{2} \times 10^{-3}, \quad \mu = 0.001, \quad w_0 = 0.2.$$

Finding the solution of the problem (1)–(5), the solution of the dimensional problem is determined by the formulas:

$$\bar{u} = \bar{u}(Lx, Ly, hz) = u_0 \cdot u(x, y, z),$$
$$\bar{v} = \bar{v}(Lx, Ly, hz) = u_0 \cdot v(x, y, z),$$
$$\bar{w} = \bar{w}(Lx, Ly, hz) = w_0 \cdot w(x, y, z), \quad w_0 = \frac{hu_0}{L},$$

where the line denotes the components of the velocity vector for the dimensional problem. Wind stress was set in accordance with the specified F. In the Western part of the integration area, the wind stress is carried out in the south, and in the Eastern one in the Northern direction. Under this wind action, the maximum speeds at the boundaries of the region were about 2.5 sm/s (Fig. 1). Taking into account the β-effect, there is an intensification of currents in the West Bank up to 7 sm/s (Fig. 2).

In this formulation, with the selected external influences, the values of the additional components are small, but they are obtained analytically and they determine the values of the vertical velocity component, since the barotropic component satisfies the continuity equation exactly automatically. In the future, the solutions obtained analytically can be compared with numerical ones (Figs. 3 and 4).

7 Conclusions

The analytical solution of the wind circulation model is constructed for its use as a test solution in the analysis of schemes and algorithms in the construction of dynamic

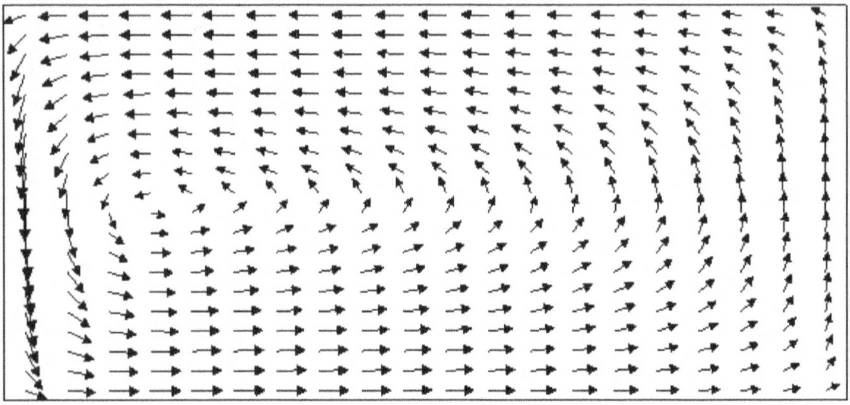

Fig. 2 Barotropic field of currents ($\beta = 0.002$)

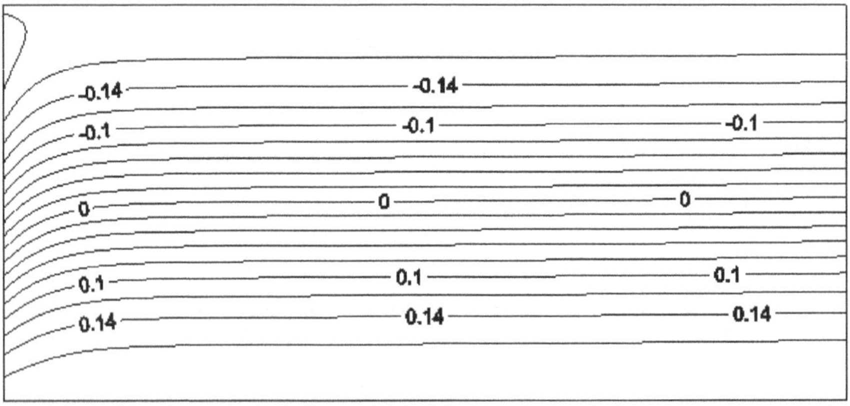

Fig. 3 The additional component of velocity $\hat{u}$ for $z = 0$

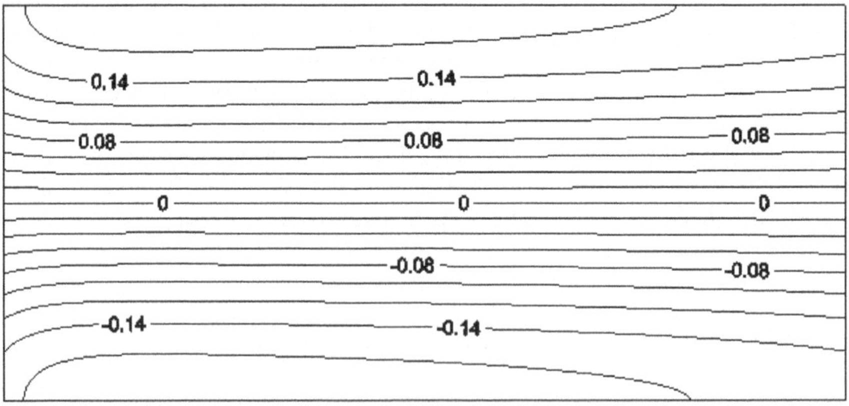

Fig. 4 The additional component of velocity $\hat{v}$ for $z = 0$

models of reservoirs. A simplified scheme of wind circulation in the Black sea was constructed using the analytical solution. The found solution allows to analyze the influence of certain input parameters on the intensity of flows in the studied basin.

Acknowledgements The work was carried out within the framework of the state task on the theme No. 0827-2018-0004 "Complex interdisciplinary studies of Oceanological processes determining the functioning and evolution of ecosystems in the coastal zones of the Black and Azov seas on the basis of modern methods of control of the marine environment and grid technologies."

References

Eremeev VN, Kochergin VP, Kochergin SV, Sklyar SN (2008) Mathematical modeling of hydrodynamics of deep-water basins. ECOSI-Gidrophisika, Sevastopol, 363p
Kochergin VP (1978) Theory and methods of ocean currents. Nauka, Moscow, 127p
Stommel H (1948) The westward intensification of wind-driven ocean currents. Trans Amer Geophys Union 29:202–206
Stommel H (1965) The gulf stream. A physical and dynamical description. University of California Press, Berkeley, CA, 227p

Determination of the Propagation Speed of Wave Perturbations in a Two-Component Field

V. V. Ocherednik and A. S. Zapevalov

Abstract The spatial-time characteristics of the wave field, representing a superposition of two components in which the waves obey different dispersion equations have analyzed. It is shown that the phase shift, determined at two points of the two-component field, depends nonlinearly on the distance between these points, which leads to errors of determining the phase velocity. The dependence of the phase velocity on the ratio of the spectra of free and bound waves on the scale of the second harmonic is constructed for the field of gravitational surface waves.

Keywords Two-component field · Phase shift · Phase velocity · Cross-spectrum

1 Introduction

The study of wave fields (surface and internal) in the World ocean is one of the most urgent tasks of modern oceanography. The solution of many fundamental and applied problems requires a detailed understanding of the structure and variability of wave fields, including the number of wave propagation velocity (Efimov et al. 1972; Kononkova and Pokazeev 1978; Litvin et al. 1992; Zapevalov et al. 2004).

The weak nonlinearity of wave fields leads to the simultaneous existence of free and coupled waves obeying different dispersion equations (Yuen and Lake 1987; Saprykina 2015). One of the main types of coupled waves is the harmonics of longer waves, which affects the spatial-time characteristics of the wave field (Zapevalov 1996). In this paper, the effect of the existence of two types of wave disturbances on the calculation of their phase velocities is analyzed.

V. V. Ocherednik
Shirshov Institute of Oceanology RAS, Moscow, Russia

A. S. Zapevalov (✉)
Marine Hydrophysical Institute RAS, Sevastopol, Russia
e-mail: sevzepter@mail.ru

C. T. Olegovna (ed.), *Processes in GeoMedia—Volume I*,
Springer Geology, https://doi.org/10.1007/978-3-030-38177-6_4

2 Spatial-Time Characteristics of the Two-Component Wave Field

The mathematical apparatus of cross-spectral analysis is used to describe the spatial-time structure of the wave field. The linear relationship between the studied parameters, determined at two points in space $X_1 = (x_1, y_1)$ and $X_2 = (x_2, y_2)$ separated by a vector $\vec{L}$, is described with the spectral phase shift and the quadratic coherence function (Jenkins and Watts 1972)

$$\varphi(\omega, \vec{L}) = arctg(-Q(\omega, \vec{L})/Co(\omega, \vec{L})) \tag{1}$$

$$R^2(\omega, \vec{L}) = \frac{Co^2(\omega, \vec{L}) + Q^2(\omega, \vec{L})}{S_1(\omega, X_1)\, S_2(\omega, X_2)},$$

where ω is circular frequency; $Co(\omega, \vec{L})$ is co-spectrum; $Q(\omega, \vec{L})$ is quadrature spectrum; $S(\omega, X_1)$ and $S(\omega, X_2)$ is spectra of the surface elevation measured at the points X_1 and X_2. If the wave field is uniform, then $S(\omega, X_1) = S(\omega, X_2) = S(\omega)$. The quadratic coherence function $R^2(\omega, \vec{L})$ is a measure of the stability of phase relations.

Let there be two types of components in a homogeneous wave field, obeying different dispersion equations and not interacting with each other. Then auto and cross-spectral characteristics can be represented as:

$$S(\omega) = \sum_{j=1}^{2} S_j(\omega),$$

$$\chi(\omega, \vec{L}) = Co(\omega, \vec{L}) - i\,Q(\omega, \vec{L}) = \sum_{j=1}^{2} \sqrt{R_j^2(\omega, \vec{L})\, S_j(\omega)}\, \exp\left(i\,\varphi_j(\omega, \vec{L})\right),$$

where j is the number of component ($j = 1, 2$).

Consider the two-dimensional problem. We assume that all components of the wave field propagate in the same direction. In this case, due to the linearity of each component, the identity $R_j^2(\omega, L) \equiv 1$ is true. For the real and imaginary parts of the cross-spectrum, we obtain that:

$$Co(\omega, L) = \sum_{j=1}^{2} S_j(\omega) \cos\left(\varphi_j(\omega, L)\right) \tag{2}$$

$$Q(\omega, L) = \sum_{j=1}^{2} S_j(\omega) \sin\left(\varphi_j(\omega, L)\right) \tag{3}$$

The magnitude of the phase shift at a distance L for each component of the wave field can be defined as $\varphi_j = 2\pi(\lambda_j(\omega)/L)$. Here $\lambda_j(\omega)$—is the wavelength of the jth component, which is determined from the equation

$$C_j = \omega / k_j,$$

where $k_j = 2\pi/\lambda_j$ is wave number.

3 A Phase Shift in the Wave Field

Without breaking the generality of the problem, for certainty we will accept that $\lambda_1 > \lambda_2$. Let's also assume that $S_1(\omega) > S_2(\omega)$. Let us consider how the phase shift changes, determined at different values of the parameter $\tilde{L} = L/\lambda_1$ in a two-component wave field. The parameter $\tilde{L}$ is a dimensionless distance between two points of the wave field. Introduction of the parameter $\tilde{L}$ allows us to eliminate an explicit dependence on frequency ω and obtain generalized characteristics of the wave field (Efimov et al. 1972).

In a linear wave field, the phase shift increases linearly with the distance increasing, i.e., φ_1 and φ_2 linearly depend on L. In the wave field, represented the superposition of two linear fields, this dependence is violated. Changes of the calculated values of the phase shift ω in the two-component wave field are shown in Fig. 1. For clarity, the values of the shift ω are normalized to the phase shift of the dominant component of the wave field (having a higher level of spectral density). The calculations are carried out for three values of the ratio of spectra $\mu = S_2/S_1$ and three values of the phase shift ratios φ_2/φ_1.

In the region of small values of the parameter $\tilde{L}$, the phase shift φ calculated from measurements at two points of the two-component wave field is equal to the

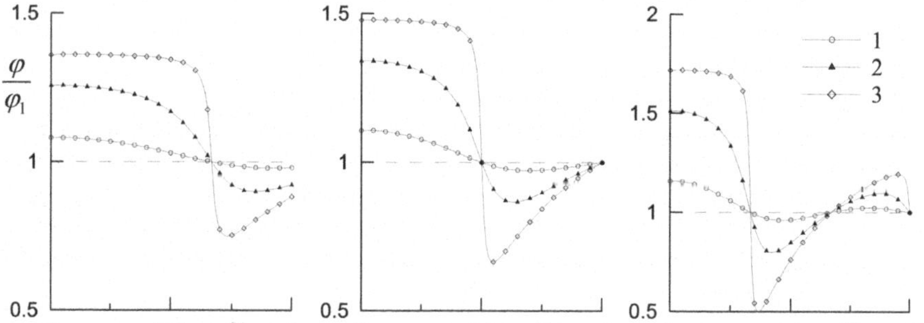

Fig. 1 The dependences of relative phase shift φ/φ_1 from distance $\tilde{L}$ under $\varphi_2/\varphi_1 > 1$. The curves 1, 2, and 3 correspond to the parameter μ values equal to 0.1, 0.3 and 0.5

average phase shift of these components with a weight proportional to the level of the corresponding spectra

$$\varphi = \frac{\varphi_1 S_1 + \varphi_2 S_2}{S_1 + S_2} \tag{4}$$

Equation (4) for small values of parameter $\tilde{L}$ can be obtained from (1)–(3), assuming that the φ_j values are small and approximations $\varphi_j \approx \sin\varphi_j$ and $\varphi_j \approx \sin\varphi_j$ can be used. The area where the expression (4) is true, as shown in Fig. 1, depends on the ratio value φ_2/φ_1.

As the parameter $\tilde{L}$ increases, the phase shift φ values approach the shift of the dominant component φ_1. Moreover, changes in φ with the $\tilde{L}$ change are non-monotonic.

One of the main characteristics of wave disturbances is their phase velocity. According to the measurements of cross-spectral characteristics at two points of the wave field, the phase velocity can be calculated using the expression

$$C(\omega) = \omega L/\varphi(\omega, L) \tag{5}$$

Here the ratio $\varphi(\omega, L)/\omega$ determines the time for which the wave passes the distances L (Efimov 1981).

It follows from (5) that the condition $\varphi_2/\varphi_1 > 1$ corresponds to the condition $C_2/C_1 < 1$. Thus, presented in Fig. 1 phase shift changes correspond to the situation when the speed of the dominant component is higher than the speed of the second component. Consider the reverse situation, i.e. the situation when the condition $\varphi_2/\varphi_1 < 1$ is satisfies. The calculation results are shown in Fig. 2. It is true the above statement that with the increase of the parameter $\tilde{L}$ the phase shift φ non-monotonically approaches the shift of the dominant component φ_1.

It should be noted that obtained in this paper results can be used in the case when in the field under study, along with wave disturbances, there are disturbances

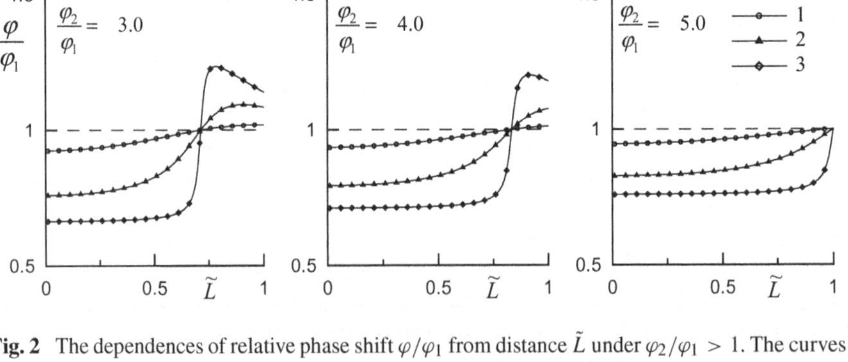

Fig. 2 The dependences of relative phase shift φ/φ_1 from distance $\tilde{L}$ under $\varphi_2/\varphi_1 > 1$. The curves 1, 2, and 3 correspond to the parameter μ values equal to 0.1, 0.3 and 0.5

generated by the flow. In this case, it can be assumed that the second component of the field under study is disturbances propagating at the flow rate. The possibility of applying the proposed approach to the calculation of phase shifts caused by the flow is determined by the extent to which the Taylor hypothesis of frozen turbulence is fulfilled for the flow field (Lamley and Panovsky 1966).

4 Phase Velocity

The results of the analysis carried out in the previous section showed that the phase shift at two points of the two-component wave field changes nonlinearly with the change of the distance between these points. As a consequence, the phase velocity $C(\omega)$ determined according to (5) would depend on the distance between the points at which the phase shift was determined. The value $C(\omega)$ also depends on the ratio of the spectrums levels of the two components of the wave field.

Experimental studies of the phase velocities of gravitational waves on the sea surface have shown that their values deviate from the values following from the dispersion equation

$$\omega^2 = gk \tag{6}$$

where g is gravitational acceleration (Yuen and Lake 1987). One of the main reasons for these deviations was the presence of harmonics of longer waves. Let us estimate the deviations, which may appear due to the second harmonic of the gravitational wave.

It follows from the dispersion Eq. (6) that the phase velocity of the gravitational wave with frequency ω is equal $C = g/\omega$. Let the frequency $2\,\omega_0$ present free waves, the phase velocity of which is equal $C_2 = g/\omega_0$, and the second harmonic the dominant wave, which propagate with the phase velocity $\mu = S_2/S_1$. The influence of the harmonic level, which can characterized by the parameter $\mu = S_2/S_1$, is shown in Fig. 3. The calculations were carried out for the situation when the distance between

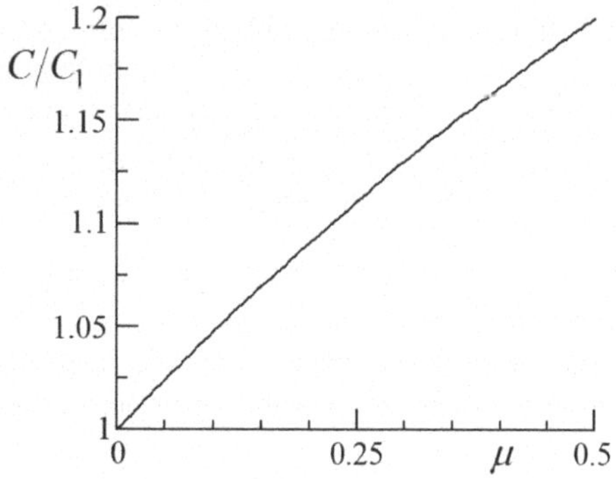

Fig. 3 Effect of the second harmonic of gravitational waves on the phase speed C calculated for a two-component wave field

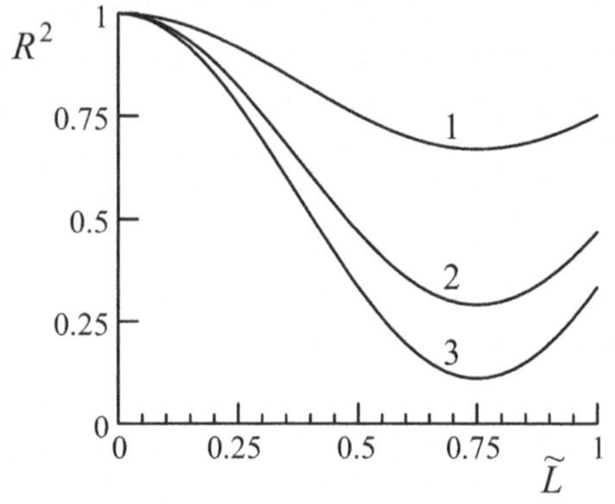

Fig. 4 The dependence of the level of coherence R^2 from the distance $\tilde{L}$ at the frequency of the second harmonic. The curves 1, 2, and 3 correspond to the parameter μ values equal to 0.1, 0.3 and 0.5

two points for which the phase shift is determined is small, and the expression (4) is true.

As the distance between the points at which the phase shift is measured increases (see Figs. 1 and 2), the influence of the second component on the magnitude of the phase shift is reduced. At the same time, as shown in Fig. 4, with increasing distance, there is a change in the level of coherence. With the decrease of coherence, confidence intervals for phase shift are extended, with zero coherence, phase shift estimates are uniformly distributed over the interval $(-\pi, \pi)$ (Jenkins and Watts 1972).

5 Conclusion

It is shown that in a wave field representing a superposition of two linear fields in which the waves obey different dispersion equations, the phase shift determined at two points of the wave field depends nonlinearly on the distance between these points. The consequence of this effect is the dependence of the phase velocity estimates on the distance between the points X_1 and X_2 for which the phase shift was determined. If the distance between the points X_1 and X_2 is small, then the phase shift is equal to the average phase shift for the two components. Moreover, averaging occurs with a weight proportional to the level of the spectrum of each component.

For the field of surface waves including gravitational waves and their harmonics, the influence of the harmonics of dominant wave on the phase velocity determination is investigated. It is shown that for gravitational surface waves, the estimates of the phase velocity at the scale of the second harmonic of the dominant wave deviate from the value corresponding to the dispersion equation by no more than 20% with the ratio of the harmonic level less than 1:2.

Acknowledgements The study was conducted with the financial support of RFBR grant No. 17-35-50030/17 mol_nr, and with state assignment Nos. 0827-2018-0003 and 0149-2018-0013.

References

Efimov VV (1981) Dynamics of wave processes in the boundary layers of the atmosphere and the ocean. Naukova dumka, Kiev, 255p

Efimov VV, Soloviev YuP, Khristoforov GN (1972) Experimental determination of the phase velocity of propagation of the spectral components of sea wind waves. Izv Acad Sci USSR Atmos Ocean Phys 8(4):435–446

Jenkins G, Watts D (1972) Spectral analysis and it's applications, vol 2. Mir, Moscow, 287p

Kononkova GE, Pokazeev KV (1978) Experimental study of the dispersion relation for the components of the frequency spectrum of wind waves. Vestn Mosk Univ Ser 3 Phys Astron 19(1):121–123

Lamley DL, Panovsky GA (1966) The structure of atmospheric turbulence. Mir, Moscow, 264p

Litvin EN, Pokazeev KV, Tuporshin VN (1992) Infragravity waves in the mouth of the Kamchatka River. Oceanology 32(2):218

Saprykina YaV (2015) Experimental investigations of role of nonlinearity in formation of infragravity waves in coastal zone. Process Geogeomed 14(1):67–74

Yuen G, Lake B (1987) Nonlinear dynamics of gravitational waves in deep water. New in foreign science. Mechanics, no 41. Mir, Moscow, 179p

Zapevalov AS (1996) On the estimation of the angular energy distribution function of dominant sea waves. Izv Atmos Ocean Phys 31(6):802–808

Zapevalov AS, Bol'shakov AN, Smolov VE (2004) Studies of the coherence level of sea surface waves. Izv Atmos Ocean Phys 40(4):483–487

A Model of a Coastal Barrier and Its Application to the Anapa Bay-Bar Coasts

I. O. Leont'yev and T. M. Akivis

Abstract A model of a coastal barrier following the sea level changes is proposed. The model is applied to the Anapa Bay-bar. Its coastline retreats landward in spite of sediment accumulation due to longshore flux discharge. Modeling shows that sea level rise generated by tectonic processes can be the cause of the phenomenon. The coastal barrier follows the sea level rise and moves landward.

Keywords Anapa Bay-bar · Coastal barrier · Sea level · Longshore sediment flux

1 Introduction

The Anapa Bay-bar is an accumulative body in the NW part of the Black Sea. It's length is about 47 km and the width varies from 100 m in the northern part to 1500 m in the south. The bay-bar is relatively closed litho-dynamic system. It is composed mainly of sand with shelly admixture. The former sources of material for the bay-bar were sand sediments of paleo Kuban-river together with biogenic shell sediments and erosion of coastal base rocks and Aeolian forms. Nowadays these sources are largely depleted. Moreover, the area undergoes significant technogenic pressure because of intensive economic activity. Observations testify increase of destructive processes on the Anapa Bay-bar coast. This paper analyzes actual sediment fluxes and probable sea level changes in order to explain the observed trends.

2 Coastal Dynamics

Modelling by LONT-2D model (Leont'yev 2014) with the recent wave climate data from Russian Maritime Register of Shipping (Lopatukhin et al. 2006) shows that

I. O. Leont'yev · T. M. Akivis (✉)
Shirshov Institute of Oceanology RAS, Moscow, Russia
e-mail: akivis@yandex.ru

I. O. Leont'yev
e-mail: igor.leontiev@gmail.com

C. T. Olegovna (ed.), *Processes in GeoMedia—Volume I*,
Springer Geology, https://doi.org/10.1007/978-3-030-38177-6_5

main sediment flux along the Anapa Bay-bar coast is directed eastward and gradually decreases along its way. The cause is a gradual turn of the coastline to the south and corresponding change of wave approaching direction (Fig. 1).

The flow capacity in the western part of the bay-bar is 150–200 thousand m^3/year. Closer to the s. Vityazevo range sediment transport fades away and further to the east, the material is transported in opposite direction. The flow capacity near s. Dzhemete approximates 100 thousand m^3/year. Thus, there is a convergence of fluxes near the s. Vityazevo range. Even for non-saturated (due to lack of material) flux it should result in predominant sediment accumulation along the coast (Fig. 1).

At the same time, observational data by Krylenko (2015) give evidence of coastal scour dominance. The most noticeable retreat is registered in Vityazevo-Dzhemete, Bugazsky Bar and Lake Solyenoe areas (Fig. 1).

This discrepancy between observations and modeling is evidently caused by some factors antagonizing accumulation. The most probable reason of the coastline retreat near Lake Solyenoe and Bugazsky and Vityazevsky Limans is a relative sea level rise. Near the Blagoveschensky outlier there is no visible coastline changes and the sea level is most likely relatively stable.

The probable relative sea level behavior can be associated only with tectonic moves of the adjacent land area. This consideration correlate with the conclusions

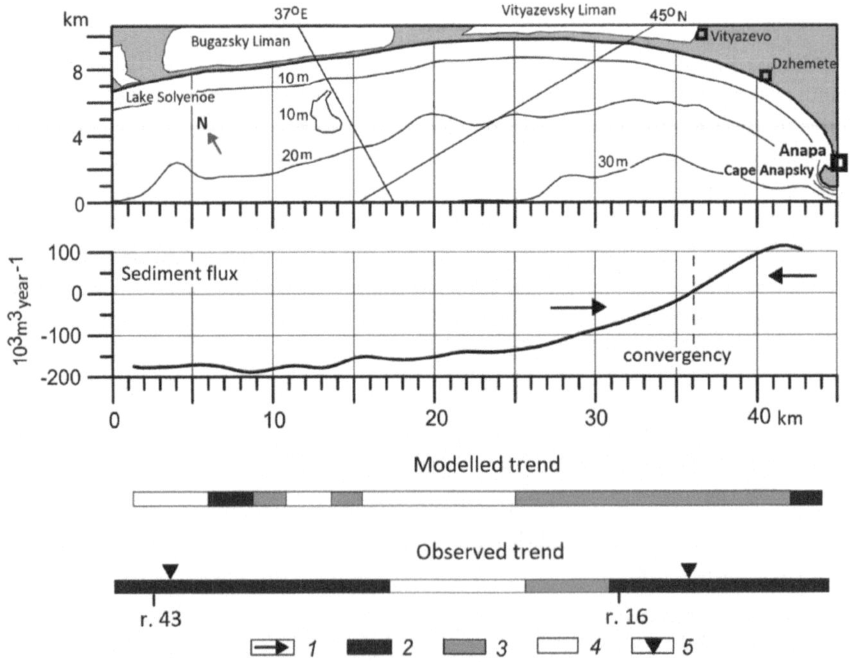

Fig. 1 A sketch of the research area, sediment fluxes and coastline change trends by modelling and observations. 1—flux direction, 2, 3 and 4—erosion, accumulation and minor changes, respectively, 5—sites of maximal observed coastline recession

by Izmailov (2005) about gradual submergence of coastal segments near Bugazsky and Vityazevsky estuaries and minor elevation of the Blagoveschensky outlier area. Kaplin et al. (1991) believe that the whole Anapa bay-bar undergoes submergence.

Further, we try to explain the observed behavior of the coasts using the model of coastal barrier development under relative sea level rise condition.

3 A Model of a Coastal Barrier

Trends of an accumulative coast development largely depends on a ratio between its profile slope $\overline{\beta}$ and the slope β_S of a base rock the accumulative body forms on (Leont'yev 2014; Cowell et al. 1995). In the context of the problem under consideration, we are interested in coastline evolution under sea level rise for two main cases, namely (a) the slopes β_S and $\overline{\beta}$ are close to each other and (b) β_S is significantly less than $\overline{\beta}$.

The schemes of corresponding configurations are shown in Fig. 2. Sea level rise causes move of the accumulative body along the base rock surface. Assume that the profile shape remains unchanged. The sediment properties and wave climate are constant too. Then in the first case (Fig. 2a) the material is transported both up and down along the slope maintaining zero sediment budget within the active profile. This is well-known Bruun type of evolution (Bruun 1988).

In the second case (Fig. 2b) we deal with the coastal barrier or bar (Kaplin et al. 1991; Zenkovich 1962) which moves landward following the sea level rise and slides along the base rock. The material is transported up the slope. The main mechanism of the movement is an overflow through the bar ridge during extreme storms.

In order to estimate the speed of the coastline displacement we use the mass conservation law that can be written as (Leont'yev 2012, 2014)

$$\frac{\partial h}{\partial t} = Er - Ac + w, \tag{1}$$

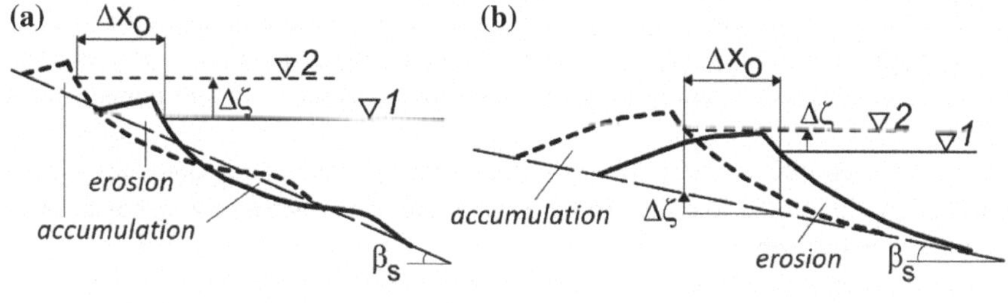

Fig. 2 A move of accumulative body profile while relative sea level rises **a** slopes β_S and $\overline{\beta}$ are close to each other, **b** β_S is significantly less than $\overline{\beta}$. 1 and 2 are successive sea level positions with the difference $\Delta\zeta$ which causes the coastline displacement to the distance Δx_0

where h is water depth, t is time (in a scale of years), Er and Ac are erosion and accumulation rates due to wave action, $\mathrm{w} = \frac{\partial \zeta}{\partial \mathrm{t}}$ is the rate of sea level ζ change. We integrate Eq. (1) over the total profile length. The boundaries are the points x_c and x_*. The former is the highest beach level z_c and the latter marks the marine end of the profile with the closure depth h_* where storm deformations of the seabed are negligible. We assume as before the shape of the profile unchanged during its move. It means the speed of all its points is the same. Therefore

$$\partial x_c/\partial t = \partial x_*/\partial t = \partial x_0/\partial t$$

The point x_0 determines the coastline position. As a result, we come to the equation (Leont'yev 2012, 2014)

$$\frac{\partial x_0}{\partial t} = \frac{B - wl_x}{z_A}, \quad z_A = h_* + z_c, \tag{2}$$

where z_A is a height of the active profile, B is a sediment budget, i.e. the difference between accumulation and erosion volumes in a time unit:

$$B = \int_0^{l_X} Acdx - \int_0^{l_X} Erdx \tag{3}$$

In case of Bruun evolution the sediment budget is balanced ($B = 0$), so the rate of the coastline move can be found from (2) as

$$\frac{\partial x_0}{\partial t} = -\frac{w}{\overline{\beta}}, \quad \overline{\beta} = \frac{z_A}{l_X}. \tag{4}$$

The relation (4) is the Bruun rule. Accordingly, the horizontal move of the coastline is directly proportional to sea level change and reversely proportional to an average profile slope $\overline{\beta}$. Sea level rise ($w > 0$) induces retreat of the coast: ($\partial x_0/\partial t < 0$).

In case of a coastal barrier, the sediment budget is not balanced. The point is a move of the form governed only by sediment transfer over the ridge to its rear part. Accordingly, the coast develops under permanent budget deficiency ($B < 0$). The balance recovers only when the sea level stabilizes and the sediment transfer from the slope stops.

If the sea level increases at $\Delta\zeta$ the profile shape can remain the same only for the barrier move at $\Delta x_0 = \Delta\zeta/\beta_S$. Therefore the rate of the barrier move has to obey the following relation

$$\frac{\partial x_0}{\partial t} = -\frac{w}{\beta_S}, \tag{5}$$

Obviously, this is an analogue of the Bruun rule (4) when the slope of a base rock is significantly less than the slope of a sand body ($\beta_S \ll \overline{\beta}$). It follows from (3) and (5) that the value *B*, namely the sediment deficiency due to the transfer over the barrier ridge, can be estimated as

$$B = z_A w \left(\frac{1}{\overline{\beta}} - \frac{1}{\beta_S} \right). \tag{6}$$

Absolute value of *B* increases with the difference between β_S and $\overline{\beta}$ and vanishes to zero when they are equal.

4 The Model Application to the Anapa Bay-Bar

Further, the model is applied to the Bugazsky and Vityazevsky Bars areas. The parameters $\overline{\beta}$ and z_A of the coastal profile are about 10^{-2} and 10 m respectively. Assume the base rock slope β_s is of order 10^{-3}. Then, according to (5), the observed rate of the coastline retreat $\partial x_0/\partial t$ of about –1 m/year can be reached if the sea level change rate *w* is about 1 mm/year. In this case, the material deficiency *B* at the marine coastal slope should be of order -10 m^3 m^{-1} year^{-1} by (6). These estimates look quite realistic.

As shown in Fig. 1, the total sediment flux falls from 120 thousand m^3/year to zero along the eastern Vityazevsky Bar. In other words, the sediment flux discharge supplies about 13 m^3 m^{-1} year^{-1} of material in case of saturated flux. Taking into account general lack of material we halve this value to 6 m^3 m^{-1} year^{-1}. It is clear that, in the case, the sediment flux discharge cannot compensate the material loss due to relative sea level rise (–10 m^3 m^{-1} year^{-1}). So the coast will retreat. It is even more relevant to the Bugazsky Bar area where the longshore gradient is minor (Fig. 1) and the flux discharge has nearly no effect on the coastal barrier retreat on account of the level rise.

There are two main mechanisms of sediment transfer from front to rear part of the barrier, namely overflow through the ridge during extreme storms and Aeolian transport.

Our focus is on the first mechanism. We model the storm effect on the retreating Bugazsky Bar coast (Fig. 1, Profile 43).

The model CROSS-P (Leont'yev 2014) was applied to 12 h long extreme western storm (average wave height $\overline{H} = 3.1$ m, period $\overline{T} = 8.5$ s) accompanied with 0.5 m level rise.

The modeling shows that storm waves cause an overflow of water through the coastal bar ridge (Fig. 3). The overflow erodes the front bar slope and transfer material to its rear part. As a result, the coastal line and the dune move landward.

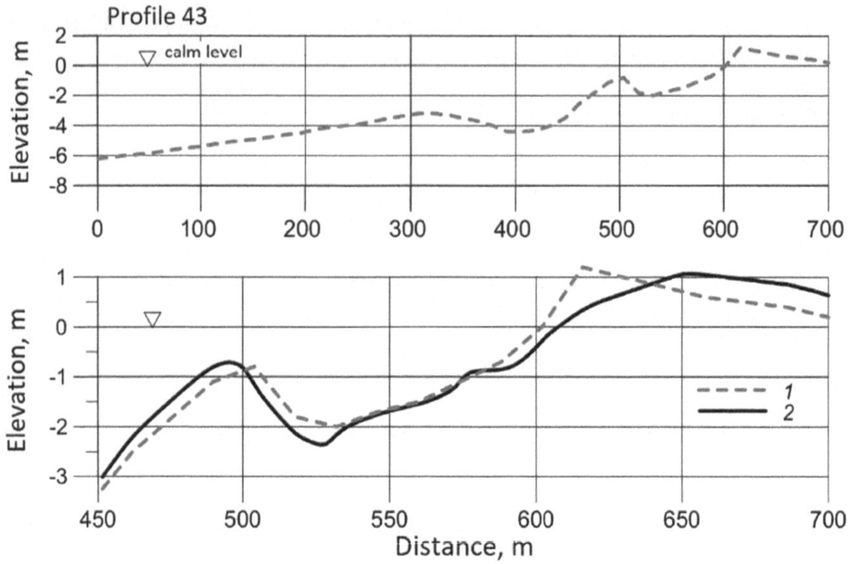

Fig. 3 Storm deformations of the Bugazsky Bar coast (profile 43) under extreme western storm action. 1 and 2 are the bed profiles before and after the storm, respectively

These results testify that the coastal barrier model based on a probable tectonic moves scenario is capable of both qualitative and quantitative explanation of the observed behavior of the Anapa Bay-bar coast.

5 Conclusions

The research shows that longshore sediment transport does not control the Anapa Bay-bar coastal dynamics. The coasts retreat despite of the longshore sediment fluxes discharge. The relative sea level rise due to tectonic processes was estimate as one of the phenomenon reasons. The model of coastal barrier following up the sea level rise explains Bugazsky and Vityazevsky coastlines retreat.

Acknowledgements This work was fulfilled in a scope of the Government assignment (Theme # 0149-2019-0005) and partially supported by RFBR (Grants ## 18-05-00741 and 18-55-3402 Cuba_t).

References

Bruun P (1988) The Bruun rule of erosion by sea-level rise: a discussion on large-scale two- and three-dimensional usages. J Coast Res 4(4):627–648

Cowell PJ, Roy PS, Jones RA (1995) Simulation of large-scale coastal change using a morphological behaviour model. Mar Geol 126:45–61

Izmailov YaA (2005) Evolutional geography of Azov and Black Seas coasts. Book 1. Anapa Bay-Bar. Lazarev Publ. House, Sochi, 174 p

Kaplin PA, Leont'yev OK, Luk'yanova SA, Nikiforov LG (1991) Coasts. Mysl', Moscow, 479 p

Krylenko VV (2015) Seashore dynamics of the Anapa bay-bar. Oceanology 55(4):742–749

Leont'yev IO (2012) Modeling beach profile evolution at centennial to millennial scales. Oceanology 52(4):550–560

Leont'yev IO (2014) Morphodynamic processes in the coastal zone of the sea. LAP LAMBERT Academic Publishing, Saarbrücken, 251 p

Lopatukhin LI, Bukhanovsky AB, Ivanov SV, Chernyshova ES (eds) (2006) Data on wind and wave regime on Baltic, North, Black and Mediterranean seas. Russian Maritime Register of Shipping, St. Petersburg, 452 p

Zenkovich VP (1962) Fundamentals of the sea coast development theory. RAS Publ. House, Moscow, 710 p

Testing a Mathematical Model of Air Movement in a Street Canyon Using the OpenFoam Package

M. V. Volik and N. S. Orlova

Abstract The paper presents the results of two-dimensional calculations of air flow in street canyons. The studies were conducted using the OpenFoam package with the support of the University Cluster program (http://www.unicluster.ru) and the Unihub Web Lab technology platform (https://unihub.ru). In order to determine which model of turbulence is appropriate to apply in further calculations, the results are compared with experimental data from different authors. It is shown that if the height of the house on the windward side of the street is less, then two-vortex structures air flow appear, at which the street ventilation significantly deteriorates.

Keywords Geoecological problems · Atmospheric pollution · Mathematical modeling · Street canyons · Aerodynamics · Turbulence · OpenFoam

At present, the ideas of environmental protection are the dominant social paradigm, which was formed under the influence of growing concern for the future of mankind. The observed environmental boom is the result of not only detrimental to human changes in the biosphere, but also the result of a certain reaction of the public consciousness, which began to approach an objective assessment of the place and purpose of man in the natural environment. The preservation of the human race is impossible without a renewed outlook and moral attitude of people not only in relation to each other, but also to nature. The consciousness of mankind is gradually coming to understand that growing industrialization without taking into account natural factors can give rise to phenomena whose destructive effect is comparable to the consequences of the use of nuclear weapons. The ecological front is at the forefront of the struggle for the survival of humanity on the same level with the settlement of regional conflicts, overcoming economic backwardness and other problems (Ivanov et al. 2016).

M. V. Volik (✉) · N. S. Orlova
Southern Mathematical Institute—the Affiliate of VSC of RAS, Vladikavkaz, Russia
e-mail: volikmv@mail.ru

M. V. Volik
Federal State-Funded Educational Institution of Higher Education "Financial University Under the Government of the Russian Federation", Moscow, Russia

C. T. Olegovna (ed.), *Processes in GeoMedia—Volume I*,
Springer Geology, https://doi.org/10.1007/978-3-030-38177-6_6

At present, air pollution is considered to be one of the main problems: geo-ecologists from various countries are engaged in the study of human exposure to the atmosphere. The reason for such interest are the largest global environmental problems ("greenhouse effect", violation of the ozone layer, acid rain). In addition, the atmosphere also performs the most important environmental function in protecting the Earth, for example, from solar radiation. In atmospheric air, global meteorological processes occur, affecting the formation of weather and climate, a large number of meteorites are delayed. At present, seriously damaged the capabilities of the natural systems are self-cleaning atmospheric air (Akimov et al. 2014; Paskhina 2013).

The importance of scientific research aimed at using a mathematical apparatus and developing software for studying the spread and accumulation of pollutants in the atmosphere of industrial cities is increasing. At present, mathematical modeling of the air flows and the spread of pollutants in urban areas in Russia and abroad is actively developing (Sukhinov et al. 2011; Kamenetsky et al. 2014; Kastner-Klein et al. 2001; Liu et al. 2004; Thangam 1991).

One of the popular applications for solving problems of continuum mechanics is an open integrable platform for numerical simulation—OpenFOAM (Open Source Field Operation And Manipulation)—a freely distributed computational fluid dynamics tool for scalar, vector and tensor field operations (OpenFOAM, http://www.openfoam.org). The tasks of mathematical modeling require a large amount of computer time. One way to overcome this problem is to use multiprocessor computing systems. The UniHUB platform via a web browser provides work with interactive software packages integrated into UniHUB, which occurs in the "cloud" infrastructure without the need to download or compile any program code. The UniHUB platform also provides access to training courses, feature articles, instructions for using programs and other necessary documentation. The ParaView package, currently integrated into OpenFOAM and UniHUB, is intended to visualize the calculation results in OpenFoam. ParaView allows you to build streamlines, graphs for one or more values, calculate average values for volume or surface, calculate pressure drop and export data to files (Resources Web-laboratory Unihub, http://desktop.weblab.cloud.unihub.ru/signin).

The main objective of this work is to find the fields of velocity and pressure during the motion of a turbulent incompressible fluid in street canyons and behind a reverse ledge. The standard solver for the unsteady incompressible turbulent flow pisoFoam was used for the calculations. In this case, LES and RANS turbulence models (kEpsilon, kOmega, and others) (OpenFOAM, http://www.openfoam.org) can be used. The sought values are: the averaged velocity U, the averaged overpressure, related to the density of the medium P and the specific kinetic energy of turbulence K.

Before starting the calculation, it is necessary to create a mesh using the blockMesh subroutine. The calculated mesh is generated from the blockMeshDict source file, which contains a scale factor, a list of vertex coordinates, a list of blocks and the number of cells in each direction, a list of areas (input, output, upper bound, lower bound).

In the computational mesh (Fig. 1), 28 vertices, 7 blocks and 12 boundary surfaces are defined. The calculated area is a street canyon. The height of the houses was

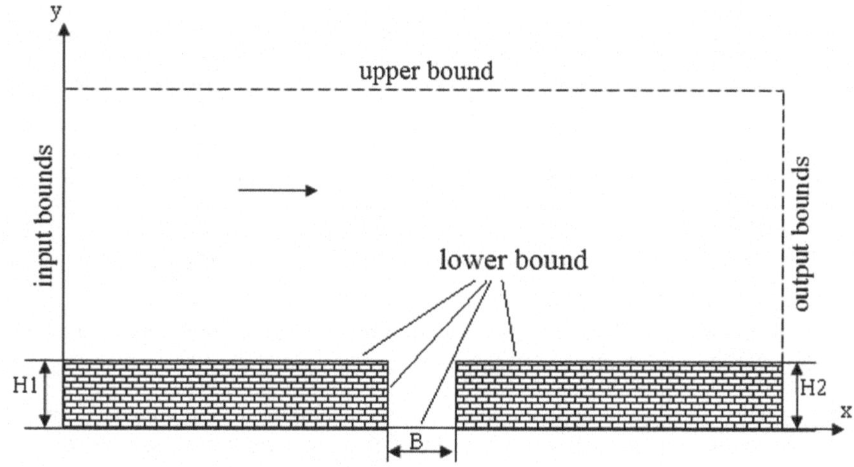

Fig. 1 Schematic image of the calculated mesh

assumed to be 20 m (H1 = H2). The distance from the entrance boundary to the street canyon and from the street canyon to the output boundary is 100 m. The width of street canyons is 20 m (B = H1). The distance from the lower boundary inside the street canyon to the upper boundary is 120 m. The calculations were carried out for a coarse mesh with a step of 1 m in the entire computational domain. A variant with a decrease in mesh steps in the lower part of the computational domain at a distance of up to 40 m from the bottom of a street canyon was investigated. The number of points in the first variant was equal to 220 × 120, and in the second variant 240 × 160. The calculations were carried out for the time interval from 0 to 1000 s with a step of 0.001 s. In addition, calculations were carried out for street canyons with a width equal to two heights of houses (B = 2 · H1) and with houses of smaller height on the windward side of the street.

It was assumed that the moving air is incompressible fluid. The system of equations included the continuity equation

$$\nabla U = 0$$

and the momentum change equation

$$\frac{\partial U}{\partial t} + \nabla \cdot (UU) - \nabla \cdot (\nu \nabla U) = -\nabla P.$$

In the calculations $K - \varepsilon$ and LES turbulence models were used.

For the standard $K - \varepsilon$ model: the equations for the kinetic energy of turbulence and its dissipation rate were solved [with the constants $\rho = 1.2$, $C_{1\varepsilon} = 1.44$, $C_{3\varepsilon} = 0.09$, $C_{2\varepsilon} = 1.92$ (OpenFOAM, http://www.openfoam.org)]:

$$\frac{\partial}{\partial t}(\rho K) + \frac{\partial}{\partial x_i}(\rho K U_i) = \frac{\partial}{\partial x_j}\left[\left(\mu + \frac{\mu_t}{\sigma_k}\right)\frac{\partial K}{\partial x_j}\right] + P_k - \rho\varepsilon$$

$$\frac{\partial}{\partial t}(\rho\varepsilon) + \frac{\partial}{\partial x_i}(\rho\varepsilon U_i) = \frac{\partial}{\partial x_j}\left[\left(\mu + \frac{\mu_t}{\sigma_\varepsilon}\right)\frac{\partial \varepsilon}{\partial x_j}\right] + C_{1\varepsilon}\frac{\varepsilon}{K}(P_k + C_{3\varepsilon}P_b) - C_{2\varepsilon 2\varepsilon}\rho\frac{\varepsilon^2}{K} + S_\varepsilon$$

For the standard LES model, the equation for the change in momentum was:

$$\rho\frac{\partial U}{\partial t} + \nabla\rho(UU) = -\nabla P + \nabla\tau; \quad \tau = 2\mu\left(S - \frac{1}{3}(\nabla U)I\right) + \frac{2}{3}KI.$$

The following boundary conditions were used (Resources Web-laboratory Unihub, http://desktop.weblab.cloud.unihub.ru/signin):

- for velocity: fixed value at the input (10 m/s); the output is zero gradient ($\frac{\partial U}{\partial n} = 0$); on the walls of houses and the bottom of the street canyon—the condition of sticking (equality of all components of velocity to zero) or near-wall functions (universal relations relating flow parameters to distance from wall); at the upper boundary, the velocity was assumed to be zero, since, on the one hand, at the distance from the bottom of the canyon to the upper boundary, it practically does not affect the flow pattern in the canyon, and on the other hand, this boundary condition is close to the conditions under physical modeling of the flow of street canyons in wind tunnels, the results of which were compared to the calculation results;
- for overpressure: at the inlet, top and bottom walls a zero gradient was taken ($\frac{\partial P}{\partial n} = 0$), the output was used a fixed value equal to zero;
- for energy turbulence: at the output, the upper and lower bounds was used a zero gradient ($\frac{\partial K}{\partial n} = 0$), at the input boundary, a fixed value of K = 0.375 was applied, which was calculated under the assumption that the turbulence at the entrance is isotropic and equal to the fluctuations of the average velocity equal to 5% of its value (OpenFOAM, http://www.openfoam.org).

As a result of calculations using the LES model of turbulence and sticking conditions on the walls, it was found that for large and small meshes inside the street canyons, main eddies and secondary eddies are formed in the lower part on both sides of the canyons. The velocity of air flow in the vortices is greater in the case of a fine mesh. If the near-wall functions are used in the calculations, only more intense main eddies are formed inside the street canyons. And in this case, the flow velocity is greater in the case of a fine mesh. The centers of the vortices are located at a height of 10 m under the condition of sticking and 13 m with the use of wall functions. The turbulence energy is higher if the near-wall functions and the fine mesh are used in the calculations.

Calculations using the $K - \varepsilon$ model of turbulence without wall functions showed that inside the street canyons there is one vortex, whose center is formed at a height of 14 m. There are no secondary eddies. The center of the vortex drops to a height of 10 m. The speed of air movement increases significantly, the difference between the results of calculations for large and small mesh becomes noticeable. The turbulence

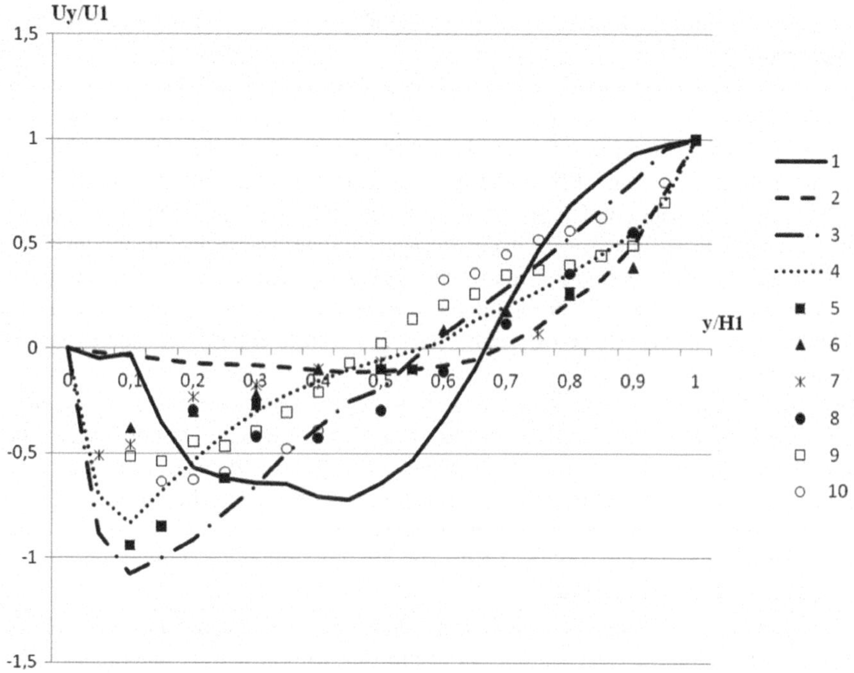

Fig. 2 The distribution of the normalized horizontal component of air velocity along the height in the centers of street canyons

energy for the K model is significantly lower if the near-wall functions are used in the calculations and little depends on the size of the mesh steps. Since a decrease in mesh steps to 0.5 m in a street canyon and directly above it affects the quantitative results of calculations, only results for a crushed mesh are given in the following.

Since the values of air velocity are obtained substantially depending on the model of turbulence and the type of boundary conditions, it is advisable to compare the obtained results with experimental data.

Figure 2 shows the distribution of the horizontal component of the air velocity along the height in the center of the street canyons. Graph 1 corresponds to the results of calculations with the LES model of turbulence without wall functions, graph 2 with the $K-\varepsilon$ model of turbulence without wall functions, graph 3 and 4, respectively, with LES and $K-\varepsilon$ models of turbulence with wall functions, and icons 5 to 10—experimental data (Kastner-Klein et al. 2001; Thangam 1991; Uehara et al. 2000; Kovar-Panskus et al. 2001, 2002; Li et al. 2008), respectively. The velocity of air at the point located in the centers of street canyons at the level of the roofs of houses was chosen as the calibre of velocity $U1$.

The large scatter of experimental data makes it difficult to choose an adequate model of turbulence and boundary conditions. But, since, for practical purposes, the most interesting is the air velocity in the lower part of the street canyon, the most acceptable results are obtained when using the $K - \varepsilon$ model of turbulence with wall functions.

A comparison is also made with experimental data (Li et al. 2008) of the results of calculations of the vertical component of air velocity across the width of a street canyon at a height equal to 0.5 of the height of houses. On the leeward side, results of calculations using the LES-model without near-wall functions correspond with the experimental data are. On the windward side, with the experimental data are in good agreement the calculations using the $K - \varepsilon$ model and near-wall functions.

In addition, a comparison was made of the calculated distribution of the horizontal component of air velocity over the height in street canyons, the width of which was assumed to be equal to two house heights ($B = 2 \cdot H1$) with experimental data (Kovar-Panskus et al. 2002). A vertical cross section located in the center of a street canyon was investigated. Note that, in the calculations for this case, one vortex is obtained, filling the street canyon, except for the case with the LES model of turbulence and the adhesion condition, which shows the formation of several vortices of different sizes. In the latter case, the picture is not completely stationary and the vortices change their size over time. In the lower part of the street canyon, the $K - \varepsilon$ model with near-wall functions gives a slightly better result, and in the upper part—the $K - \varepsilon$ model with a sticking condition.

In general, the results of calculations show that, despite the large scatter of experimental data, the $K - \varepsilon$ model of turbulence with near-wall functions can be considered more acceptable for street canyons with houses of the same height.

The work involved calculations in which the height of houses on the windward side of the streets (H2) is 0.75, 0.5 and 0 of the height of houses on the leeward side.

For comparison with experimental data (Thangam 1991) in the variant with "reverse ledge" ($H2 = 0$), a computational mesh similar to the size of the experimental setup was used: the distance from the input boundary to the edge of the ledge was two ledge heights, the distance from the edge of the ledge to the exit boundary was ten heights, the distance from the lower boundary of the computational domain to the upper is three heights. The calculations were carried out using the models of turbulence and boundary conditions given earlier. Only in the case of using the $K - \varepsilon$ model of turbulence and near-wall functions, a steady flow over a step is obtained with the formation of one vortex with a size equal to 5.75 from the height of the step. The experiment showed that behind the "reverse ledge" a vortex is formed, the dimensions of which are seven ledge heights.

If the houses are lower on the windward side of the street canyons, two modes of air flow are possible with the formation of one or two vortices (Fig. 3).

The transition from a single-vortex pattern air flow to a two-vortex one occurs when the height of the house on the windward side of the street decreases. Depending on the choice of the turbulence model and the boundary conditions, the height of these houses, at which the transition occurs, is different. When using the $K - \varepsilon$ model of turbulence and of the near-wall functions, the two-vortex structure occurs only when $H2 = 0.5 \cdot H1$. Without the near-wall functions, the two-vortex structure does not arise at all with such a model of turbulence. When using the LES model of turbulence, a two-vortex structure occurs under any boundary conditions already at

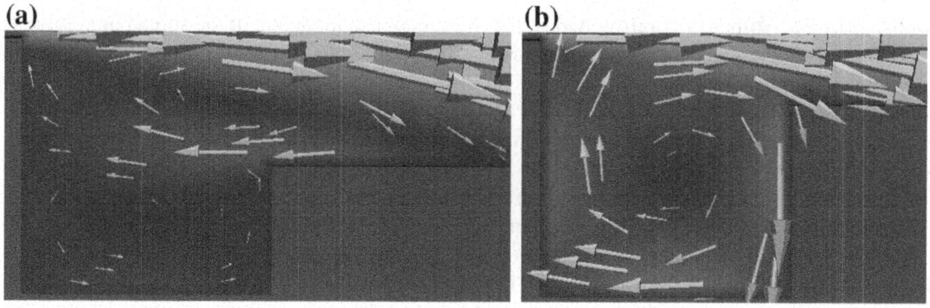

Fig. 3 A fragment of the air flow with the formation of a two-vortex (**a**) and single-vortex (**b**) structure

$H2 = 0.75 \cdot H1$. The distance from the street canyon to the point of attachment of the upper vortex essentially depends on both the boundary conditions and the turbulence model. At $H2 = 0.75 \cdot H1$, the transition from the wall functions to the sticking condition for the LES model of turbulence leads to a change in this distance from 10 to 70 m. If $H2 = 0.5 \cdot H1$, then the corresponding change in the boundary conditions for the same model of turbulence changes the distance to the sticking point from 50 to 100 m. $K - \varepsilon$ model with near-wall functions reduces the distance to the sticking point to 20 m.

The air flow has a significant effect on the distribution and accumulation of gaseous pollutants. The study of the aerodynamics of buildings of different configurations will determine the most unfavorable areas of street canyons in terms of the accumulation of pollutants (Volik 2016; Volik et al. 2016).

As a result of computational experiments in a two-dimensional formulation, a two-vortex air flow regime, not described in the literature, was found when flowing around a street canyon with lower houses on the windward side, for streets whose width is not less than the height of houses. This mode is observed regardless of the choice of the model of turbulence when using the near-wall functions. In the case of sticking conditions on the walls, such a structure appears only in the LES-model of turbulence. Due to a significant deterioration in the ventilation of a street canyon in the case of a two-swirl structure, it seems necessary to conduct experimental studies of the flow around a street canyon with a smaller height of houses on the windward side to determine the moment of transition to the two-swirl flow structure.

References

Akimov LM, Vinogradov PM, Akimov EL (2014) Comprehensive assessment of the environmental situation, taking into account the state of the atmosphere and the functional planning structure of the city. Vestnik VSU, Series: Geography. Geoecology 4:57–67

Ivanov VA, Katunina EV, Sovga EE (2016) Estimates of anthropogenic impacts on the ecosystem of the waters of the Herakleian Peninsula in the area of the location of deep sewers. Process Geomedia 1(5):62–68

Kamenetsky ES, Volik MV, Tagirov AM (2014) Mathematical modeling of the spread of pollutants emitted by motor transport. News of the Kabardino-Balkarian Scientific Center of the Russian Academy of Sciences 6(62):23–32

Kastner-Klein P, Fedorovich E, Rotach MW (2001) A wind tunnel study of organised and turbulent air motions in urban street canyons. J Wind Eng Ind Aerodyn 89:849–861

Kovar-Panskus A, Louka P, Sini J-F, Savory E, Czech M, Abdelqar A, Mestayer PG, Toy N (2001) Influence of geometry on the flow and turbulence characteristics within urban street canyons—intercomparison of wind tunnel experiments and numerical simulations. In: Proceedings of the 3rd international conference on urban air quality, Loutraki, Greece

Kovar-Panskus A, Louka P, Sini J-F, Savory E, Czech M, Abdelqari A, Mestayer PG, Toy N (2002) Influence of geometry on the mean flow within urban street canyons—a comparison of wind tunnel experiments and numerical simulations. J Water Air Soil Pollut Focus 2:365–380

Li X-X, Dennis Y, Leung C, Liu C, Lam KM (2008) Physical modeling of flow field inside urban street canyons. J Appl Meteorol Climatol 47:2058–2067

Liu CH, Barth MC, Leung DYC (2004) Large-eddy simulation of flow and pollutant transport in street canyons of different building-heigth-to-street-width ratios. J Appl Meteorol 43:1410–1422

OpenFOAM—UserGuide 1.7.1 [Electronic resource]. Access mode: http://www.openfoam.org

Paskhina MV (2013) Geoecological assessment of urbanized areas for urban planning purposes: Diss. … Cand. geographer. Sciences: 25.00.36. Moscow, 153 p.: ill

Resources Web-laboratory Unihub [Electronic resource]. Access mode: http://desktop.weblab.cloud.unihub.ru/signin

Sukhinov AI, Chistyakov AE, Khachunts DS (2011) Mathematical modeling of the movement of multicomponent air and transport of pollutants. News of the Southern Federal University. Theoretical and applied questions of mathematical modeling 8(121):73–79

Thangam S (1991) Analysis of two-equation turbulence models for recirculating flows. http://www.dtic.mil/dtic/tr/fulltext/u2/a240683.pdf

Uehara K, Murakami S, Oikava S, Wakamatsu S (2000) Wind tunnel experiments on how thermal stratification affects flow in and above urban street canyon. Atmos Environ 34:1533–1562

Volik MV (2016) Numerical modeling of the spread of pollutants emitted from low-lying sources inside streets. News of the Kabardino-Balkarian Scientific Center of the Russian Academy of Sciences 1(69):20–27

Volik MV, Kamenetsky ES, Kusraev AG, Orlova NS, Khubezhty ShS (2016) Study of the influence of street length and height of houses on the movement of air. Geology and Geophysics of South Russia 4:31–38

The Manifestation of Caldera-Forming Volcanism in the Formation of the Coast (On Example of Iturup Island of the Great Kuril Ridge)

V. V. Afanasiev, N. N. Dunaev, A. O. Gorbunov, and A. V. Uba

Abstract The results of geological and geomorphological studies of the central part of the Iturup island of the Great Kuril ridge are considered. The study area is represented by the Vetrovoy Isthmus, emphasizing the structural-tectonic saddle of the island. Composed of volcanic pumice stones, the isthmus is a natural model of the origin of the marine coastal zone as a result of the manifestation of underwater caldera-forming volcanism. The received data indicate several major emissions pumiceous pyroclastic material in late Pleistocene–Holocene. Modern sea shore retreat velocities of the isthmus are estimated to exceed 0.5–1 m/year.

Keywords Coast · Sea level · Volcanism · Caldera

1 Introduction

The islands of the Great Kuril Ridge (GKR) are characterized by intensive modern geological processes, which is expressed in the high-amplitude movements of the Earth's surface along the faults, manifestations of intense volcanism, increased seismicity and other phenomena. Such island systems are the key structural elements of the geodynamic concept of plate tectonics, the areas of active interaction between the matter of the Earth's crust and mantle. They are also interesting in practical terms, as these areas are associated with a variety of mineral deposits. Therefore, much attention has been paid to the study of island arcs, but many aspects of their intrinsic nature and functioning remain unclear. This also applies to their coastal morphosystems. In the presented paper features of a geomorphological structure and development of sea coast area of an initial volcanotectonic origin in Holocene are considered.

V. V. Afanasiev (✉) · A. O. Gorbunov · A. V. Uba
Institute of Marine Geology and Geophysics, Far Eastern Branch Russian Academy of Sciences, Yuzhno-Sakhalinsk, Russia
e-mail: vvasand@mail.ru

N. N. Dunaev
Shirshov Institute of Oceanology, Russian Academy of Sciences, Moscow, Russia

C. T. Olegovna (ed.), *Processes in GeoMedia—Volume I*,
Springer Geology, https://doi.org/10.1007/978-3-030-38177-6_7

2 Iturup's Location Characteristic

The largest in the Kuril Arc System, Iturup Island is a part of GKR formed by Neogene and Quaternary-age formations representing the island-arc structural tier. This tier include volcanogenic, volcanogenic-sedimentary and sedimentary deposits broken through by numerous relatively small extrusive and subvolcanic bodies and dikes of a wide petrographic range—from basalts and dolerites to rhyolites and granites.

In the relevant literature, GKR is considered as a fragment of the *Kuril-Kamchatka Volcanic Arc* formed as a lithodynamic type of volcanic-magmatic activity, confined to the seismofocal zone of the Pacific lithospheric plate feat under the Eurasian plate (Fig. 1) and superimposed on the Kuril Island Arc with a "granite" layer of the continental crust in its bedding (Ermakov 1997; Ermakov 2009). The rocks of the pre-island-arc complex in Big Kuril Area are not expressed (Atlas of the Kuril Islands 2009). It is possible that the formation of volcanic mountains here in the Quaternary period occurred on the ancient East Asian landmass submerged under the sea level, formed by Paleozoic and Mesozoic rocks.

In the Miocene-Pliocene as the earliest period of conditions for this area formation, the latter was a sea basin with intense hydro-volcanic eruptions. From the second half of the Pliocene, block tectonics is activated and individual blocks are raised above sea level. At the end of Pliocene-Early Neo-Pleistocene, the general rise of the area continues. In Eopleistocene-beginning of Middle Neopleistocene subaqueous conditions prevail. Subaerial conditions are progressing from the Middle Neopleistocene to the present day, and active volcanism is taking place with weakening of the effusive and strengthening of the explosive forms, especially in the last 50 thousand years (Atlas of the Kuril Islands 2009; Bazanova et al. 2016). At this time, the amplitude of vertical tectonic movements increases (Bulgakov 1994).

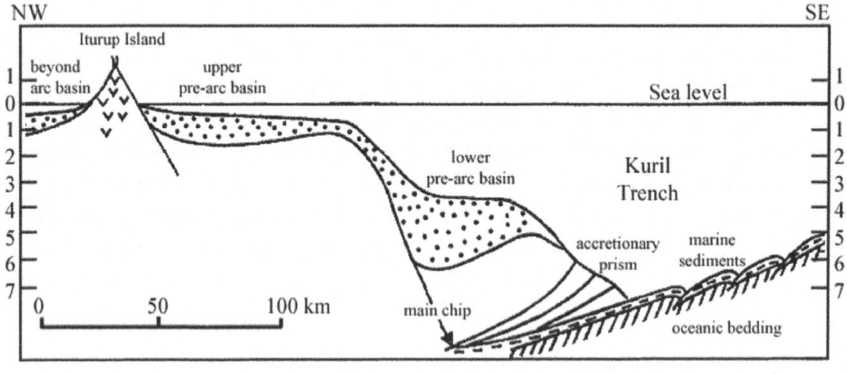

Fig. 1 Cut through the Kuril Island Arc (Zonenshain and Kuzmin 1993)

3 Geological Structure of the Iturup Island

Iturup Island is composed of the entities of the abovementioned island-arc structural tier. Neogene sediments are dominated by tuffogenic and effusive pyroclastic deposits of acid composition, and the quaternary system is characterized by basaltoid and andesite complexes, as well as products of their exogenous processing. Volcanogenic-sedimentary and sedimentary material has formed aeolian, deluvial, alluvial, lake, coastal-marine, proluvial, eluvial and biogenic varieties. At the same time, stratigraphy, genesis and chronology of Holocene sediments about. The Iturup Island has been studied rather poorly and are based mainly on a series of C^{14} datings from the deposits on the shores of Kasatka Bay and Olya Bay (Bulgakov 1996). The formation of the main structures of the GKR is due to the mobility of tectonic blocks (Kostenko et al. 1998), which is expressed along the longitudinal and sectional Iturup Island faults. Kinematically large number of dump-shifts are important.

4 Characteristics of the Research Area

The area of our direct study is represented by the intermountain saddle, the central part of which is expressed by the accumulation Vetrovoy (Wind) Isthmus with the area of just over 20 km^2 (Fig. 2). The isthmus is predetermined by a NW graben with width of about 12 km, in the middle part of which an underwater pyroclastic volcanic eruption took place, accompanied by a discharge of pumice material and a collapse calder of 6–7 km in diameter (Gorshkov 1967). Currently it is the narrowest (6.4–10 km) and lowest, covered with a layer of pumice with a capacity of more

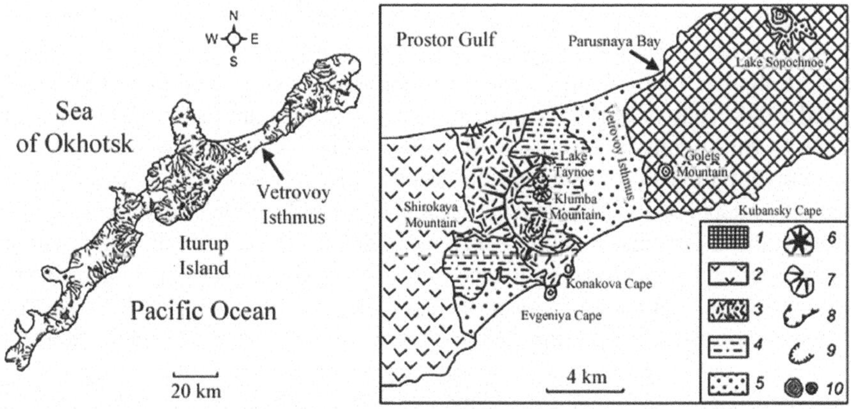

Fig. 2 Location and geological scheme of Vetrovoy Isthmus (Smirnov et al. 2017): 1—fundament; 2—effusives of quaternary volcanoes; 3—pyroclastic sediments; 4—sediments of sea terrace of 80–120 m; 5—sediments of low sea terrace and beaches; 6—volcanic structures; 7—volcanic structures strongly dissected; 8—caldera ledges; 9—erosion ledges; 10—extrusion domes

than 200 m, territory of the island connects its northern and southern megablocks. Geomorphologically, the isthmus is represented by a late Neopleistocene terrace up to 12–15 m high (Grabkov and Ishchenko 1982).

The age of the pumice, which formes the layered thickness called *the Rokovskaya*, is determined by C^{14} as later Neopleistocene within 38,500 ± 500 years (GIN-7092) and 5350 ± 50 years (GIN-7094). According to other data, age is estimated at about 20,000 years on basis thermoluminescence analysis (Bulgakov 1994). This rocks were formed as a result of pyroclastic volcanism, and the internal structure is the result of metabasite melting. The composition of pyroclastics corresponds to moderately aluminous dacites and riodacites of normal series (Smirnov et al. 2017).

According to preliminary estimates, the volume of erupted rocks of the Vetrovoy Isthmus caldera is about 100 km^3 (Melekestsev et al. 1988). Probably, a part of the pumice strata is introduced during the formation of the caldera-forming eruption in the Prostor Bay, framing the isthmus of the Sea of Okhotsk (Melekestsev 2005).

5 Paleogeographic Situation of the Region Under Study

The contemporary evolution of coasts and littoral area is primarily associated to fluctuations of sea level. Regional features are determined by the tectonic regime and the geological-geomorphological structure of the coastal area, the climate, the balance and the dynamics of load in the near-zero zone of the sea. For the area under study, sea level is the most controversial issue. Thus, some researchers claim that there are no traces of ancient (early-middle Pleistocene) coastlines, while others identify and attribute to this age a number of marine terraces with marks from 20–30 m to 200–250 m. Repeated attempts have also been made to identify Late Pleistocene elevations of marine and coastal-marine complexes, but disagreements remain. Thus, a terrace 25–30 m high is considered to be either Late Neopleistocene (Gorshkov 1967) or an Middle Pleistocene (Razzhigaeva et al. 2003). In the most substantiated actual work (Bulgakov 1996), there are traces of four transgressions on the coast of Iturup Island and the coast of Middle Neopleistocene (*Kargin* warming), the height of which increases from south to north from 10 to 60 m (Iturupskaya terrace) is confidently diagnosed. At the same time, the coastal accumulative forms of this age are located in the sea at a depth of 20–40 m, and the age and origin of Late Pleistocene terrace-like levels on the coast of the islands are incorrectly defined (Kaplin 2010). The situation is similar as for the sea level in Holocene: either it exceeded the current one up to 3.5 m or more (Korotky et al. 2000), or its excess is denied (Bulgakov 1994). No traces of quaternary glaciers—presumably of a semi-coverage scale of Middle (Early?) Pleistocene and a limited mountain-valley late Neopleistocene—have been found. There is also no common understanding of the latest vertical tectonic movements in the assessment of the manifestation of the newest vertical tectonic movements. For example, Markov (2009) is convinced that there are no signs of vertical displacements since Pliocene-Pleistocene time, and other researchers find confirmation of moves in the presence of elevated sea terraces and wave-surface niches above the sea level corresponding to their age (Korotky 2004).

The validity of the identification and correlation of ancient coastlines, as well as the morphology and dynamics of modern coastlines, is one of the most important problems in the study of the GKR's coastlines (Bulgakov 1994; Korotky et al. 2000; Alexandrova 1971; Kulakov 1973; Afanasiev 1992; Braitseva et al. 1995; Melekestsev et al. 1990; Korsunskaya 1958; Lutaenko et al. 2007; Tsukada 1986, etc.). In our studies, define the coastal zone (near-seazone) is the land adjacent to the shore as a frontal part of the seaside (seaside-zone), the exogenous specificity of which is determined, above all, by the microclimate, commonly referred to as the marine one. Here the peculiarities of atmospheric characteristics, micro-relief, vegetation, soil formation, some forms of the animal world are clearly evident. In this context, the Vetrovoy Isthmus is fully represented by the near-sea zone.

6 Overview of the Problem

It is known that for the development of the coast of the Kuril Islands, along with the factors considered above, large-scale caldera-forming volcanic eruptions were of great importance, which led to the arrival of millions of cubic meters of pumice pyroclastic material and tefra to the zone of wave processing of Late Neo-Pleistocene–Holocene (Bazanova et al. 2016; Braitseva et al. 1995; Melekestsev et al. 1990). Thus, on Iturup Island in the Late Neo-Pleistocene there were four large collapse calders, the formation of which, according to Melekestsev (2005), was erupted, 450 km^3 mainly dacite pyroclastics weighing 600×10^9 tons. The formation of the subaerial Vetrovoy Isthmus is explained by the accumulation of such material in the shallow strait between adjacent islands with powerful explosive caldera-forming eruptions (Korsunskaya 1958). As a result, a single island, Iturup, was formed.

The modern geomorphological appearance of the isthmus is associated with subsequent wave and aeolian processing of pyroclastics with accompanying appearance of transgressive sea and lagoon terraces.

Thus, taking into account the peculiarities of the volcanic and tectonic development of the GKR with reference to the problem of pyroclastic processing by coastal processes, we turn to the late Quaternary history, first of all, to the modern Holocene period of the island coast development by the example of the Vetrovoy Isthmus. As applied to this time, transgressive phases of sedimentation have been identified on the island, which are comparable with the Atlantic, Subboreal and Sub-Atlantic periods of the Holocene with a maximum sea level rise up to 3.5 m above the current level during its Atlantic warming epoch (Korotky et al. 2000). The epoch stands out for the region in the range of 6500–5000 thousand years (BP) (Korotky et al. 2000; Lutaenko et al. 2007), or 7000–4000 BP (Tsukada 1986) with an optimum of 5350 ± 50 years (Bulgakov 1996). This is consistent with the rhythms of coastal-sea deposition in the Holocene established earlier for the shores of the Far East seas (Melekestsev et al. 1990). However, the conclusions regarding aeolian sedimentation during periods of low sea level cannot be extended to dune formation regions, which is associated with the flow of estuary sediments into the coastal zone due to increased erosion of the coastline as a result of sea-level rise.

On the current geomorphological map of the Iturup Island (Grabkov and Ishchenko 1982) area of the Vetrovoy Isthmus is designated as the territory of modern cliffs and terraces of the Middle Pleistocene with heights of 12–15 and 30–40 m. This paper presents new data on the geomorphology of Vetrovoy Isthmus area, based on the results of field research in 2017. The aim of the work is to analyze the conditions of formation of the sea coast and the coast of the investigated territory in the late Neo-Pleistocene-Golocene from the point of view of the geomorphology of the sea coast depending on the regional conditions of the geological structure, rates of sedimentation, sea level fluctuations, peculiarities of tectonic shifts and volcanotectonic manifestations.

7 Research Methodology

Cartometric studies of the coastal-sea relief forms are carried out at the Vetrovoy Isthmus. The analysis of the Vetrovoy Isthmus relief is carried out through the visualization of data from the Aster Global Digital Elevation Model (GDEMV2), a product of the Ministry of Economy, Trade and Industry (METI) and NASA. Information is available in the United States Geological Survey's Earth Explorer system. WGS84 coordinate data system, pixel size ~20 × 30 m, standard error (RMSE) in height <10 m (https://lpdaac.usgs.gov/dataset_discovery/aster/aster_products_table/aster_gdem_version_2_validation). To visualize the levels, 3D maps of heights have been built with color separation of altitude ranges from 0 to 120 m with 10 m pitch and shaded relief image for higher levels. The profiles performed in the Baltic elevation system (BSV-77) show a more accurate picture in comparison with the previous (Grabkov and Ishchenko 1982; Kulakov 1973) studies of this direction. To substantiate the genesis and age of these surfaces, it is necessary to carefully search for coastal-marine sediments and to study both lithological and social features and organic residues. Nevertheless, the analysis of the elevation field allows to state the presence of several abrasion and abrasion-accumulative terrace levels in the pyroclastic sediments of the late Neo-Pleistocene, covered by coastal-sea and slope sediments.

8 The Results

In addition to the terrace levels, crest-like ridges up to 50 m high, oriented to the northwest are also mapped. Their origin and age are not quite clear, but the ridge closest to the Okhotsk Sea coast is a typical avandune. The base age of one of these dunes is defined as 1280 ± 110 years (Razjigaeva et al. 2002). It has also been established that the Iturupe Terrace, which was previously singled out, unites all terraced Late Pleistocene surfaces. The pebble cover of the terrace is covered by soil and pyroclastic sediments, the age of which is 31,400 ± 800 years according to C^{14} (GIN-7097). In the estuary zone of the Pemzovy Creek, cut into the sea terrace on depth of the contact with quaternary volcanoes' effusives and overlapping pumice pyroclastic deposits, the river terrace complex, morphometric parameters of which are shown in Fig. 3, was studied. The level of 22.7 m, determined by us, corresponds to the previously dated (Bulgakov 1996; Braitseva and Melekestsev 1974) 16–25 m terrace to the West of the creek with the age of 5350 ± 50 years (GIN-7094). Therefore, the three terraces of the Pemzovy Creek located below the floodplain should also be of Holocene age. On the south-western border of the Okhotsk Sea coast of the Vetrovoy Isthmus under the aeolian deposits up to 15–20 m thick, forming the avandune, the structure of a typical bank ledge is mapped (Fig. 4). In the middle part of the allocated aeolian sediments forming the coastal avandune, an ash interbed represented by white alevrite is revealed.

The formation of this layer according to C^{14} dating (Razgigaeva and Hansei 2006; Lyashchevskaya and Hansei 2011) occurred probably in the interval from 250 (250 ± 45, AA-20942) up to 800 years ago (800 ± 80, AA-20941; 880 ± 40, GIN-7344). Such a young age of the dune is also evidenced by a fragment of a tree trunk protruding from the pumice layer in the same coastal section at a height of 2.7 m. The angle of inclination of the pumice surface (supposed sea terrace) covered by aeolian sediments is 4°–5°. The peculiarity of the structure of the pumice of the bank ledge is the presence of the soil horizon covered with aeolian deposits with the capacity of up to 20 cm, which fixes the erosion cut along the pumice. Calibrated (calendar) according to C^{14} age of its base according to the sample selected by us is 1050 ± 70 years (LU-8685). The age value is given on the basis of the calibration

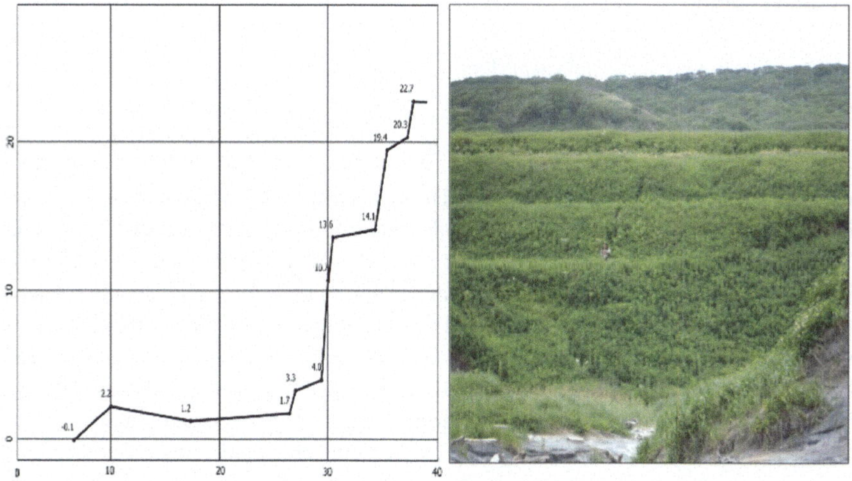

Fig. 3 River terraces at the estuary of the Pemzovy Creek

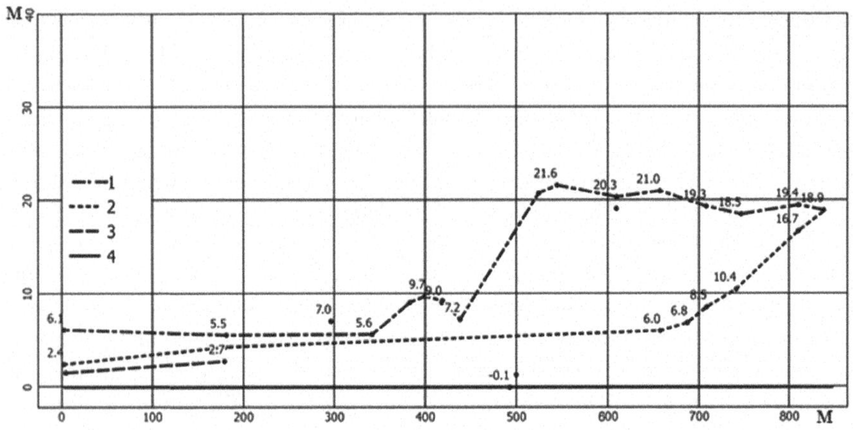

Fig. 4 Schematic view of the shore shelf structure at the South-Western border of the Okhotsk Sea coast of the Vetrovoy Isthmus: 1—aeolian deposits surface, 2—pumice bench surface, 3—position of the rear edge of the beach, 4—sea level (BSV-77)

program "OxCal 4.2" (Calibration curve "IntCal 13", http://c14arch.ox.ac.uk). In 1100 m to the South-West on the surface of a terrace-like ledge about 20 m wide at an altitude of 18–20 m, aeolian deposits were also found. The angle of inclination of the pumice surface, also the proposed sea terrace, under the aeolian deposits here has approximately the same values and is 5°–6°. This testifies to similar conditions of formation of the Vetrovoy Isthmus' Okhotsk coast.

9 Discussion

The data obtained indicate several large emissions of piroclastic materials near Vetrovoy Isthmus.

At the same time, the explosions of the early Holocene occurred at sea level, which at this time did not exceed the marks that were 40–50 m below its present location (Umitsu 1991; Siddall 2003). The pyroclastic sediments of late neopleistocene considered above were formed under conditions when the sea level was about 70–80 m lower than the present one (Nakada et al. 1991). Volcano-tectonic deformations associated with eruptions near the central part of the isthmus, where the remains of a large caldera are observed, have led to the collapse of the oldest pumice packs with a 2–3 m layer of beach material at their base. Layer incidence angles exceed 80° for azimuths falling towards the middle part of the Vetrovoy Isthmus (Fig. 5). This can be explained by the fact that during the collapse of the caldera there was a rise in the tectonic block of the adjacent area of the soma caldera by several tens of meters. The pumice, overlapping the ancient horizon, may also be associated with the formation of an underwater caldera in the Prostor Gulf (Melekestsev 2005).

Fig. 5 Subvertically occurring layers of pumice pyroclastic deposits on the shore of the Vetrovoy Isthmus

At least, the angles of slope of the pyroclastic flows surface indicate a higher relief in the northern part of the extensive pumice pyroclastic cover, the remains of which are currently observed at the Vetrovoy Isthmus (Smirnov et al. 2017). Average pumice fracture rates established by comparing the profile of the basement surface of the terrace facing the axial part of the isthmus and the surface of the terrace currently being abraded are significant. They can probably exceed 0.5–1.0 m/year. It is quite possible that the dislocated ancient pumice cover had a capacity of several hundred meters. Taking into account that the terraces of Middle-Holocene age in the southern part of the island are 2–3 m lower than the corresponding levels in the area of the Vetrovoy Isthmus, one can assume that the tendency of increasing the surface elevation of the marine Late Pleistocene terraces in the direction from south to north is also characteristic of the Holocene levels.

Acknowledgements The work was performed in accordance with the theme of State Assignment No. 0149-2018-0015 and with the support of the Russian Foundation for Basic Research, project No. 16-05-00364.

References

Afanasiev V (1992) Evolution of the Coast of the Far Eastern Seas in the Holocene. In: Evolution of the shores in conditions of raising the ocean level. MSU, Moscow, pp 166–174 (in Russian)

Alexandrova AN (1971) Some issues of the stratigraphy of quaternary sediments on. Iturup. Izv Sakhal ord VGO of the USSR 2:74–81 (in Russian)

Atlas of the Kuril Islands (2009) Moscow-Vladivostok: CPI “DIC”. 516 p (in Russian)

Bazanova LI, Melekessev IV, Ponomareva VV, Dirksen OV, Dirksen VG (2016) Volcanic catastrophes of the late Pleistocene–Holocene in Kamchatka and the Kuril Islands. Part 1. Types and classes of catastrophic eruptions—the main components of volcanic catastrophism. Volcanol Seismol 3:3–21 (in Russian)

Braitseva OA, Melekestsev IV, Ponomareva VV, Sulerzhitsky LD (1995) Ages of calderas, large explosive craters and active volcanoes in the Kuril-Kamchatka region, Russia. Bul Volcanol 57(6):383–402

Braitseva OA, Melekestsev IV (1974) The volume, sequence and questions of the methodology for studying the newest period of relief development. Kamchatka, Kuril Islands and the Commander Islands, pp 83–99 (in Russian)

Bulgakov RF (1994) The history of the development of the southern islands of the Great Kurile chain in the Pleistocene: Avtoref Dis Cand Geogr Sci. MSU, Moscow, 20 p (in Russian)

Bulgakov R (1996) Reconstruction of quaternary history of southern Kuril Islands. J Coastal Res 12(4):930–939

Calibration curve "IntCal 13". Christopher Bronk Ramsey. http://c14arch.ox.ac.uk (Date of appeal 19.03.2018)

Ermakov VA (1997) The island arc is a special tectonotype in the structures of the western frame of the Pacific Ocean (on the example of the Kurile–Kamchatka arc). In: Tectonics of Asia: materials of the XXX tectonic meeting. GEOS, Moscow, pp 92–94 (in Russian)

Ermakov VA (2009) Island and volcanic arcs. Problems of correlation and typification of arcs (on the example of the Kurile–Kamchatka and other regions). In: Vulcanism and geodynamics: materials IV All–Russian symposium on volcanology and paleovolcanology. Petropavlovsk–Kamchatsky, vol 1, pp 61–64 (in Russian)

Gorshkov GS (1967) Volcanism of the Kuril island arc. Moscow, Nauka, 183 p (in Russian)

Grabkov VK, Ishchenko AA (1982) Morphogenetic types of the relief of the Kurile Islands. In: Relief and volcanism of the Kurile island–arc system. DVNTS AN SSSR, Vladivostok, pp 13–24 (in Russian)

https://lpdaac.usgs.gov/dataset_discovery/aster/aster_products_table/aster_gdem_version_2_validation

Kaplin PA (2010) Questions of geomorphology and paleogeography of sea coasts and the shelf. MSU, Moscow, 617 p (in Russian)

Korotky AM, Razzhigaeva NG, Grebennikova TA, Mokhova LM, Bazarova VB, Sulerzhitsky LD, Lutayenko KA (2000) Sedimentation and paleolandscapes of the late Pleistocene-Holocene basin Kurilki (Iturup Island, Kuril Islands). Pacific Geol 19(5):61–77 (in Russian)

Korotky AM (2004) The sea terraces of the Kuril Islands: the influence of tectonics and glacio–elvetics (critical analysis of the state of the problem). In: Relief-forming processes: theory, practice, research methods. Institute of Geology and Geophysics SB RAS, Novosibirsk, pp 145–147 (in Russian)

Korsunskaya GV (1958) Kuril island ridge. Geografgiz, Moscow, 224 p (in Russian)

Kostenko NP, Korchuganova NI, Dunaev NN (1998) Modern geological structure of the Kurile–Kamchatka region. OCEANOL 38(2):293–304 (in Russian)

Kulakov AP (1973) Quaternary coastlines of the Sea of Okhotsk and the Sea of Japan. Nauka, Novosibirsk, 183 p (in Russian)

Lutaenko KA, Zhushchikhovskaya IS, Mikishin YA, Popov AN (2007) Mid–Holocene climatic changes and cultural dynamics in the basin of the Sea of Japan and adjacent areas. In: Anderson DG, Maasch KA, Sandweiss DH (eds) Climate change and cultural dynamics: a global perspective on Mid–Holocene transitions, pp 331–406

Lyashchevskaya MS, Hansei KS (2011) Development of the natural environment of the central part of Iturup Island in the middle-late Holocene (Kuril Islands). Vestnik KRAUNTS. Ser.: Earth Sci 1:120–130

Markov AY (2009) Geomorphological structure and geodynamics of the South Kurile island arc and forecast of oil and gas potential. In: Materials of the XVIII international scientific conference (school) on marine geology. GEOS, Moscow, pp 94–99 (in Russian)

Melekestsev IV, Braitseva OA, Ponomareva VV, Sulerzhitsky LD (1990) Ages and dynamics of development of the active volcanoes of the Kuril-Kamchatka region. Int Geol Rev 32(5):436–448

Melekestsev IV, Braitseva OA, Sulerzhitskii LD (1988) Catastrophic explosive eruptions of the Kurilo–Kamchatka region in the end of pleistocene–beginning of Holocene. Dokl AN SSSR 300(1):175–181 (in Russian)

Melekestsev IV (2005) The newest (N_2^2–Q_4) surface and underwater volcanism of the Kuril island arc. In: Novel and modern volcanism in Russia. Nauka, Moscow, pp 233–335 (in Russian)

Nakada M, Yonekura N, Lambeck K (1991) Late pleistocene and halocene sea-level changes in Japan: implications for tectonic histories and mantle rheology. Palaeogeogr Palaeoclimatol Palaeoecol 85(1–2):107–122

Razgigaeva NG, Hansei LA (2006) Conditions of sedimentation of island territories in the Pleistocene-Holocene. Dal'nauka, Vladivostok, 365 p

Razjigaeva NG, Korotky AM, Grebennikova TA, Ganzey LA, Mokhova LM, Bazarova VB, Sulerzhitsky LD, Lutaenko KA (2002) Holocene climatic changes and environmental history of Iturup Island, Kurile Islands, northwestern Pacific. The Holocene 12(4):469–480

Razzhigaeva NG, Grebennikova TA, Mokhova LM (2003) Middle Pleistocene coastal deposits of. Iturup, Kuril Islands. Pacific Geol 22(4):48–58 (in Russian)

Siddall M (2003) Sea-level fluctuations during the last glacial cycle. Nature 423(6942):853–868

Smirnov SZ, Rybin AV, Sokolova EN, Kuzmin DV, Degterev AV, Timina TY (2017) Acid magmas of caldera eruptions in the island of Iturup: the first results of the investigation of melt inclusions in phenocrysts of pumice caldera Lion's Maw and the Isthmus of isthmus. Pacific Geol 36(1):52–69 (in Russian)

Tsukada M (1986) Vegetation in prehistoric Japan: the last 20000 years. In: Pearson RJ (ed) Windows on the Japanese past: prehistory. Center for Japanese Studies, University of Michigan, Ann Arbor, pp 11–56

Umitsu M (1991) Holocene sea-level changes and coastal evolution in Japan. Quat Res (Daiyonki-Kenkyu) 30(3):187–196

Zonenshain LP, Kuzmin MI (1993) Paleogeodynamics. Nauka, Moscow, 192 p (in Russian)

Comparative Evaluation of Accuracy of Numerical Wave Models Based on Satellite Altimetry Data

V. G. Polnikov, N. S. Zilitinkevich, F. A. Pogarskii, and A. A. Kubryakov

Abstract We developed a technique of selection of along-track data of altimeters in accordance with their binding to the data of buoy measurements and data of two wave models: WAM4 and its modified version—WAM4-M. We estimated the root-mean-square deviations (RMSD) of the model data of the wind wave heights from the calibrated altimetric data in the Indian Ocean for the period 2013–2015. The RMSD were shown to have significant variability over space. We established advantages of WAM4-M over WAM4 model in terms of the standard and corrected RMSD for the different zones of the Indian Ocean and the entire ocean. The reasons for the variations of the RMSD of the two study models are discussed in dependence on the zone of the Indian Ocean.

Keywords Altimetry · Along-track data · Calibration · Verification of the wave model

1 Introduction

To assess the validity of numerical wave models on the basis of satellite data, we used along track altimetry data (Young et al. 2011, 2012, 2013).

For the entire Indian Ocean, we used modeling data of wind wave heights obtained for comparative verification of two numerical wave models: the original European version of WAM4 (Gunter et al. 1992) and its modification WAM4-M (Polnikov 2005; Polnikov et al. 2007).

Our objectives were: (1) to assess the significance of the spatial and temporal limits used to select altimetry data (A-data) in accordance with their binding to the buoy data (B-data) and data obtained from numerical modeling (M-data) (collocation);

V. G. Polnikov (✉) · N. S. Zilitinkevich · F. A. Pogarskii
Obukhov Institute of Atmospheric Physics of the RAS, Moscow, Russia
e-mail: polnikov@mail.ru

A. A. Kubryakov
Marine Hydrophysical Institute of the RAS, Sevastopol, Russia

Saint Petersburg State University, St. Petersburg, Russia

C. T. Olegovna (ed.), *Processes in GeoMedia—Volume I*,
Springer Geology, https://doi.org/10.1007/978-3-030-38177-6_8

(2) to perform two options for calibrating A-data with B-data: for each satellite individually and for all satellites together; (3) to estimate discrepancies between M-data and calibrated A-data (model verification) in the Indian Ocean for two wave models: WAM4 and its modification WAM4-M.

Our goal was to obtain objective information about the validity of the models; to find the advantages of the modified numerical model over the original one in terms of the statistical parameters in various zones of the Indian Ocean and the ocean as a whole; to analyze changes in statistical parameters obtained for these models in dependence on the Indian Ocean zone.

2 Materials and Methods

The most accurate error estimates of wind wave modeling data were found using the triple collocation method when all three types of data (A-, B- and M-data) were available and corresponded in time and space (Janssen et al. 2007). In our case, such type of correspondence was in A- and M-data for the period 2013-2015 and for A- and B-data for the period 2013–2016. Besides, all three types of data corresponded in time.

AVISO track data of the second processing level (http://www.aviso.altimetry.fr), including observations from four satellites over the period from March 2013 to December 2016, were used as altimetry data according to the significant wave height (H_S).

For the Indian Ocean, the modeling data of significant wave height obtained for our objectives were used. The calculations were performed for two versions of the numerical model: the original version of WAM4 (Gunter et al. 1992) and its modification—WAM4-M (Polnikov 2005).

The modeling was carried out on the basis of the European wind reanalysis, ERA-Interim (http://www.ecmwf.int/research/era/do/get/era-interim) for the period 2013–2015. The wind speed was taken with a resolution of 6 h in time and $1.5° \times 1.5°$ in space. The significant wave height was calculated with a resolution of 1 h in time and $1.5° \times 1.5°$ in space.

It was previously established that the entire Indian Ocean (IO) can be divided into six zones (the boundaries of the zones are shown in Fig. 1 and Table 1). This allowed us to conduct independent analysis of wave heights in such zones.

For calibration of the A-data, we selected 40 buoys in the Pacific and Atlantic oceans, located over 80 km away from the coast, with the data on hourly measurements of significant wave heights for the period 2013–2016 taken from the NDBC Buoy Database (http://www.ndbc.noaa.gov/historical_data.shtml). Additionally, we used data of the Australian buoy SOFS-4 located in the Southern Ocean, kindly provided by A. Babanin. Figure 2 shows a map of the locations of all the selected buoys.

The calibration process included 2 stages: (1) selection of A-data corresponding to the B-data in time and space within the specified spatial and temporal limits;

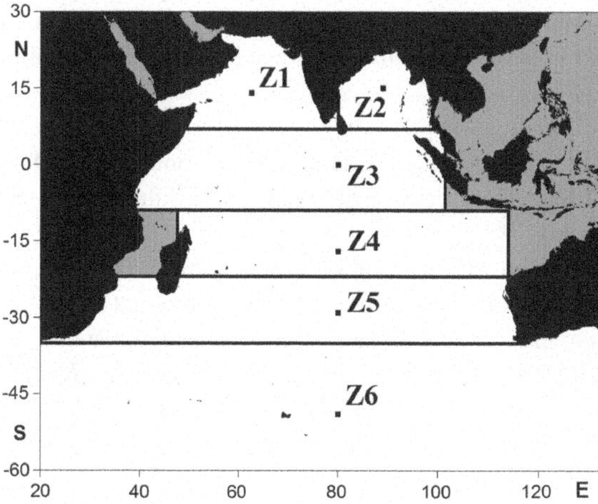

Fig. 1 Diagram of the spatial division of the entire water area of IO into 6 zones

Table 1 The boundaries of the zones in the Indian Ocean

No. zone	Boundaries of zones degrees of latitude; longitude
Z1	25 N–07 N; 40 E–80 E
Z2	25 N–07 N; 80 E–100 E
Z3	07 N–09 S; 35 E–105 E
Z4	09 S–22 S; 35 E–115 E
Z5	22 S–35 S; 20 E–115 E
Z6	35 S–60 S: 20 E–130 E

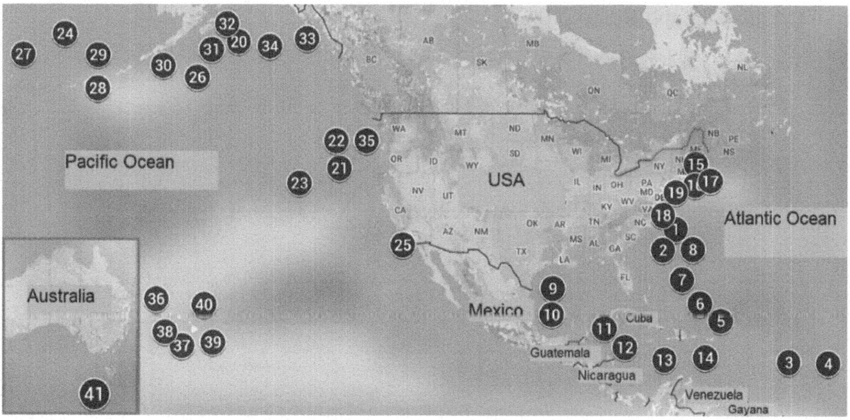

Fig. 2 Map of the buoys used for altimeter's calibration. Circles indicate the serial number of the buoy

(2) the matchup data of A-data were used to carry out the regression calibration against B-data. A range of the significant wave height used for calibration was $0 \leq H_S \leq H_{Smax}$.

The final model verification procedure was performed in 2 stages. First, similarly to the binding of the A-data to the B-data, we performed the binding of the A-data to the M-data. Then A-data calibrated based on B-data were compared with the corresponding M-data.

The assessment of the validity of the model, or its verification, included evaluation of common statistical characteristics of the significant wave height parameter. We used the following statistical parameters: bias of the means—*B*, root-mean-square deviation (RMSD)—*R* and the correlation coefficient—*K*. We also calculated corrected root-mean-square deviation—R_C (Hanna and Heinold 1985).

It was shown that the root-mean-square deviation may have a tendency to decrease for modeling results that are subject to negative bias (Mentaschi and Besio 2013). Using the geometric decomposition of the RMSD in its components (bias and scattering component), we can see that the RMSD is equal to the scattering component under zero bias, otherwise the bias affects the scattering component by increasing or decreasing the RMSD values (Mentaschi and Besio 2013). Smaller values of the RMSD do not always indicate that numerical modeling gives the good parameters (Mentaschi and Besio 2013). For this reason, the corrected RMSD was applied (Hanna and Heinold 1985). The lowest values of the corrected RMSD can be obtained for the modeling results that are not subject to bias. That means that the corrected RMSD have minimum values at zero bias and they increase with increasing bias. Therefore, this indicator provides more reliable information about the accuracy of the results of numerical models (Mentaschi and Besio 2013).

The formulas for calculating the statistical parameters of the variables were taken from (Mentaschi and Besio 2013):

$$\text{bias:} \quad B = \frac{1}{N}\sum_{i=1}^{i=N}(M_i - A_i) = \overline{M} - \overline{A} \tag{1}$$

$$\text{root-mean-square deviation:} \quad R = \left(\sum_{i=1}^{N}(M_i - A_i)^2/N\right)^{1/2} \tag{2}$$

$$\text{correlation coefficient:} \quad K = \frac{\left(\sum_{i=1}^{N}(M_i - \overline{M})(A_i - \overline{A})\right)}{\left[\sum_{i=1}^{N}(M_i - \overline{M})^2 \cdot \sum_{i=1}^{N}(A_i - \overline{A})^2\right]^{1/2}} \tag{3}$$

$$\text{corrected root - mean - square deviation:} \quad R_C = \left(\sum_{i=1}^{N}(M_i - A_i)^2 \Big/ \sum_{i=1}^{N} M_i A_i\right)^{1/2} \tag{4}$$

M_i and A_i mean *i* member of the rows of M- and A-data, N—the number of row members. The bar at the top of the variable indicates mean value.

3 Calculation Results

The selection of spatial-temporal ranges of conformity of A- and B-data (collocation).

We analyzed the spatial intervals from 0 to 99 km where the sections along the distance were such that each subsequent section increased by 3 km (3, 6, 9, …, 99). The time ranges were considered from −30 to +30 min where the time sections increased so that each subsequent section was larger by 6 min (0, 6, 12, …, 30). The rows of A- and B-data were compared calculating RMSD for each section along the distance and the time. Then, according to the criterion of the minimum of RMSD between the rows of A- and B-data, the optimal spatial-temporal limits were determined (Fig. 3).

Figure 3 shows the dependence of the RMSD on the time ΔT and distance ΔX differences between A and B data. On the basis of the RMSD, we selected the limits

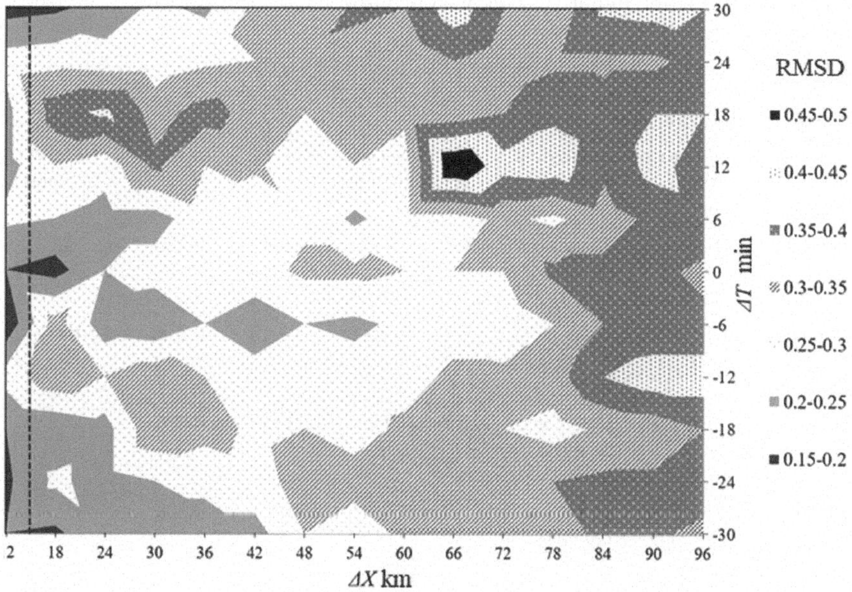

Fig. 3 Dependence of the RMSD between A- and B-data on the size of the spatial-temporal ranges: on the time difference ΔT and on the distance ΔX. The dashed line separates the limits of the most accurate spatial-temporal ranges

of the most accurate spatial-temporal ranges: $\Delta X = 15$ km, $\Delta T = 30$ min. These restrictions were accepted for altimeter calibrations and model verification.

Calibration

Table 2 includes regression formulas of relations of A-data on B-data (calibration equations) and statistical parameters for individual altimeters and for all altimeters together. Note. H_{SA*} and H_{SA} mean H_S value calibrated and measured with an altimeter. These calibration equations for each altimeter were compared with those obtained elsewhere (Young et al. 2017; Liu et al. 2016; Zieger et al. 2009). However, the calibration equations obtained for the same water areas and for time periods differed within three years are similar. Differences in the calibration equations of relations of A-data on B-data enhance with increasing inconsistency between time periods (over 3 years) and calibration areas.

Verification

Table 3 presents statistical parameters on the validity of the two analyzed models calculated using calibration of individual altimeters and all the altimeters together. The values for the WAM4 | WAM4-M models are shown respectively.

Figure 4a–c shows regression graphs of A-data in dependence on M-data for several zones in the Indian Ocean which allow us to illustrate their statistical correspondence.

Table 2 Results of altimeter calibration

Name of the satellite	Calibration period, years	Regression equation	Coefficient of determination, R^2	Bias B, m	RMSD R, m	Number of points
Jason-1	2013	$H_{SA*} = 1.067H_{SA} - 0.125$	$R^2 = 0.969$	−0.019	0.220	262
Jason-2	2013–2016	$H_{SA*} = 1.018H_{SA} - 0.061$	$R^2 = 0.963$	0.018	0.231	2044
H2Y	2013–2016	$H_{SA*} = 0.969H_{SA} - 0.060$	$R^2 = 0.956$	0.140	0.296	353
Altika	2013–2016	$H_{SA*} = 0.994H_{SA} - 0.047$	$R^2 = 0.970$	0.061	0.202	951
Totally	2013–2016	$H_{SA*} = 1.008H_{SA} - 0.058$	$R^2 = 0.963$	0.038	0.230	3610

Table 3 Statistical characteristics of the significant wave height parameter calculated for two models

No. zone of IO	Calibration of individual altimeters (WAM4 I WAM4-M)			Calibration of all altimeters together (WAM4 I WAM4-M)		
	B, m	*R*, m	*K*	*B*, m	*R*, m	*K*
1	0.133 I 0.073	0.414 I 0.345	0.953 I 0.964	0.121 I 0.061	0.419 I 0.345	0.952 I 0.963
2	0.119 I−0.078	0.367 I 0.336	0.896 I 0.907	0.119 I−0.079	0.366 I 0.335	0.897 I 0.908
3	0.101 I−0.027	0.352 I 0.319	0.848 I 0.868	0.081 I−0.040	0.354 I 0.326	0.846 I 0.865
4	−0.041 I −0.072	0.346 I 0.341	0.886 I 0.899	−0.059 I −0.090	0.355 I 0.352	0.886 I 0.896
5	−0.172 I −0.110	0.480 I 0.440	0.888 I 0.897	−0.190 I −0.128	0.496 I 0.453	0.886 I 0.895
6	−0.175 I 0.294	0.602 I 0.705	0.920 I 0.921	−0.199 I 0.270	0.619 I 0.702	0.919 I 0.919
Entire ocean	−0.096 I 0.112	0.51 I 0.56	0.943 I 0.947	−0.116 I 0.092	0.523 I 0.560	0.942 I 0.945

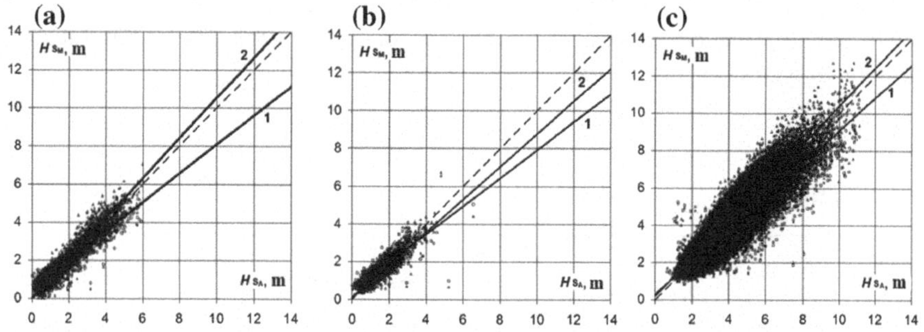

Fig. 4 Regression relations between M- and A-data on the significant wave height (H_S) obtained with separate calibration of A-data for two wave models: WAM4 (squares), straight line 1 and WAM4-M (triangles), straight line 2: **a** zone Z1, **b** zone Z2, **c** zone Z6. Dashed line—regression relation 1:1

4 Results

We have made separate calibration of altimeters and calibration for all altimeters together. The statistical parameters of models' verification showed that the way of calibration doesn't have a major impact on the results of verification (Table 3).

Table 3 demonstrated that in the modified model WAM4-M, the bias of the mean H_S is commonly less than that in the original model WAM4 which testifies in favor of the modified version. Meanwhile, in zones with heavy swell (zones 2, 3, 4, 5)

(Vethamony et al. 2006), the WAM4-M model underestimates H_S, which is explained by excessive attenuation of the swell in the WAM4-M model version.

Data in Table 3 suggest that in all the zones, except zone 6, the modified model WAM4-M has lower RMSD values than WAM4, which indicates the advantage of WAM4-M in terms of calculating the significant wave height.

This advantage is provided by the better description of physical processes incorporated in the modified version of WAM4-M (Polnikov et al. 2007). Zones 1 and 2 are of great practical significance in terms of shipping and marine activities in the territorial waters of India.

Altimeters significantly underestimate the heights of high waves (Fig. 5). For this reason, in zone 6, WAM4-M results which on the contrary overestimate altimetry, may be more accurate than altimetry data calibrated for the entire range of wave heights. On the contrary, WAM4 underestimates altimetry in zone 6. Due to the fact that altimeters overestimate low waves and underestimate high waves (Fig. 5), altimeters should be calibrated separately for the wave heights under 4 m and above 4 m.

Since we conducted statistics for comparing M-data with A-data calibrated for the whole range of wave heights, both models showed the highest RMSD in zone 6 relative to the other zones. In zone 6 there was also the biggest bias, therefore the usage of RMSD as an indicator of accuracy in this case is doubtful. Thus, we suggest that the fact that WAM4-M model gave worse estimates than WAM4 model in this high-wave zone can be ignored. These large errors in zone 6 are reflected in statistics throughout the ocean as a whole.

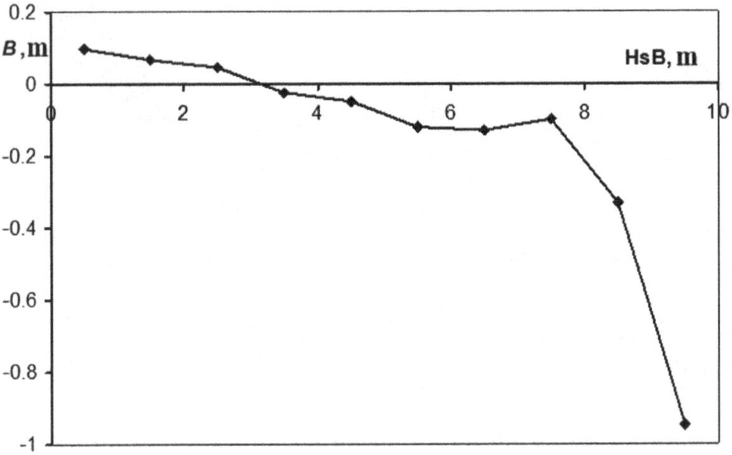

Fig. 5 The deviation of the means of A-data from the means of B-data in dependence on H_S. Significance of the bias values (B) at H_S (H_SB) > 8 m is small due to the few records of A-data within this range of wave heights

Table 3 documented that for both models in all the zones and in the ocean as a whole the correlation coefficient between M-data and calibrated A-data was over 0.85. This fact has already been highlighted earlier (Polnikov et al. 2015).

RMSD (R) and corrected RMSD (R_C) are shifted provided there is a bias. However, at a bias, R can either decrease or increase while R_C only increases. Therefore, R_C more reliably characterizes model accuracy than R (Mentaschi and Besio 2013).

Table 4 contains R and R_C values and corresponding biases obtained for the two models using A-data based on separate calibration of altimeters. At different positive and negative biases, both models suggested that $R > R_C$ by 1.5–2 times in zones 1–3, by 2.5–3 times in zones 4 and 5, and by 3.5–4.5 times in zone 6 and throughout the ocean (Table 4).

WAM4-M model in all the zones gives better estimates in terms of R_C than WAM4, except for zone 6, which is associated with the above-mentioned influence of altimeter calibration for the entire range of wave heights.

Figure 6 shows the distribution of R and R_C and the corresponding biases among the zones and throughout the Indian Ocean calculated for the two models, WAM4 and WAM4-M.

According to the both models in all the zones, including the whole Indian Ocean, R and R_C are distributed differently, especially in zone 6 and throughout the Indian Ocean. Both models indicated that R exhibited insignificant variability in zones 1–4, showed intensive growth in zones 4–6, and had high values throughout the Indian Ocean. Both models also evidenced that R_C showed insignificant variability in zones 1–4 with a maximum value in zone 2 and a minimum value in zone 4. R_C slightly increased in zones 4–5 while in zones 5–6 and throughout the Indian Ocean, in contrast to R, this parameter stabilized remaining within the same range as in zones 1–4 (Fig. 6).

Figure 6 shows that both models did not reveal any trends in connections of the increase or decrease of biases with corresponding changes of R and R_C in 1–4 zones. Both models indicated that in 4–6 zones, R increased with a significant increase in the bias. In contrast, R_C did not demonstrate such a trend in 4–6 zones changing identically at the large and small biases. To conclude, verification of the two models in terms of R and R_C appeared identical.

5 Conclusions

(1) We determined the limits of the most accurate spatial-temporal ranges for the binding of A-, B- and M-data: ΔX (space limit) = 15 km and ΔT (time limit) = 30 min. Therefore, the accuracy of the previously assessed spatial-temporal ranges (Zieger et al. 2009) increases with a decrease of the spatial limits.

(2) A-data were calibrated based on the B-data during survey period from 2013 to 2016. Calibrations obtained in the current study were compared to the calibrations published elsewhere (Young et al. 2017; Liu et al. 2016; Zieger et al. 2009). We found that our calibrations are close to those for the same altimeters,

Table 4 R and R_C at different biases are given for two models

No. zone	*B*WAM4, (m)	*B*WAM4-M, (m)	*R*WAM4	*R*WAM4-M	*Rc*WAM4	*Rc*WAM4-M	*R*WAM4/*Rc*WAM4	*R*WAM4-M/*Rc*WAM4-M
1	0.13	0.07	0.414	0.345	0.212	0.170	2.0	2.0
2	0.12	−0.08	0.367	0.336	0.216	0.207	1.7	1.6
3	0.10	−0.03	0.352	0.319	0.188	0.174	1.9	1.8
4	−0.04	−0.07	0.346	0.341	0.136	0.134	2.5	2.5
5	−0.17	−0.11	0.480	0.440	0.174	0.157	2.8	2.8
6	−0.18	0.29	0.602	0.706	0.147	0.162	4.1	4.4
IO	−0.10	0.11	0.510	0.559	0.153	0.160	3.3	3.5

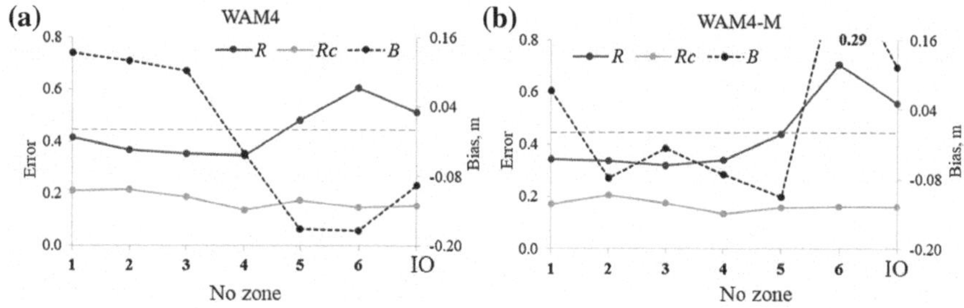

Fig. 6 Distribution of the errors R and R_C and the corresponding biases B in the individual zones and throughout the Indian Ocean calculated for the two models, **a** WAM4 and **b** WAM4-M. The dashed line indicates the zero bias

for closely located water areas and time periods within an interval of 3 years between calibrations.

(3) Based on data of the Indian Ocean for the period 2013-2015, we verified two numerical models: the European model of wind waves WAM4 (Gunter et al. 1992) and its modification WAM4-M (Polnikov 2005) using along-track altimeter data. We found the differences between statistical parameters of the both models obtained for 6 zones of the Indian Ocean and for the ocean as a whole. WAM4-M in 5 zones gives more accurate results than WAM4.

    However, in the zone of high waves (zone 6), due to the strong bias of A-data relative to B-data, the statistic parameters are not significant. In this case, separate calibration of A-data according to B-data is required for the ranges of H_S below 4 m and above 4 m.

(4) For estimation of the accuracy of numerical modeling of waving, we used corrected RMSD (Hanna and Heinold 1985). The distribution of this indicator was analyzed in dependence on the RMSD and the bias for each zone and for the ocean as a whole.

Acknowledgements The studies were performed with the support from the Russian Science Foundation (Grant No.: 15-17-20020) (calibration of altimeter data and comparative analysis of satellite measurements), from the Russian Foundation for Basic Research (Grant 14-05-62692-Ind-a and 18-05-00161) (wave modeling and comparison with altimeter data).

References

Gunter H, Hasselmann S, Janssen PA (1992) Technical report #4. DKRZ WAM4 Model Documentation. Hamburg, 101p

Hanna S, Heinold D (1985) Development and application of a simple method for evaluating air quality. In: API Pub. no 4409, Washington, USA

http://www.aviso.altimetry.fr

http://www.ecmwf.int/research/era/do/get/era-interim

http://www.ndbc.noaa.gov/historical_data.shtml
Janssen PA, Abdalla S, Hersbach H (2007) Error estimation of buoy, satellite and model wave height data. J Atm Ocean Tech 24(9):1665–1677
Liu Q, Babanin AV, Guan C (2016) Calibration and validation of HY-2 altimeter wave height. J Atmos Ocean Technol 33(3):919–936
Mentaschi L, Besio G (2013) Problems in RMSE-based wave model validations. Ocean Model 72(1):53–58
Polnikov VG (2005) Model of wind waves with an optimized source function. In: Izv RAS Physics of the atmosphere and the ocean, vol 41(5), pp 655–672
Polnikov VG, Dymov VI, Pasechnik TA (2007) Advantages of a wind wave model with an optimized source function. Rep Russ Acad Sci 417(9):1375–1379
Polnikov VG, Kubryakov AA, Pogarsky FA, Stanichny SV (2015) Comparison of numerical and satellite data on wave fields. Process Geoenviron 2(2):56–68
Vethamony P, Sudheesh K, Rupal SP (2006) Wave modelling for the north Indian Ocean using MSMR analysed winds. Int J Remote Sens 27(18):3767–3780
Young IR, Zieger S, Babanin AV (2011) Global trends in wind speed and wave height. Science 332(6028):451–455. https://doi.org/10.1126/science.1197219
Young Y, Vinoth J, Zieger S, Babanin AV (2012) Investigation of trends in extreme value wave height and wind speed. J Geophys Res 117. #C11. C00J06. https://doi.org/10.1029/2011jc007753
Young IR, Babanin AV, Zieger S (2013) The decay rate of ocean swell observed by altimeter. J Phys Oceanogr 43:2322–2333
Young IR, Sanina E, Babanin AV (2017) Calibration and cross-validation of a global wind and wave database of altimeter, radiometer and scatterometer measurements. J Atm Ocean Technol 34:1285–1306
Zieger S, Vinoth J, Young IR (2009) Joint calibration of multiplatform altimeter measurements of wind speed and wave height over the past 20 years. J Atmos Ocean Technol 26(12):2549–2564 (Faculty of Engineering and Industrial Sciences)

Study of Falling Rocks Using Discrete Element Method

N. S. Orlova and M. V. Volik

Abstract Investigation of rockfalls motion to estimate affected area and hazard effects is very important. The importance of studying of such phenomena is determined by a number of reasons, among them the danger to human life in the relevant territories. Mathematical and computer modeling methods allow to estimate the influence of the slope geometry (the angle of the slope and the height of the slope) on the distance of the run after rockfalls. This paper presents the results of modeling of the rocks motion along the slope that is attached to a horizontal section. We obtained the results of modeling using the discrete element method for different values of the angle of the slope and the height of the slope. The results of modeling were compared with the experimental data. In general, the model, based on the discrete element method, satisfactorily describes the experiments of the falling dolomite rock in the investigated range of the angle of the slope and the height of the slope. The results of calculations of the distance of the run after rockfalls are overestimated in comparison with the experimental data. The distance of the run after rockfall increases almost linearly with the angle of the slope. The difference between the results of calculations and experiments increases with the increasing of the height of the slope. The difference also increases with the increasing of the angle of the slope in the case of the high value of the height of the slope. We concluded, that the model, based on the discrete element method, can be used to simulate rockfalls.

Keywords Rockfall · Angle of a slope · Height of a slope · Distance of the run · Mathematical modeling · Falling dolomite rock · Discrete element method

1 Introduction

Investigation of rockfalls motion to estimate affected area and hazard effects is very important. The importance of studying of such phenomena is determined by a number of reasons, among them the danger to human life in the relevant territories; the prevalence, increasing in connection with the development of mountain and

N. S. Orlova (✉) · M. V. Volik
Southern Mathematical Institute—The Affiliate of VSC of RAS, Vladikavkaz, Russia
e-mail: norlova.umi.vnc@gmail.com

C. T. Olegovna (ed.), *Processes in GeoMedia—Volume I*,
Springer Geology, https://doi.org/10.1007/978-3-030-38177-6_9

foothill territories; the unexpectedness of their occurrence; participation in shaping and changing the terrain, etc. Mathematical and computer modeling methods allow to estimate the influence of the slope geometry (the angle of the slope and the height of the slope) on the distance of the run after rockfalls. The height of the slope is the height, on which the rocks is located at the initial moment of time. There are many models, which describe rockfalls. These models are based on a discrete approach in the main. For example, the results of rockfalls modelling using a discrete approach are presented in (Chen et al. 2013; Luuk 2003; Shi and Goodman 1985; Rammer et al. 2010; Hungr 1995; Wei et al. 2003; Kusraev et al. 2016).

There are also the models, which are based on a continuum approach (Orlova and Volik 2016, 2017; Orlova and Kamenetskii 2018). It should be noted that the models, based on the discrete approach, are more physical to simulate falling rocks than the models, based on the continuum approach. It can be explained as follows. The models, based on the continuum approach, describe the motion of a falling flow of rocks as the motion of a continuous medium. The models, based on the discrete approach, describe the motion of individual particles interacting with each other. In this regard, the implementation of discrete models requires the use of sufficiently powerful computing resources.

This paper presents the results of modeling of the rocks motion along the slope that is attached to a horizontal section. We obtained the results of modeling using the Discrete Element Method. The Discrete Element Method is a powerful numerical method used to investigate deformations and dynamics of particulate matter (Hashemnia and Pourandi 2018). In this method, the equations of motion of all particles are solved together in each time step after calculating the contact forces between the particles knowing the current positions and velocities of particles in contact (Hashemnia and Pourandi 2018; Xiang et al. 2010).

The influence of the slope geometry on the affected area is investigated. We used the results of an experimental study obtained under laboratory conditions to verify the model based on the discrete element method (Kusraev et al. 2016). We used a free program code LIGGGHTS, in which the discrete element method was implemented (LIGGGHTS, http://www.liggghts.com).

2 Discrete Element Method

A free program code LIGGGHTS was used to calculate the falling dolomite rock (LIGGGHTS, http://www.liggghts.com). The discrete element method was implemented in this program. The main equations used in the discrete element method are the Newton's and Euler's equations (Eqs. 1 and 2). These equations allow to calculate new positions, velocities and accelerations of all particles (Hashemnia and Pourandi 2018). m is the mass of each particle, I_p is the centroidal mass moment of inertia of each particle, v is the translational velocity of each particle, ω is the rotational velocity of each particle, g is the gravitational acceleration. F_c and T_p

are the resultant contact force and couple applied to each particle (Hashemnia and Pourandi 2018).

$$m\frac{d\overline{v}}{dt} = \overline{F}_C - m\overline{g}; \tag{1}$$

$$I_p\frac{d\overline{\omega}}{dt} = \overline{T}_p. \tag{2}$$

There are different contact force models used in the discrete element method. These models are based on different assumptions about the elasticity of the contacting particles. For example, elastic models (the linear-spring and Hertz models), visco-elastic models (the linear-spring/dashpot (LSD) and Langston models) and elasto-plastic models (the Hertz-Mindlin, Thornton and Walton-Braun models) (Hashemnia and Pourandi 2018; Norouzi et al. 2016; Crowe et al. 2012; Di Renzo and Di Maio 2004; Stevens and Hrenya 2005). As the amount of plastic deformation of the dolomite spheres was negligible, the Hertz-Mindlin model was used. This model considers the elastic and plastic deformations of the colliding particles in normal and tangential directions and the coulomb friction force between the particles in the tangential direction (Hashemnia and Pourandi 2018). The contact forces between contacting particles and the walls, are calculated by the normal and the tangential overlaps between the bodies. The normal overlap of two spherical particles in contact δ_n (see Fig. 1) is defined as the amount of relative displacement of the particles i and j in the normal direction, and the tangential overlap δ_t is defined as the relative

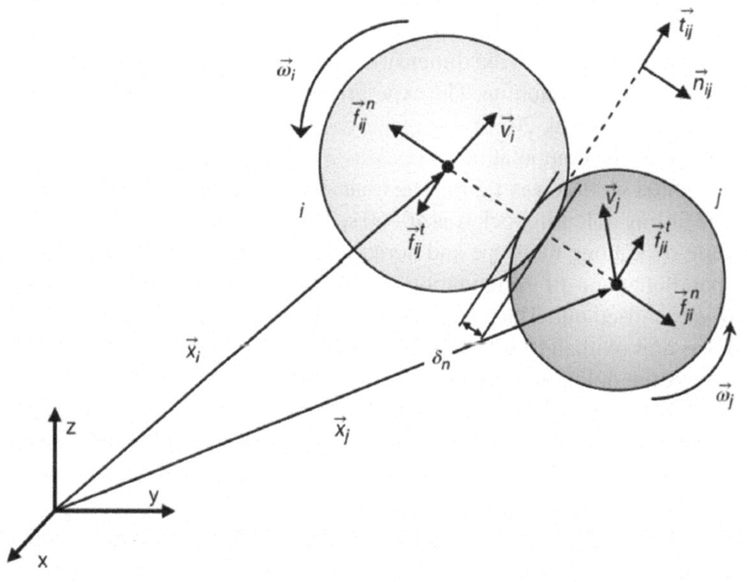

Fig. 1 Two contacting particles and the corresponding interaction forces

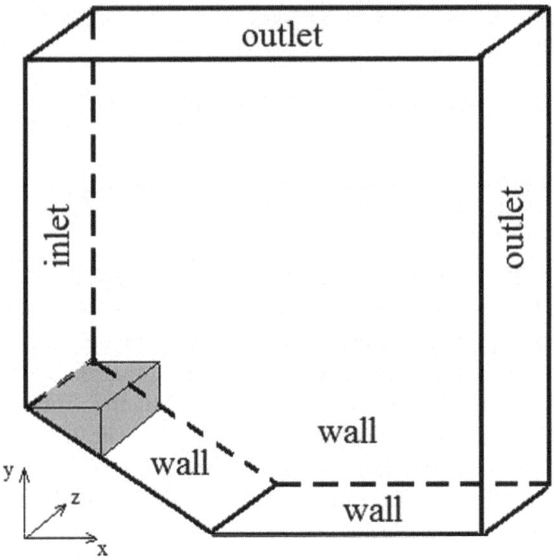

Fig. 2 The computational area

displacement of the particles in the tangential direction (the rolling contribution is not taken into account) (Hashemnia and Pourandi 2018; Norouzi et al. 2016). The formulas for calculating the overlaps and contact forces are presented in (Hashemnia and Pourandi 2018). A more detailed description of the discrete element method is given in (Kusraev et al. 2016; Hashemnia and Pourandi 2018; Norouzi et al. 2016; Crowe et al. 2012; Di Renzo and Di Maio 2004; Stevens and Hrenya 2005).

We performed 3D modeling of the falling rock. The computational area (Fig. 2) was built in accordance with the dimensions of the laboratory equipment. The rocks are often composed of dolomite. The experimental study of the falling dolomite rock is given in (Kusraev et al. 2016).

The height of the computational area was 3 m, the width was 0.45 m, the length of the horizontal section was 1.5 m. The time step was equal to 0.005 s. The average time of the falling dolomite rock was 40–60 s. This allowed us to consider the motion of dolomite rock along the slope and horizontal section until it stopped.

The parallelization of computations on 4 cores processor with a frequency of 3.1 GHz was carried out. The computation time for one case of the discrete element method was approximately 120 min.

There was the dolomite rock in the form of a triangular prism with a volume of approximately 0.0012 m^3 in the upper part of the slope at the initial moment of time (Fig. 1). The dolomite rock was represented as a complex of spherical solid particles with a diameter of 0.005 m. The diameter value approximately corresponds to the average particle size of dolomite in experiments.

It should be noted that we used the value of the coefficient of friction 0.7 (Smith and Faulkner 2010). Young's modulus for dolomite was equal 3×10^5 kg/cm^2,

Poisson's ratio was equal to 0.14 (Makse et al. 2004), density of dolomite was equal 2600 kg/m^3. A more detailed description of the parameters is given in (Kusraev et al. 2016).

3 Analysis of the Modeling Results

We obtained the results of modeling using the discrete element method for different values of the angle of the slope (from 35° to 53°) and the height of the slope (from 43 to 56 cm). We presented snapshots of particle distribution (Fig. 3) after rockfall. The results of modeling were performed for the angle of the slope 35° and the height

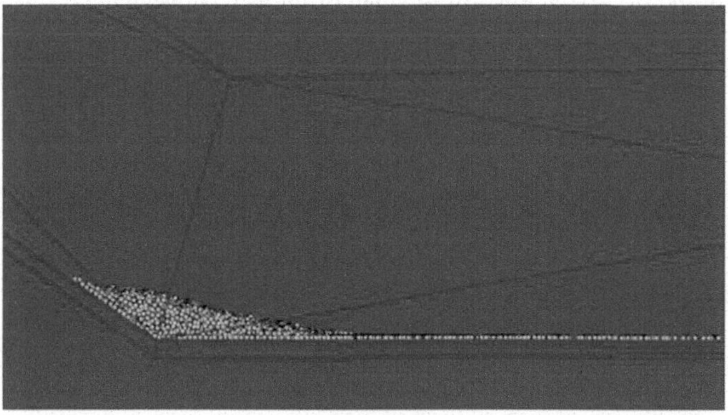

side view

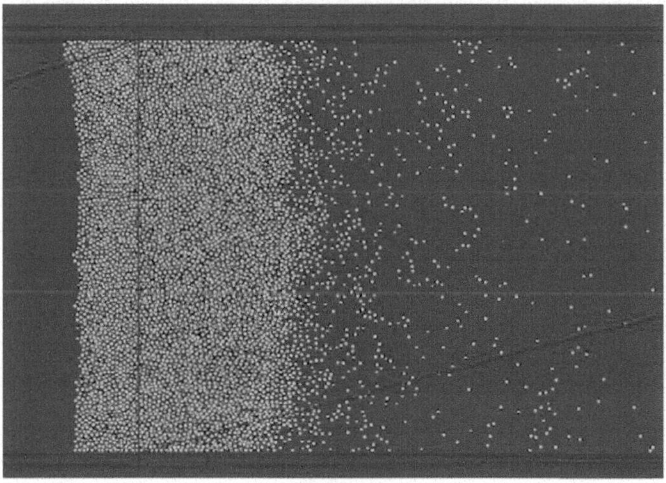

top view

Fig. 3 Snapshots of particle distribution after rockfall of the slope 50 cm. It can be seen that the continuum part of the fallen dolomite mass is separated from the part in which the mass of dolomite is loose.

We compared the calculated distance of the run after rockfall with the experiments. We presented the results of modeling for the value of the height of the slope 43 cm (see Fig. 4a), the value of the height of the slope 50 cm (see Fig. 4b) and the value of the height of the slope 56 cm (see Fig. 4c). The procedure for determining the distance of the run is described in (Kusraev et al. 2016).

In general, the results of calculations satisfactorily describe the experiments. The distance of the run after rockfall increases almost linearly with the angle of the slope. The results of calculations are overestimated in comparison with the experimental data (except Fig. 4b). The difference between the results of calculations and experiments increases with the increasing of the height of the slope. The difference also increases with the increasing of the angle of the slope in the case of the high value of the height of the slope (see Fig. 4c). A small difference between the results of calculations and experiments can be explained by the insufficient accuracy in determining

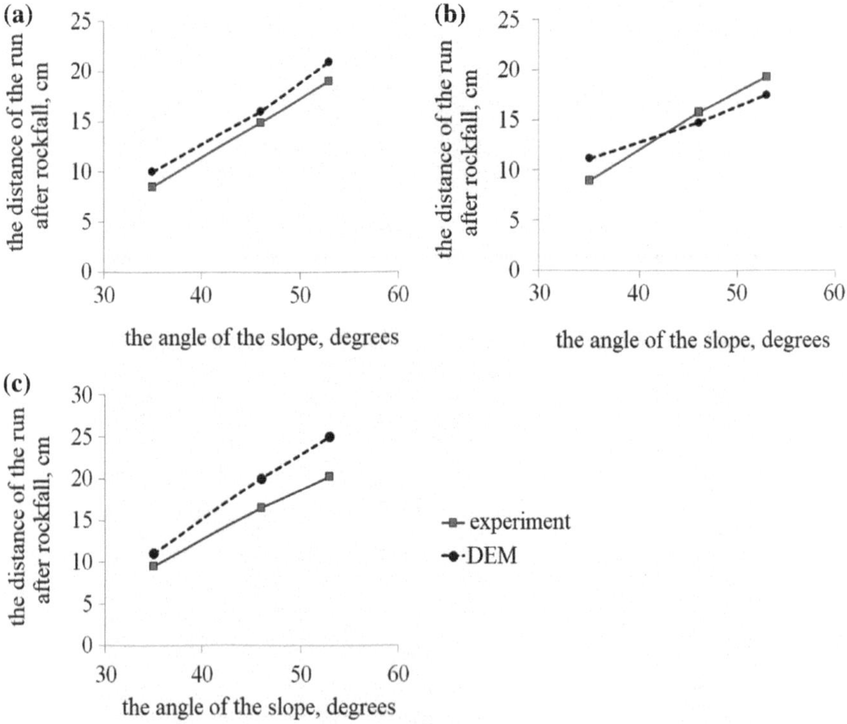

Fig. 4 Dependence of the distance of the run on the angle of the slope. **a** The height of the slope 43 cm. **b** The height of the slope 50 cm. **c** The height of the slope 56 cm

the distance of the run both in experiments and in the calculations. Moreover, we used spherical particles (particle diameter 5 mm) in the calculations. In fact, dolomite particles had irregular shape in experiments, and their sizes were varied from 3 to 6 mm.

4 Conclusions

In general, the discrete element method satisfactorily describes the experiments of the falling dolomite rock in the investigated range of the angle of the slope and the height of the slope. The results of calculations of the distance of the run after rockfalls are overestimated in comparison with the experimental data. The distance of the run after rockfall increases almost linearly with the angle of the slope. The difference between the results of calculations and experiments increases with the increasing of the height of the slope. The difference also increases with the increasing of the angle of the slope in the case of the high value of the height of the slope. A small difference between the results of calculations and experiments can be explained by the insufficient accuracy in determining the distance of the run both in experiments and in the calculations. So we concluded, that the discrete element method can be used to simulate the falling rock.

References

Chen G, Zheng L, Zhang Y, Wu J (2013) Numerical simulation in rockfall analysis: a close comparison of 2-D and 3-D DDA. Rock Mech Rock Eng 46:527–541

Crowe CT, Schwarzkopf JD, Sommerfeld M, Tsuji Y (2012) Multiphase flows with droplets and particles, 2nd edn. CRC Press, Taylor and Francis Group, New York

Di Renzo A, Di Maio FP (2004) Comparison of contact-force models for the simulation of collisions in DEM-based granular flow codes. Chem Eng Sci 59:525–541

Hashemnia K, Pourandi S (2018) Study the effect of vibration frequency and amplitude on the quality of fluidization of a vibrated granular flow using discrete element method. Powder Technol 327:335–345

Hungr O (1995) A model for the runout analysis of rapid flow slides, debris flows and avalanches. Can Geotech J 32(4):610–623

Kusraev A, Minasjan D, Orlova N, Pantileev D, Hubezhty Sh (2016) Verification of the model of rockfall using the discrete element method. Geol Geophys South Russia 4:83–93

LIGGGHTS Open Source Discrete Element Method Particle Simulation Code, http://www.liggghts.com, last accessed 2018/07/13

Luuk KA (2003) Dorren: a review of rockfall mechanics and modeling approaches. Prog Phys Geogr 27(1):69–87

Makse H, Gland N, Johnson D, Schwartz L (2004) Granular packings: nonlinear elasticity, sound propagation, and collective relaxation dynamics. Phys Rev E 70(6):061302

Norouzi H, Zarghami R, Sotudeh-Gharebagh R, Mostoufi N (2016) Coupled CFD-DEM modeling. Wiley, West Sussex, United Kingdom

Orlova N, Kamenetskii E (2018) Verification of rock falls model using the continuum approach. Sustain Dev Mt Territ 10(1):7–13

Orlova N, Volik M (2016) Mathematical modeling of the motion of rockfall using the continuum approach. Proceedings of the universities. North Caucasus region. Nat Sci 3:20–24

Orlova N, Volik M (2017) Investigation of the influence of restitution coefficient on the results of the rockfall modeling. Process Geo-environ 4:693–699

Rammer W, Brauner M, Dorren L, Berger F, Lexer M (2010) Evaluation of a 3-D rockfall module within a forest patch model. Nat Hazards Earth Syst Sci 10:669–711

Shi GH, Goodman RE (1985) Two-dimensional discontinuous deformation analysis. Int J Numer Anal Meth Geomech 9(6):541–556

Smith S, Faulkner D (2010) Laboratory measurements of the frictional properties of the Zuccale low-angle normal fault, Elba Island, Italy. J Geophys Res 115:B02407. https://doi.org/10.1029/2008JB006274

Stevens AB, Hrenya CM (2005) Comparison of soft-sphere models to measurements of collision properties during normal impacts. Powder Technol 154:99–109

Wei F, Kaiheng H, Lopez J, Peng C (2003) Method and its application of the momentum model for debris flow risk zoning. Chin Sci Bull 48(6):594–598

Xiang L, Shuyan W, Huilin L, Goudong L, Juhui C, Yikun L (2010) Numerical simulation of particle motion in vibrated fluidized beds. Powder Technol 197:25–35

The Application of Mathematical Modelling to Assess Ecological Safety of the Coastal Area of the International Resort of Varadero (Cuba)

N. N. Dunaev, I. O. Leont'yev, T. Y. Repkina, and J. L. Juanes Marti

Abstract Geological and geomorphological location is reviewed for the seaside-resort Varadero which is recreational area of international importance. The outlook on the natural scenario of this area evolution is of great concern. The analysis of coastal morpho- and lithodynamics is implemented. The interpretation of the term « coastal zone » is given along with the approach to the identification of its boundaries and components. The materials obtained formed the basis of a mathematical model characterizing the lithodynamics component of the studied area evolution. Specific measures to preserve beaches from erosion by storm waves are proposed.

Keywords Near-Shore coastal zone · Ecology · Modeling · Artificial beach

1 Introduction

The concept of "ecological safety" is interpreted differently practically from the moment of the term "ecology" occurrence (from ancient Greek οἶκος—dwelling, habitation, and λόγος—concept, doctrine, science), proposed by the German biologist E. Geckel in 1866 in his work "General morphology of organisms." According to the Federal Law of the Russian Federation of March 5, 1992, N 2446-1 "On safety" is a state of protection from danger. The level of the facility's ultimate endurance in relation to the damaging impacts may be the basic criterion for environmental safety. In the case of coastal zones of the sea, this concept can be used to define the acceptable level of negative impact of natural and anthropogenic factors on these

N. N. Dunaev (✉) · I. O. Leont'yev
Shirshov Institute of Oceanology, Russian Academy of Sciences, Moscow, Russia
e-mail: dunaev@ocean.ru

T. Y. Repkina
M.V. Lomonosov Moscow State University, Moscow, Russia

J. L. J. Marti
Institute of Marine Sciences (ICIMAR), Habana, Cuba

C. T. Olegovna (ed.), *Processes in GeoMedia—Volume I*,
Springer Geology, https://doi.org/10.1007/978-3-030-38177-6_10

zones, as a result of which they remain stable, capable of self-regulation and maintain the quality of the environment. It seems that environmental safety presupposes sustainable development of a natural object in a state of comfort for existing biogeocenosis. The main objective of the research in this field is to study the regional environment in order to preserve its vital resources. While the notion of the environment developed at least in the V-IV centuries BC and was introduced into the science of "ecology" by the term "umwelt" in the second half of the nineteenth century by the German biologist Jacob Ikskyl (1864–1944), it is reflected in the Federal Law of the Russian Federation "On Environmental Protection" dated January 10, 2002 No. 7-FZ: "Environment—a set of components of the natural milieu, natural and natural-anthropogenic objects, as well as anthropogenic objects. The term "environment" comes from Old French and is translated approximately as "to surround". Therefore, the environment is what surrounds us. In a broad sense, environment includes all factors of external influence on an individual's development (Odum 1975; Watt 1971; Watt 1973; Reimers 1994; Rosenberg et al. 1998). The attention is paid by nations to the problem of ecological safety and this attention results in formation and control over execution of ecological doctrines based on fundamental scientific knowledge in the field of ecology and related sciences, assessment of the current state of the environment, taking into account global and regional peculiarities of human-nature interaction. As an example, the Ecological Doctrine of the Russian Federation from August 31, 2002 N 1225-r can serve.

The environmental situation within the coastal zones of the sea areas has now become much more pronounced. In order to predict the condition of such areas, the study of their features in different geographical zones and sectors is an urgent problem. The object of the proposed research is the area of Varadero, located within the western coast of the peninsula Hicacos (Republic of Cuba), which has received international recognition as an excellent beach and swimming recreational area with a commercial component. Its sandy beaches, stretching for about 10 km, are among the top five beaches in the world. At present, their natural dynamics is a cause for concern for modern coastal scientists.

The study of any natural object will always be more productive if it is based on a normative or at least routinely accepted definition of the object that reflects the most significant or vivid aspects of condition, origin and development. Still R. Descartes (1596–1650) and later R. Tagore (1861–1941) called to clarify the concepts to rid the world of misconceptions. These thoughts were reinforced by A. Lavoisier (1743–1794), who reminds that no matter how reliable the facts were and how correct their representations were, they would be perceived erroneously without precise expressions for their transmission. The study of a natural object begins with its contouring, the components allocation, emphasizing among them the most important for the solution of the problem, the establishment of links between object parts in terms of energy distribution, and the role of controlling factors.

At the moment, the terminology for the land–sea interface is still under development (Gogoberidze 2008). Against this background, the formation and development of this boundary strip is proposed to be considered as a result of the processes occurring in a single zone—an integral relatively autonomous natural–territorial complex (the geosystem—according to Sochava 1963), consisting of interconnected direct energy exchange of nearshore (coastal) subaerial and subaquatic segments. It should be noted that there is a certain peculiarity of exogenous processes in relation to the main components of this geosystem—the shore and its coastal parts, which Longinov (1956) called "coastal zone of the sea" and "coastal zone of the land".

According to the authors of the paper, the coastal zone (nearshore zone) should be determined by the territory within which the morpho-lithodynamic processes take place at the direct interaction of the nearshore land and water area at a given mean multiyear level and be defined easily enough on the ground. Using the landscape approach, the term "coastal zone of the sea" (SCZ), i.e. the territory adjacent to the shore, can be used to describe the land–sea interface strip, within which there is a direct energy exchange between the water area and the adjacent land, where, depending on the region, different types of sea coast are formed, especially the subaerial and subaquatic subzone.

In this context, there is no disagreement on the definition of "shore". The ground part can be identified with the concept of "near–sea zone". This is a land area adjacent to the shore in the form of a seaside zone, the exogenous specificity of which is determined primarily by the microclimate, commonly referred to as the sea climate.

Here the peculiarities of atmospheric characteristics, micro–relief, vegetation, soil formation, some forms of the animal world are clearly shown. The sea part (near–land zone) will be defined by the territory within which there is a direct energy exchange between water area and coast under the influence of wind sea waves as the most significant exogenous factor of its condition. The internal boundary of this coastal zone component is traditionally carried out along the line of maximum surfacing wave flow, while the external boundary remains debatable. It appears to be appropriate to carry it out along the line of noticeable impact of the strongest storm waves on the seabed, which, in particular, is realized in the movement of loose geological beaching, commonly referred to as sediments. In terms of energy exchange between the coast and the water area, the lower (outer) boundary of the coastal zone will be determined cross-shore distance where the sediments carried out from shore under a given storm conditions are deposited area due to the action of storm waves of this provision. In the first approximation, it corresponds to the closing contour of the most active exchange of matter between the shore and the sea floor, the seaward of which of which the deformation amplitude of the latter, fading with the distance from the shore, becomes comparable with the error of the depth of the bottom measurement. Based on the practice of studies of fine-grained sediments carried away from the shore, it equals 5–10 cm. The depth of water where the circuit occurs to correspond the doubled height of the waves of the given coverage.

Its maximum value approximately coincides with the doubled height of the waves of the strongest storm registered once a year. Waves practically cease to experience deformation due to contact with the bottom located deeper and to be universal regulator of the distribution of bottom sediments and microrelief forms. The evolution of CZS according to the natural scenario is mainly conditioned by the hydrokynematics of the World Ocean level, the climate and such features of the land and water area interface as the geological structure, the latest tectonics and their derivatives of

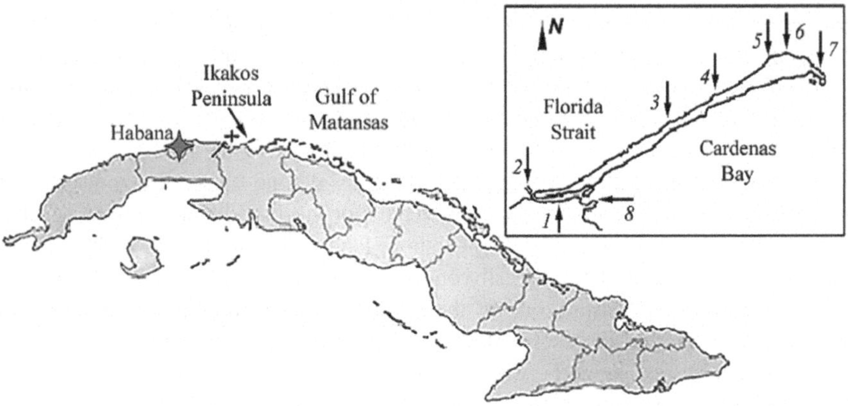

Fig. 1 Location of the Hicacos Peninsula: Cuba Island; in box: *1*—Paso Malo Lagoon, *2*—channel Paso Malo, *3*—Cape Bernardino, *4*—Cape Chapelin, *5*—Cape Frances, *6*—Cape Hicacos, *7*—Cape Molas, *8*—Kawama Channel

the initial relief and hydrological regime of the respective basin. Among the many factors influencing the evolution of CZS are three top factors: sea level fluctuations, sediment budgets and the geological structure of the coastline (Cowell and Thom 1994).

The Hicacos Peninsula is located in the north of Cuba (Fig. 1). It stretches northeast (70°) for a distance of 22 km with a base width of 0.5 km and up to 1.2–2.5 km as it approaches the distal. Since 1956, Hicacos has been practically made an island as a result of the construction of the Paso–Malo channel at its base, which allows the passage of small vessels from Cardenas Bay to the Strait of Florida (Fig. 1).

The shape of the Hicacos peninsula resembles a slanting, accumulative, Azov-type coastal form. Its relief is represented by a low (average 10 m) abrasion–accumulative plain with several remains of carbonate bedrock, the maximum mark of which is 27 m. The eastern coast of the peninsula is the lowest (1–1.5 m) and is distinguished by the development of bogs and mangrove vegetation. The West Bank is characterised by sandy beaches 20–30 m wide and in some places up to 50 m wide. At three sites, bedrock limestone outcrops come to the water's edge, forming capes with well-defined cliffs up to 8–12 m high. They have niches 1–1.5 m deep and up to 0.5 m high. The relief of the Hicacos Peninsula is young, not older than the late neopleistocene (Avelio Suarez et al. 1975; Santana and González 2002). The geological structure of the Hicacos Peninsula is basically represented by several Miocene limestone sandstone (calcarenite) remains, which during the Sangamon transgression were joined by sandy ridges that formed the eolianites. As a result, the peninsula got a close modern look.

The authors' studies are focused on the western coast of the Hicacos Peninsula. This is due to the fact that the safety of the beaches located here in the Varadero resort area causes great concern. To maintain them in recent decades, sand has been periodically dumped mostly south of Chapelin for approximately 12 km. Thus, since

1987, about 3 million m^3 of sand has been dumped here at the cost of about 5 Cuban pesos per 1 m^3. The current dynamics of the coastline is assessed by direct study and also by comparing space images of December 31, 2003 and December 15, 2013, posted in Google Earth. Varadero beach is represented by marine organogenic sand, which includes benthic foraminifera (often more than 50%), shell detritus (20–35%), halimede algae scales (15–30%) and reef components (1–3%). Sand sorting index $S_0 = 1.1\text{–}1.43$, median diameter $M_d = 0.24\text{–}0.55$.

Hydrometeorological conditions in the chosen area are largely determined by northeastern trade winds with a velocity of 2.5–3.5 m/s, which create sea waves 0.5–1.0 m high, and northwestern anticyclonic winds, which velocity on stormy days (usually 2–3 days) reaches 15–20 m/s, and the height of generated sea waves, which have a destructive impact on the beaches of the peninsula, exceeds 2 m. In case of moderate unrest, the beaches are rebuilt up to half their width. Rare hurricanes with velocities of 40–55 and up to 73 m/s cause catastrophic effects on the shores when the beaches are completely eroded and the redevelopment of the underwater profile can be traced back to a bottom depth of 10 m. On average, sand losses from the beaches are approaching 50,000 m^3/year. This is facilitated by the specificity of the regional beach–forming material, which is characterized by rapid abrasion and increased flotation. The western shores of the Hicacos Peninsula are most susceptible to the destructive effects of northwestern tropical cyclones. In the Varadero area, extreme storms usually occur against the background of high storm surges, causing partial or total flooding of the beach. Direct wave effects on the beach and the adjacent avandune result in scouring and the removal of significant amounts of sand on the underwater slope. Part of the sand then moves beyond the shelf edge, which is close enough to the shore, and is irrecoverably lost to the shore.

The adjacent western shelf of the peninsula is shallow. The bend to the island slope occurs at an average depth of 25–40 m, sometimes slightly less or more. It extends from 0.5 km at the beginning of the peninsula to 7 km in its marginal part. Correspondingly, the slopes of the coastal shoal vary from 0.011–0.014 to 0.005–0.008. Two generations of submarine sand rolls of slightly undulating shape stretching sub-parallel to the shore are clearly distinguished. The first shaft is 80–160 m away from the cut at a depth of about 2 m. Its relative height is 1 m, and the water depth above it is 0.3–0.7 m. The second shaft is larger with a relative height of 2–2.5 m. It is formed at the bottom depth of 5–6 m and is located at a distance of 200–250 to 700 m and more from the cut–off point. The surface of the shelf is complicated by three generations of reef structures, forming intermittent ridges, moving away from the shore in the direction of the distal peninsula at a distance of 0.7 to 5 km. Reefs close to the shore, located at a depth of 10–15 m at a distance of 0.7–2.5 km from the shore, are relict, apparently, of Sangamon age. On average, the water depth to the reef surface is within 7–28 m (Medvedev and Juanes 1981).

The current kinematics of sea level in the region is controversial. According to the observation posts established along the coasts of Cuba, its relative fluctuations vary from negative to mostly ambiguous positive values. This is due to the unaccounted for influence of the newest tectonics of the locations of observation posts. Therefore, for example, over the period 1966–2005, the rate of increase in the level was within the range of 0.06–1.46–2.14 mm/year. According to the Institute of Oceanology of Cuba, the level has been increasing at a rate of 2 mm/year in recent decades, and according to the Cuban Environmental Protection Agency, it increased by 6.74 cm between 1966 and 2015, i.e. its average speed was 1.35 mm/year.

There is currently an unsaturated flow of sediment along the peninsula due to the depletion of debris in the offshore segment of the coastal zone. This means that local blurring centers caused by direct storm exposure can expand. With a constant relative sea level, and even more so if it rises, Varadero beach will gradually lose material and degrade (Juanes Marti 1997). To preserve it, periodic artificial recharge is required. Episodic sand dumping improves the situation for a short time, but does not solve the problem in the long term (Santana and González 2002). The construction of artificial beaches is now considered one of the most promising ways of coastal protection. Their parameters can be calculated, for example, on the basis of the model (Leont'yev and Akivis 2017) created specially for conditions of significant storm surges. The morphology of coastal profiles composed of biogenic or terrigenous sand does not show significant differences. In both cases, in particular, the development of submarine berm systems with similar parameters is noted, and the observed slopes of beaches in the Varadero area are of the same order as those on the sandy shores of the Baltic Sea, with appropriate particle sizes. The analogy of lithodynamic processes can therefore be assumed to be analogous, and it can be assumed that, in a very rough approximation, the dynamics of a beach composed of biogenic material can be described within the framework of traditional sediment transport models.

Our calculations take wave climate data from (http://wisuki.com/statistics/849/varadero). Roses of different wave gradations of significant heights are shown in Fig. 2. As can be seen, for the most significant storm waves with a height of more than 2 m, the prevailing direction is the West-North-Western direction.

The estimates of the storm deformations of Varadero beach under extreme storms and overtaking of different heights are made on the basis of the CROSS–P model (Leont'yev and Akivis 2017; Leont'yev et al. 2015) for a fairly typical shore profile (Fig. 3a), composed of medium–grained sand with an average size of 0.3 mm. The storm impact is characterized by an average wave height of 3 m, an average period of 8 s, and a 30° approach angle (taking into account the position of the normal shoreline in relation to the most severe north–western storms). The duration of exposure is assumed to be 6 h, which is typical enough for a storm peak when its parameters remain at their maximum values.

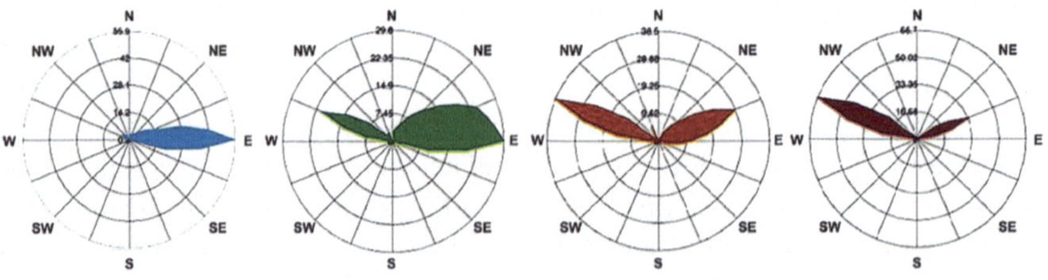

Fig. 2 Rose waves for different wave heights: 0–1, 1–2, 2–3, 3+ m

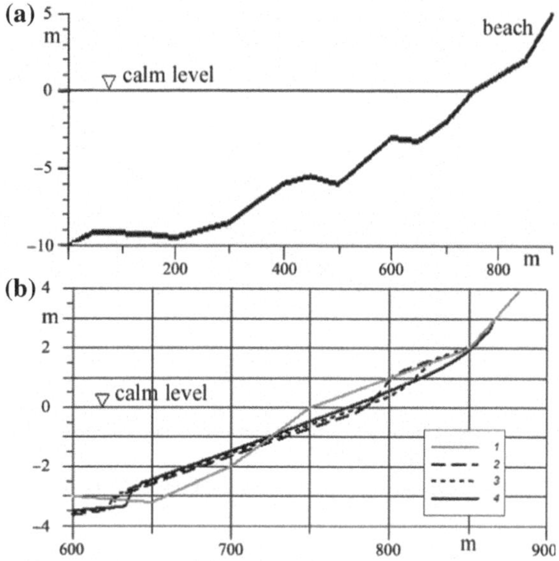

Fig. 3 Typical coastal slope profile in the Varadero area (**a**) and beach storm deformation from modelling (**b**): *1*—initial shore profile, *2*, *3* and *4*—post-storm profiles at storm surge heights of 1, 1.5 and 2 m, respectively

The results are presented in Fig. 3b, which shows the coastal part of the profile, including the beach. Its erosion occurs in all cases, and as the height of the storm surge increases, the erosion zone expands towards the shore. At an storm surge of 1 m it has a length of about 50 m, and at an surge of 2 m it more than doubles. The surface of the beach is lowered by up to 0.7 m. During the storm, 20–40 m^3 of sand is carried away from the beach per meter of shoreline (the volume of the removal increases with the height of the surge). Moderate waves (1–1.5 m) with surges of up to 0.5 m are calculated not to cause significant changes to the beach profile. During periods of low wave activityl, observations indicate that partial beach recovery is possible, but limited to the irrecoverable loss of material described above.

The problem of coastal protection is proposed to be solved by creating an artificial beach, the characteristics of which are estimated on the basis of the model (Leont'yev and Akivis 2017), taking into account the above mentioned parameters of an extreme storm. Input data for calculations are given in the left part of the Table 1, where d_g—average sand size, and $\overline{H}$ $\overline{T}$—average height and wave period, η—beach elevation t. The beach profile is shown in Fig. 4.

The main indicators of the artificial beach (in addition to X) are its maximum elevation above the calm level z_m and the width of the berm (Fig. 4), as well as the volume of material used V per unit of coast length. These indicators are reflected in the right side of Table 1. It is planned to extend the shoreline at a distance of $X =$

Table 1 Input data for calculations and artificial beach profile parameters

d_g (mm)	$\overline{H}$ (m)	$\overline{T}$	η (m)	X (m)	z_m (m)	l_a (m)	V (m^3/m)
0.5	3.0	8.3	2.0	20	3.2	73	288

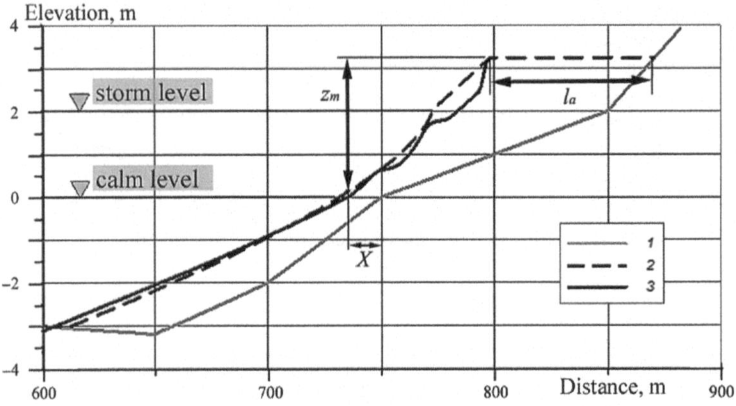

Fig. 4 Profile of an artificial beach and its deformation in an extreme storm: *1*—natural profile, *2*—artificial profile (sand size 0.5 mm), *3*—post–storm profile of artificial beach, with z_m as beach elevation, l_a—construction berm width, *X*—extension of the bank

20 m, which will provide the necessary stock of material even in case of catastrophic erosion of 0.5–0.7 m. As can be seen, the construction of a 1 km long beach will require about 300,000 m^3 of sand.

To test the stability of the artificial beach, an extreme storm impact modelling was carried out with the parameters specified in the table (storm duration, as before, 6 h). The post–storm profile is shown in the Fig. 4. As can be seen, erosion in the upper part of the profile is much less than in the current state of the coast (Fig. 3). Obviously, the project profile can withstand more than one extreme storm attack. At other duration of the storm, its strength and parameters of sea waves, the process of modeling is correlated with the introduction of new initial data.

As a result of longshore sediment transport, the volume of beach material will decrease over time. According to R. G. Dean's calculations (Dean 2002), the relative losses increase with increasing sediment flow capacity and decrease with increasing beach length along the coast. Based on the available wave data (http://wisuki.com/statistics/849/varadero) and using the calculation methodology (Leont'yev 2014a, b), it is possible to estimate the scale of the longshore sediment flux in the Varadero area, which is about 50,000 m^3/year. Then it turns out that with the given parameters of the beach even after 20 years can be preserved about half of its original volume. Of course, periodic backfilling of the material will help to prolong the existence of the beach. Additional erection of specially selected and space–oriented wave–quenching structures in the storm wave deformation band directly behind the usual storm wave destruction zone will weaken their impact, which will increase the stability of the beach.

In order to check the effectiveness of the recommendations, it is proposed to select a 1 km pilot plot in the eroded part of the Varadero coast and to carry out a beach fill based on the results of the calculations. The calculations can be easily repeated with other input data, such as for other sand sizes or for other storm parameters, if they are further refined.

In conclusion to the modelling section, it should be emphasised once again that the parameters of the coastal protection beach are designed for terrigenous medium–grained sand, whereas only biogenic sand is available in the Varadero area. The extent to which the recommendations are applicable to such material is still open and requires further research. The procedure sometimes used in practice to introduce correction factors, such as hydraulic particle size, density, porosity and other characteristics of bottom sediments, is not always and universally justified. Therefore, monitoring of the beach established in the pilot site could provide very valuable information of both scientific and practical interest.

The performed research expand the knowledge of coastal geosystems in different geological–geographical conditions. Their practical significance is largely related to the optimal solution of the problems of recreational direction. The case of the Varadero region has been used as an example to consider the development of the tropical latitude coastal zone. As a practical recommendation it is proposed to protect the most vulnerable parts of the Varadero coast with the help of an artificial beach, the parameters of which can be determined on the basis of the proposed original mathematical model.

In the near future, the morpholithodynamics of the coastal zone of Varadero in the twenty-first century will be determined mainly by the kinematics of sea level and sand reserves in the primitive zone of the shelf. Leaving the issue of possible significant sea–level rise and global warming out of the debate, as authors believe there is little reason to assume that these factors will change significantly, at least at an engineering time-scale. In the context of reduced bioproductivity and increased flotation of beach–forming material, its amount will be reduced. In order to maintain the beaches, it will therefore be necessary to carry out competent and timely sand backfilling and to provide for measures to dissipate storm wave energy at a distance from shore. The latter is prompted by nature itself—the frozen ice of the Chukchi Sea at a water depth of 10–15 m at a distance of 3–4 km from the edge protect the coastal zone of the shelf and shore from severe erosion. The forecast of the development of the coastal zone in question, taking into account the known regularities of the region's evolution in late Quaternary time, corresponds to the natural scenario. The predictability limit implies possible errors, which should not exceed the average magnitude variations of predicted values (Zakharov 1990). The chosen engineering time-scale corresponds to this postulate.

Acknowledgements The work is supported by the Russian Foundation for Basic Research (RFBR), project "Cuba_t" No. 18-55-34002 and the state assignment of IO RAS, theme No. 0149-2019-0005.

References

Avelio Suarez O, Ionin AS, Medvedev VS, Nevessky EN, Pavlidis YA (1975) Geomorphology and history of Hicacos (Cuba). Tropical ocean island shelves. IOAN USSR, Moscow, 1, pp 99–113
Cowell PJ, Thom BG (1994) Morphodynamics of coastal evolution. In: Carter RWG, Woodroffe CD (eds) Coastal evolution: late quaternary shoreline morphodynamics. Cambridge University Press, UK, pp 33–86
Dean RG (2002) Beach nourishment: theory and practice. World Scientific Inc., 399p
Gogoberidze GG (2008) Coastal zone: basics of the conceptual apparatus and principles of the geostrategic development (Russia, Saint-Petersburg). Probl Mod Econ 3(27):384–388
http://wisuki.com/statistics/849/varadero
Juanes Marti, JL (1997) Erosion on the beaches of Cuba. Alternatives for its control. PhD thesis in Geographic Sciences. University of Havana, Faculty of Geography. Havana. 101 h
Leont'yev IO (2014a) Morphodynamic processes in the coastal zone of the sea. LAP LAMBERT Academic Publishing, Saarbrücken, 251p
Leont'yev IO (2014b) On calculation of longshore sediment transport. Oceanology 54(2):226–232
Leont'yev IO, Akivis TM (2017) An artificial beach as a means for sea coast protection from storm surges (by the example of the Eastern Gulf of Finland). In: Proceedings of international conference "managing risks to coastal regions and communities in changing world" (EMECS' 11—Sea Coasts XXVI), pp 73–81
Leont'yev IO, Ryabchuk DV, Sergeev AY (2015) Modeling of storm deformations of the sandy shore (on the example of the eastern part of the Gulf of Finland). Oceanology 55(1):147–158
Longinov VV (1956) Some methodical questions of studying the dynamics of the coastal zone. In: Labor of the institute of oceanology of the USSR academy of sciences, vol XIX, pp 144–155
Medvedev BC, Juanes JL (1981) Morpholithodynamic research in the coastal zone and on the shelf of the northern coast of Cuba—continental and island shelves. Landscape and rainfall. M.: "Science" pp 229–251
Odum Y (1975) Basics of ecology. Mir, Moscow 741p
Reimers NF (1994) Ecology (theories, laws, rules of principles and hypotheses) M.: Magazine "Russia Molodaya", 367p
Rosenberg GS, Chernikova SA, Krasnoschekov GP, Krylov YM, Gelashvili DB (1998) Myths and the reality of sustainable development. Togliatti, IEBB RAS, pp 130–154
Santana JRH, González RR (2002) Varadero Beach, Hicacos Península, Cuba: relief genesis and evolution, and environmental experience linked to its artificial regeneration. Investigaciones Geográfica, Universidad Nacional Autónoma de México vol 49, pp 43–56
Sochava VB (1963) Definition of some concepts and terms of physical Geography. Dokl. Siberia and Daln Geography Institute of the East. 3, pp 50–59
Watt K (1971) Ecology and natural resource management. Mir, Moscow, 463p
Watt KEF (1973) Principles of environmental science. McGraw-Hill Book Company, New York, 319p
Zakharov VF (1990) Marine forecasts. L.: Hydrometeoizdat, 319p

Anisotropy Study of Statistical Characteristics of Wind Waves Under the Influence of Hydrodynamic Perturbations in Laser-Reflective Method

V. N. Nosov, S. G. Ivanov, V. I. Timonin, and S. B. Kaledin

Abstract The results of experimental studies of the anisotropy of the statistical characteristics of wind waves under the influence of hydrodynamic disturbances by the laser-glare method are considered. The experiments were carried out in a non-flowing basin under controlled conditions of wind and hydrodynamic action. During the experiment, hydrodynamic disturbances were created by the submerged propeller of the electric motor. As a source of wind exposure, an air fan with an adjustable operating mode is used. Processing of recorded signals is carried out using special algorithms and programs. It has been determined the next statistical characteristics of informative parameters: the radii of curvature of the water surface, time intervals between pulses and phase angles, defining the slopes of the waves.

Keywords Hydrodynamic disturbance · Wind disturbance · Anisotropy · Laser-Glare method

1 Introduction

The study of hydrodynamic processes in the World Ocean is actual problem now. Consideration of processes occurring in the water mass as well as on the sea surface is not only of scientific value, but also of great importance for applications. The urgent issue is to study the influence of various deep-water hydrodynamic processes on the near-surface water layers and on the sea surface directly.

Among the other research techniques, the optical remote methods are of special note because of certain advantages. For instance, laser-glare method based on the laser beam sea surface scanning appear to be highly effective in the solving such problems (Wu et al. 1981; Zubkov et al. 1997; Esipov et al. 1986; Chandra and Ghosh 2008; Shilin and Nokakov 2004; Phillips 1980; Monitor 1983; Shifrin 1981; Bakhanov et al. 2007; Nosov et al. 2017).

V. N. Nosov (✉) · S. G. Ivanov
Vernadsky Institute of Geochemistry and Analytical Chemistry of the RAS, Moscow, Russia
e-mail: viktor_nosov@mail.ru

V. I. Timonin · S. B. Kaledin
N.E. Bauman Moscow State Technical University, Moscow, Russia

C. T. Olegovna (ed.), *Processes in GeoMedia—Volume I*,
Springer Geology, https://doi.org/10.1007/978-3-030-38177-6_11

In actual field conditions, marine environment depends significantly on hydrometeorological parameters, namely: wind direction and force, temperature, humidity, underwater flows, etc. In particular, the spatial correlation of directions of the wind and underwater flows caused by natural or artificial sources of hydrodynamic disturbances (HDD) is of great importance for laser-glare method.

Full-scale marine experiment under controlled conditions is very difficult technical challenge, because it is practically impossible to provide stability of main parameters, such as: wind force and direction, HDD intensity, etc. Therefore, in our opinion, the laboratory simulation is possible to solve the problem under discussion.

The paper is mainly devoted to experimental laboratory studies of HDD influence on the wind wave statistical characteristics, which are detected by laser-glare method taking anisotropy into account.

2 Experimenting

The purpose of experimental research is to study the anisotropy of wind wave statistical characteristics of the water surface under HDD influence by laser-glare method.

The laser-glare method is implemented using scanning laser locator (SLL) model. SLL is a laser optoelectronic device based on well-known principle of scanning a water surface with a narrow laser beam and recording reflected glare signals. The SLL scheme is shown in Fig. 1.

SLL uses a scheme with coincided optical axes of the laser illumination channel and the photodetector channel. In illumination channel, a semiconductor laser with continuous emission of 80 mW power and 0.65 μm wavelength is used. To form a narrow laser beam a special optical system (FOS) is used, the system creates a spot not exceeding 8 mm in diameter on the water surface.

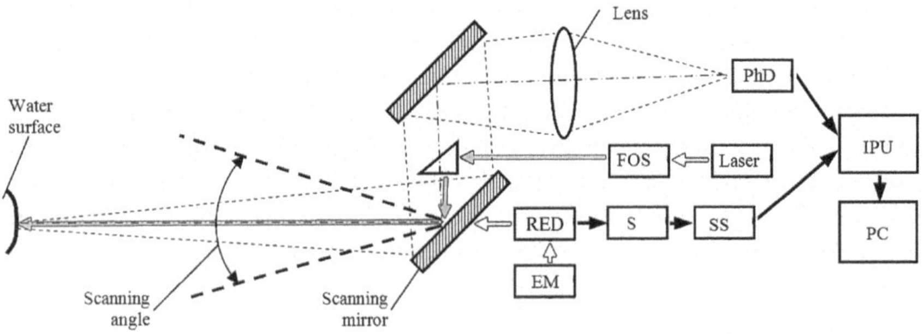

Fig. 1 Scheme of the scanning laser locator: FOS—forming optical system; EM—electric motor; RED—reducer; PhD—photodetector; S—synchronization sensor; SS—synchronization system; IPU—information processing unit; PC—personal computer

Photodetector channel consists of lens and photodetector (PhD). At the PhD input point diaphragm and interference filter, fitting to the wavelength of laser radiation, are installed. Photoelectric multiplier (PhEM) is used as photodetector. The narrow instantaneous angular field of photodetector channel (about 4 mrad) provides high spatial resolution and noise immunity to background radiation. The optical axes of the channels are coincided on the scanning mirror. The scanning angle is ±11° to the vertical. The water surface is scanned at 7 Hz frequency along the straight path.

Synchronization system (SS) is used for the subsequent processing of SLL signals. Reference signal is generated there and depends on angular position of the scanning mirror. At the unit of information processing (IPU) signals are digitized and transmitted to the personal computer (PC).

The main feature of the mirror-reflected method is generation of short pulse signals when the optical axis coincides with the normal to the water surface. There is analytical relation between equivalent curvature of water surface and amplitude of pulse signal, as well as between wave slope and moment of pulse occurrence. Thus, statistical characteristics, namely: radius of surface curvature, time interval between pulses and phase angle defining the glare position relative to vertical, are to be informative. Figure 2 schematically shows the basic informative parameters registered in the course of experiment.

Experimental studies were carried out in the indoor water pool of 10 × 5 m and 4 m deep filled with fresh water. The screw shaft of electric motor submerged to the depth of 25 cm was used as a source of HDD. It was located at distance of 1.7 m from the SSL scanning zone. The water temperature was about 16 °C.

As the source of the wind exposure the air fan with an adjustable operating mode was used. The rotational velocity of the fan varied discretely and had three modes: slow (V_1), medium (V_2) and maximum (V_3). The position of the fan could be fixed in accordance with the scheme in Fig. 3.

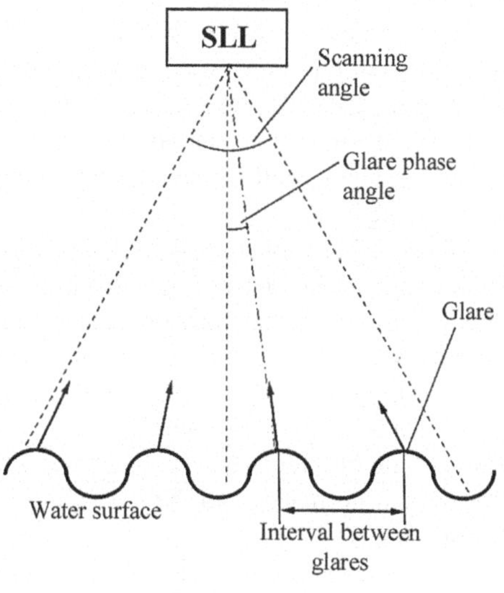

Fig. 2 Scheme of registration of wind wave parameters

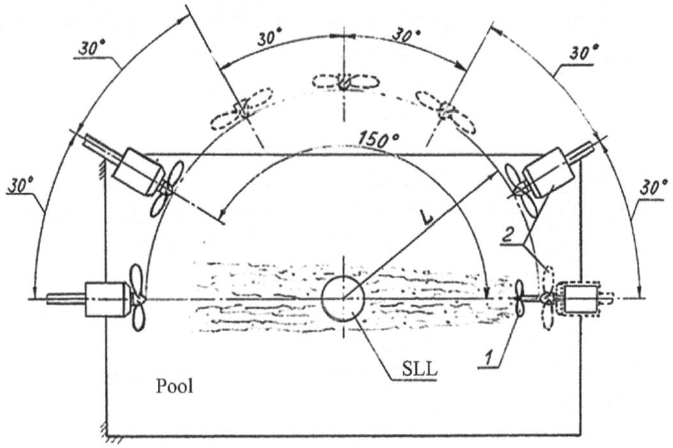

Fig. 3 Scheme of the experiment in water pool: 1—underwater HDD-source; 2—wind fan

The model of the scanning laser locator (SLL) was suspended above the center of the pool at a height of 2.3 m from the water surface. The scanning line of the laser beam passed along the axis of the pool and the direction of the under-water stream from the HDD-source (screw shaft). The length of SLL scanning line on the water surface is about 80 cm.

3 Procedure and Results of the Experiment

Experimental studies were carried out at two values of the wind velocity, V_1 and V_3. The position of the air fan has been changed during the experiment along the circular line with radius $L = 2.5$ m, with step of 30°. The origin of the angles (0°) coincided with the position of the propeller axis. The wind velocity values, V_1 and V_3, at this distance were equal to: 0.8 m/s and 1.6 m/s, respectively.

Each experiment was started at the stable state of the water surface. After switching on the fan wind swell formed during 1 min, and then electric motor with propeller was switched on.

Signals of SLL photodetector channel and synchronization system are stored in PC memory and processed by software using special algorithms. In processing the effect of low-frequency interference is eliminated, and the noise level of PhD is also estimated, and, finally, the "useful" signals from the glares are identified.

In processing peak values of pulse amplitudes, pulse duration, time intervals between adjacent pulses, counting rate of pulse (number of pulse per time unit), and phase angle of the glare appearance (i.e. relative deviation of the optical axis from the vertical normalized to the maximum scanning angle) are used as informative parameters.

Table 1 Results of the experiments

Wind velocity, angle α, deg	Average phase angle	Average time intervals between pulses, μs	Average pulse amplitude, mV	Average pulse duration, μs	Intensity, imp/s
V_1, 300	0.055	–	1772	144	129
V_1, 1200	0.064	–	2387	150	144
V_1, 1800	0.05	6672	1536	117	137.3
V_3, 00	0.06	8277	1953	132	118.9
V_3, 900	0.07	3994	2204	109	254.8
V_3, 1800	0.14	2033	2531	114	441.5

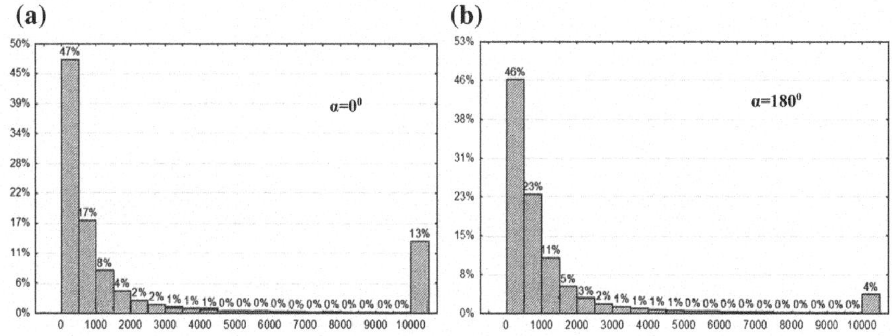

Fig. 4 Distribution of time intervals between pulses (μs)

As statistical estimates of informative parameters mentioned above, their mean values are used as well as histograms of pulse number divided by cells, constructed considering possible changes within dynamic range.

Due to the fact that different numbers of pulses are recorded in different experiments, histograms show results as percentage (relative) numbers of pulses in the cells.

As example, Table 1 shows some results of processing the signals at different fan rotational velocities V_1 and V_3.

Figures 4, 5 and 6 show the histograms of distributions of calculated pulse parameters at the fan rotational velocity $V3$.

Histograms illustrate the difference in the distributions of pulse parameters under different experimental conditions.

The anisotropy of statistical parameters is shown more clearly by graphs of dependence of the mean values of pulse parameters on various experimental conditions (angle of the fan location) (Figs. 7 and 8).

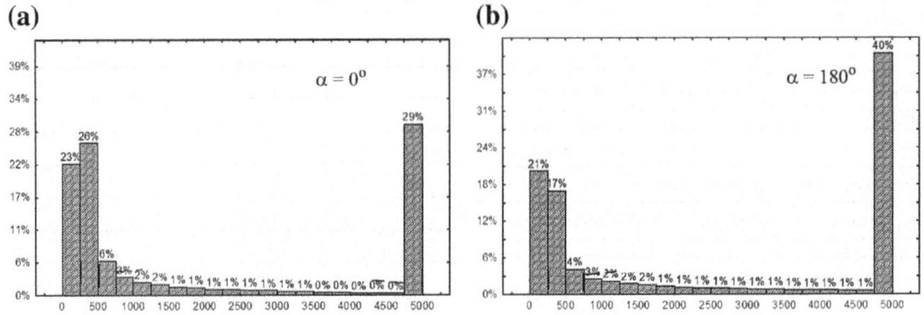

Fig. 5 Distribution of pulse amplitudes (mV)

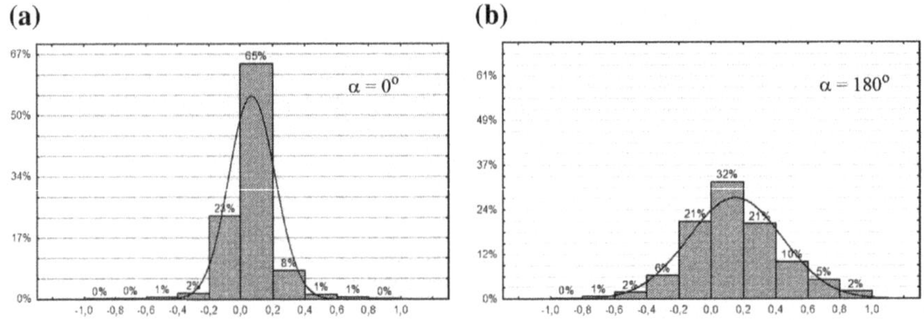

Fig. 6 Distribution of phase angles of pulse reflection (conventional units)

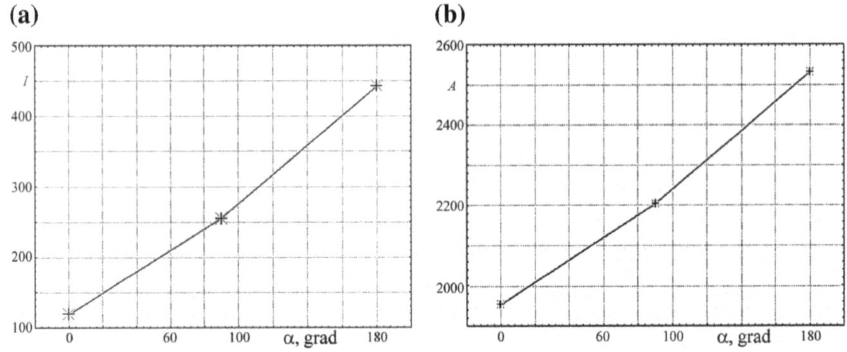

Fig. 7 Dependence of counting rate of pulse **a** and of mean value of the pulse amplitude **b** on the fan location angle α (degree)

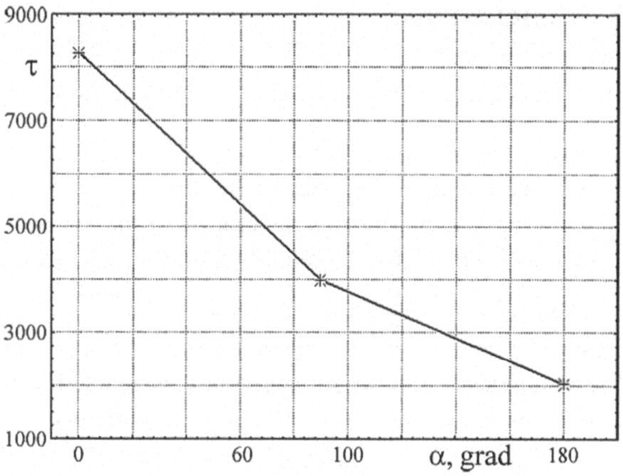

Fig. 8 Dependence of average interval value (the time between pulses) on the fan location angle α (degree)

4 Results of Processing and Discussion

Comparison of the histograms in Figs. 3, 4 and 5 shows that for different values of the angle α change in the wave spectrum is manifested in redistribution of the pulse amplitudes statistics as well as in reduction of the time intervals between the pulses. It is due to the fact that if the wind direction is opposite under-water stream shorter waves appear. The signal amplitude statistics also changes, but maximum values of signals are observed due to limitations of the dynamic range.

It should be mentioned that correlation between the signal amplitude at glare reflection (V) and the equivalent radius of the water surface curvature (R) is determined by quadratic dependence written as: $V = CR_2$, where C is a constant depending on the parameters of the photometric system (laser power, light-collecting power of the lens, distance, etc.). This allows to estimate the limit values of the curvature radius of the water surface registered within the linear range of PhD. Within the range from the noise level of PhD (about 10 mV) up to saturation level (5 V) the value of dynamic range equals to 500, which corresponds to the radius range of about 22. Assuming the minimum radius is equal to 5 mm the maximum value of radius is of 110 mm. Waves with less curvature (and larger radius) will cause signals with amplitudes corresponding to the saturation area. In this case, waves will not be taken into account in statistical processing. The statistics of phase angle change at wind velocity V_3 also depends on the direction of the wind in relation to the water flow.

When the fan rotation velocity is minimal, it's not possible to obtain so clear results. It may be caused by the fact that wind velocity value V_1 is not enough to create wind ripples with necessary intensity.

The graphs in Figs. 6, 7 and 8 clearly show the appearance of the significant anisotropy of statistical parameters of the pulses recorded by the laser-glare method.

5 Conclusions

Thus, the analysis of the experimental results shows the following:

firstly, data obtained by the laser locator are sensitive to changes caused by HDD influence on disturbed water surface;
secondly, the algorithms developed to mark out SLL pulses and estimate their characteristics allow to reliably identify these changes.

The results of experimental studies in laboratory controlled conditions show the presence of anisotropy of statistical parameters of wind waves under HDD influence. The obtained results, in our opinion, can be used in real-life conditions to identify sources of underwater hydrodynamic disturbances. Of course, this issue requires more detailed and in-depth study in order to collect sufficient data to create the model description, which will be the subject of further work.

References

Bakhanov VV, Goryachkin YuN, Korchagin NN, Repina IA (2007) Local manifestations of deep processes on the sea surface and in the near-water layer of the atmosphere. Dokl. RAS 414(1):111–115 (in Russia)
Chandra AM, Ghosh SK (2008) Remote sensing and geographic information systems. Technosphere, Moscow (in Russia)
Esipov IB, Naugolnykh KA, Nosov VN, Pashin SY (1986) Measurement of the probable distribution of the curvature radii of the sea surface. Izv. Academy of Sciences. Ser. Physics of the atmosphere and ocean, 10:1115–1117. (in Russia)
Monitor AS (ed) (1983) Optics of the ocean: applied optics of the ocean, 2nd edn. Nauka, Moscow (1983). (in Russia)
Nosov VN, Ivanov SG, Kaledin SB, Savin AS (2017) Registration of manifestation of deep processes in the near-surface layers of sea water and atmosphere. Processy v geosredah 2:522–528 (in Russia)
Phillips OM (1980) Dynamics of the upper layer of the ocean. L.: Gidrometeoizdat, 320p (in Russia)
Shifrin KS (ed) (1981) Optics of the ocean and the atmosphere. Nauka, Moscow. (in Russia)
Shilin BV, Nokakov VV (2004) Video spectral aerial survey—the leading direction of remote sensing in the optical range. Optical J. 71(3):55–58 (in Russsia)
Wu J, Heimbach SP, Hsu YuL (1981) Scanner for measuring fine sea-surfase structures. Rev Sci Instrum 52(8):1246–1251
Zubkov EV, Lyubimov YuK, Shamayev SI (1997) A method for recognizing the anomaly in the marine environment by the signals of laser aerosensing. Atm. and Ocean Optics 10(9):1093–1102 (in Russia)

Comparative Analysis of Eutrophication Level of the Sevastopol Bay Areas Based on the Results of E-Trix Index Numerical Modeling

K. A. Slepchuk

Abstract E-TRIX index gives opportunity to conduct a comparative study of different water areas' ecological state by the trophicity level. To estimate the index in the areas of the Sevastopol Bay the annual dynamics of the biogenic elements was calculated using one-dimensional variant of the water quantity model calibrated for those areas. Comparison of the E-TRIX index annual variations for different bay areas is made. It is shown that the Yuzhnaya Bay is the most contaminated one during any season of the year. The trophicity level of the rest areas depends on the season. Generally it increases in autumn-winter period when the nutrients concentrations are maximal.

Keywords Eutrophication level · E-TRIX index · Water quality model · The Sevastopol bay · The Yuzhnaya bay

1 Introduction

Serious environmental problems, for instance a significant decrease in the water quality in gulfs and bays, can be the results of the increasing anthropogenic load. It influences the ecological state of semi-enclosed shallow-water bodies. The water area (basin) trophicity level increase is one of the negative consequences of human impact on the environment. Their current state and future development. However, a standard method for assessing the sea water trophicity level does not exist. In practice, for each water area, the calculation of the ecological state is determined by a set of measured parameters, as well as the marine environment indicators (Ivanov et al. 2003).

The water trophic index E-TRIX (Vollenweider et al. 1998) is an integrated complex indicator, as a function of the deviation from 100% water saturation with oxygen, associate. The cause for the saturation of basins with nutrients, accompanied by an increase in biological productivity, may be river runoff and industrial and domestic wastewater, the effect of which is local. The assessment of eutrophication

K. A. Slepchuk (✉)
Marine Hydrophysical Institute, Russian Academy of Sciences, Sevastopol, Russia
e-mail: skira@mhi-ras.ru

C. T. Olegovna (ed.), *Processes in GeoMedia—Volume I*,
Springer Geology, https://doi.org/10.1007/978-3-030-38177-6_12

level of marine ecosystems provides the refinement ofed with the characteristics of phytoplankton primary production (content of photosynthetic pigments, mainly chlorophyll-a) and the concentration of nutrients. E-TRIX advantage over many other indicators, by which water quality can also be assessed, is that the standard characteristics of hydrochemical and hydrobiological monitoring are used for calculations. This provides a correct performance of the water ecological state comparative analysis for different marine areas in terms of their trophicity (Slepchuk et al. 2017).

The main goal of this work is to perform ecological state comparative analysis of the waters in different areas of the Sevastopol Bay by E-TRIX trophicity index using biogeochemical parameters calculated by the water quality model.

The Sevastopol Bay is oriented in latitudinal direction from east to west. It is a shallow and semi-enclosed water area. The water exchange with the open part of the sea is impeded in the bay.

The bay is under intensive anthropogenic loading since it is regularly used in economic purposes. Domestic sewage, as well as industrial and storm waters flow to the bay. The capacity of these emissions varies from 10 to 15 thousands m^3 per day. These are untreated or conditionally pure waters which contain variety of contaminants. Generally, such substances' concentrations far surpass critical standards (Ivanov et al. 2006). The Chernaya river flow also worsen the ecological state of the bay (Fig. 1). Consistent severe contamination areas appear in the Sevastopol Bay (for instance, the Yuzhnaya Bay) along with relatively "pure" areas. It is caused by locations of pollution source, hydrometeorological conditions and morphometry of the bay. Earlier in the study (Ivanov et al. 2006) the water area of the Sevastopol Bay was sliced by contamination level which is shown in Fig. 1.

In Fig. 2 diagrams of the surface layer of the Sevastopol Bay water areas are shown. They are obtained from Marine Hydrophysical Institute Oceanographic Database covering 1998–2011. In the presented figure it is shown that the most contaminated area of the Sevastopol Bay is the Yuzhnaya Bay, which is proven by maximum

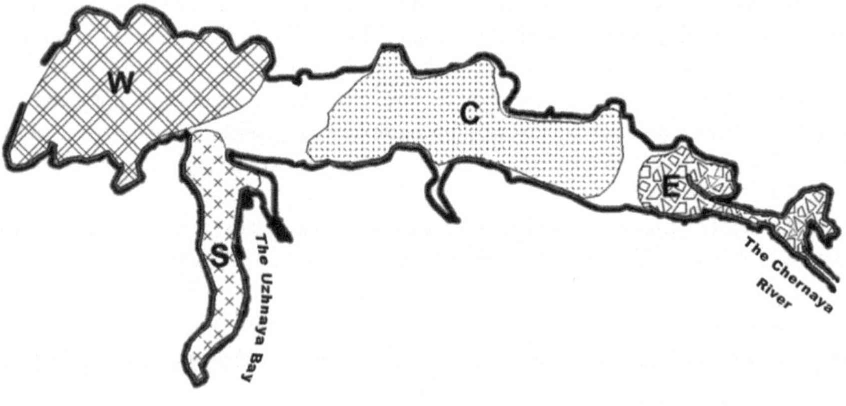

Fig. 1 The Sevastopol Bay water area contamination level (acc. to Ivanov et al. 2006): W—the areas of mild, E—moderate, C—severe and S—extremely severe contamination

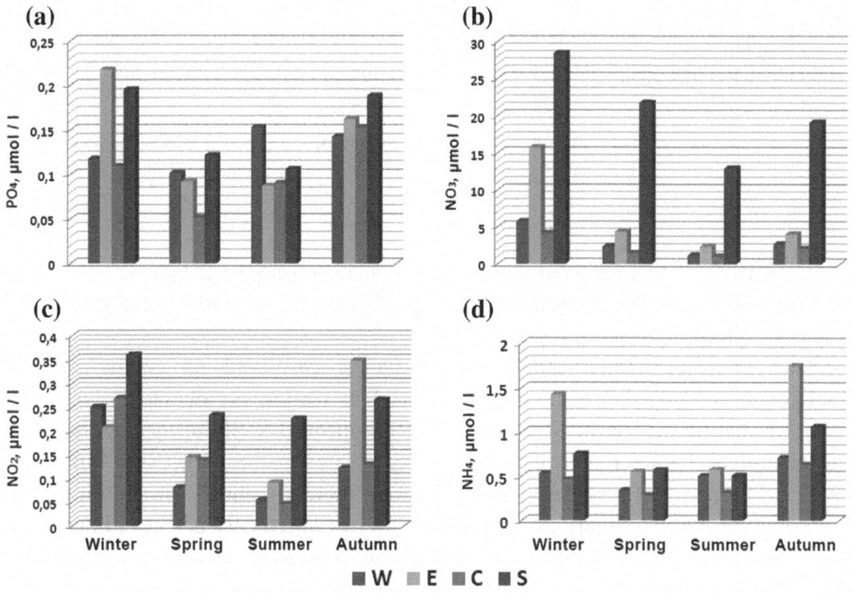

Fig. 2 Content (μmol/l) of phosphates **a**, nitrates **b**, nitrites **c**, ammonia **d** in the surface layer of the Sevastopol Bay water areas (acc. to Ivanov et al. 2006) in different seasons of the year in 1998–2011

concentration of mineral nitrogen. Also seasonal dependence is hardly in evidence which indicates the human origin of biogenic elements in the bay (Sovga et al. 2014).

2 Research Materials and Methods

E-TRIX index is used to estimate the ecological state of the bay waters. Water trophic index is the function of dissolved oxygen content and chlorophyll-a, nitrates, nitrites and ammonia, organic phosphor and phosphates concentration. The primary function of phytoplankton is characterized by chlorophyll-a. According to the study (Vollenweider et al. 1998), the trophic index is specified by formula

$$\text{E-TRIX} = \frac{(\lg[\text{Ch} \cdot D\%O \cdot N \cdot P] + 1.5)}{1.2}$$

where Ch—chlorophyll-a concentration, μg/l; $D\%O$—excursion in absolute values of dissolved oxygen content from 100% saturation; N—nitrates, nitrites and ammonia concentration, μg/l; P—organic phosphor and phosphates concentration, μg/l.

Trophic levels depending on E-TRIX values are presented in Table 1.

It is worth noting that when E-TRIX < 4 the waters have high transparency and low concentration of major biogenic elements, well aerating across the whole width.

Table 1 Trophic categories and water quality state depending on E-TRIX values

E-TRIX values	Trophic level	Water quality
<4	Low	High
4–5	Mean	Good
5–6	High	Satisfactory
6–10	Very high	Bad

When E-TRIX > 6 there is a risk of hypoxia in near-bottom water layers. The waters have low transparency and significant content of biogenic substances (Moncheva and Doncheva 2000).

Using one-dimensional quality water model case and its ecological part (Tuchkovenko 2006), data necessary for E-TRIX estimation is calculated. Earlier the model was calibrated for the Sevastopol Bay. It approved itself during calculation of hydrochemical characteristics of the whole bay and its separate parts (Slepchuk 2014; Sovga 2015). As input parameters defined are the meteorological data (humidity and cloud amount, wind velocity and direction, photosynthetically active radiation, air temperature) and annual water budget and dissolved substances run-off of the rivers which flow into the water area. In addition, the annual cycle of transparency, sea-water temperature values, salinity, phytoplankton concentration, biogenic elements, oxygen, organic phosphor and organic nitrogen defined on January 1 of the target year are used.

3 Research Results and Their Analysis

In order to assess the water trophicity level in the Sevastopol Bay, the annual variation of chemical-biological water quality parameters applied in the E-TRIX index simulation was calculated (Fig. 3).

It can be seen from the figure that during almost the entire reference year the Yuzhnaya Bay is the most polluted part of the water area (E-TRIXmean = 4.49). The Yuzhnaya Bay water quality can be described as good with an average trophicity level. The minimum value of the trophicity index (E-TRIXmin = 3.23) is observed in August, the maximum (E-TRIXmax = 5.55)—in October, which coincides with the autumn peak of phytoplankton bloom. The waters of the rest of the Sevastopol Bay have approximately the same trophicity index. The water quality of the central (E-TRIXmean = 3.63), eastern (E-TRIXmean = 3.81) and western (E-TRIXmean = 3.75) areas of the Sevastopol Bay is high with low eutrophication level. The values of trophicity indicators of the Sevastopol Bay areas are shown in Table 2.

We divide the reference year into 3 time intervals (from 1 to 120 days, from 121 to 240 days and from 241 to 365 days), then in the first third of the year, the Yuzhnaya Bay (E-TRIXmean = 4.51) can be attributed to the average eutrophication level, the western region (E-TRIXmean = 4.02) to the transitional level from medium to low, eastern (E-TRIXmean = 3.63) and central (E-TRIXmean = 3.36) regions to

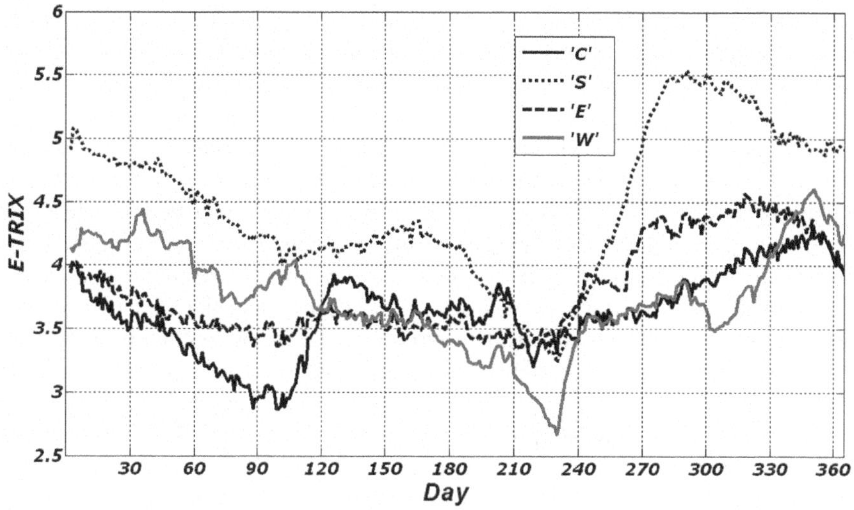

Fig. 3 E-TRIX index annual variation for the central (C), eastern (E), western (W) areas and the Yuzhnaya Bay (S)

Table 2 The values of trophicity indicators of the Sevastopol Bay areas

	"S"	"C"	"E"	"W"
$E\text{-}TRIX_{mean}$	4.49	3.63	3.81	3.75
$E\text{-}TRIX_{min}$	3.23	2.86	3.34	2.67
$E\text{-}TRIX_{max}$	5.55	4.30	4.57	4.60

loweutrophication level. In the second third of the estimated year, all areas of the Sevastopol bay have a low trophicity level (E-TRIXmean < 4), but in the third third of the year, the Yuzhnaya Bay is the most polluted part (E-TRIXmean = 5.00), then the eastern region (E-TRIXmean = 4.24) follows, central (E-TRIXmean = 3.89) and western (E-TRIXmean = 3.88) areas. The values of the average indices of the Sevastopol Bay trophicity areas in time intervals are shown in Table 3.

From the obtained results it follows that the Yuzhnaya Bay area (S) is undoubtedly the most polluted one, as indicated by the field data (Fig. 2) and the calculated trophic

Table 3 The values of trophicity indicators of the Sevastopol Bay areas in time intervals

	1–120 Computational day	121–240 Computational day	241–365 Computational day
"S"	4.51	3.93	5.00
"C"	3.36	3.63	3.89
"E"	3.63	3.52	4.24
"W"	4.02	3.35	3.88

index. However, the eutrophication level of other areas depends on the season, which is not reflected in (Ivanov et al. 2006). If in the first third of the reference year the Yuzhnaya Bay is most polluted part, then follows the western region, eastern and central, then in the second third the zoning coincides with (Ivanov et al. 2006). But in the last four months of the reference year, the eastern bay is followed by the eastern region, then the central and western ones. It should be noted that the maximums of the trophicity index of each region fall in the autumn-winter period, when the concentrations of nutrients are high, especially the mineral forms of nitrogen. They are the main factor determining the eutrophication level of the Sevastopol Baywaters. This was previously shown in (Slepchuk et al. 2017; Slepchuk and Sovga 2018). So, the western region, which is considered to be a zone of low pollution in (Ivanov et al. 2006), during the winter months can be attributed to the average eutrophication level ($4 < \text{E-TRIXmean} < 5$), which indicates a decrease in phytoplankton biomass concentration and an increase in the concentrations of nutrients. The central region, noted in (Ivanov et al. 2006) by the region of high pollution, has only an average eutrophication level from mid-October to the end of the reference year, and a low level on other days. It should also be noted that the zoning in (Ivanov et al. 2006) was carried out according to the field data of 1998–2000, and in this work the data for a longer period of 1998–2011 were used.

4 Conclusions

Modeling of the annual course of biogeochemical parameters of the Sevastopol Bay areas and further calculation of E-TRIX index showed that, on average, the Yuzhnaya Bay (E-TRIXmean. = 4.49) has a higher eutrophication level with a maximum in October. The waters of the rest of the Sevastopol Bay have a low trophicity level. Separately, it is possible to distinguish situations when E-TRIX index of the western (1–59 and 329–365 computational day), eastern (263–365 computational day) and central (315–365 computational day) regions is average, as well as the Yuzhnaya Bay (270–345 computational day) is high. This is due to the high concentrations of phosphates and, in particular, the mineral forms of nitrogen, which have the largest relative contribution to the index computational formula. The central region traditionally considered as a one of high pollution, on the average by the value of E-TRIX index, turned out to be clean.

Acknowledgements The study was carried out within the frame work of the State Order No. 0827-2019-0004 Complex interdisciplinary research of oceanographic processes determining the functioning and evolution of ecosystems of the Black Sea and the Sea of Azov coastal zones, as well as with a partial financial support of RFBR and the city of Sevastopol within the framework of scientific project No. 18-45-920002 p_a.

References

Ivanov VA, Tuchkovenko, YS (2006) Applied mathematical modeling of water quality of shelf Marine Ecosystems. Sevastopol, p 368

Ivanov VA, Mikhailova EN, Repetin LN, Shapiro NB (2003) The Sevastopol Bay model. Reconstruction of vertical structure of salinity and temperature fields in 1997–1999. Morskoy Gidrofizicheskiy Zhurnal. 4:15–35

Ivanov VA, Ovsyanyy EI, Repetin LN, Romanov AS, Ignat'eva OG (2006) Hydrological and hydrochemical regime of the Sevastopol Bay and its changing under influence of climatic and antropogenic factors. Sevastopol: ECOSI-Gidrofizika, 90 p

Ivanov VA, Mezentceva IV, Sovga, EE, Slepchuk KA, Khmara TV (2015) Assessment of self-purification ability of the Sevastopol Bay ecosystem in relation to inorganic forms of nitrogen. Processes in GeoMedia 2(2):55–65

Moncheva S, Doncheva V (2000) Eutrophication index ((E) TRIX)—an operational tool for the Black Sea coastal water ecological quality assessment and monitoring. In: International symposium "The Black Sea ecological problems". Odessa: SCSEIO, pp 178–185

Slepchuk KA (2014) Simulation of annual dynamics of phytoplankton and biogenic elements in the Sevastopol Bay using the optimization method to calibrate the biogeochemical model. Ecological safety of coastal and shelf zones and comprehensive use of shelf resources. Sevastopol: ECOSI-Gidrofizika 28:231–236

Slepchuk KA, Sovga EE (2018) Eutrophication level of the Sevastopol Bay eastern region by the results of E-TRIX index numerical modeling. Ecological safety of coastal and shelf zones and comprehensive use of shelf resources. Sevastopol 2:5359

Slepchuk KA, Khmara TV, Man'kovskaya EV (2017) Comparative assessment of the trophic level of the Sevastopol and Yuzhnaya Bays Using E-TRIX Index. Physical Oceanography 5:67–78

Sovga EE, Mezentseva IV, Khmara TV, Slepchuk KA (2014) On the prospects and opportunity to assess the self-purifying capacity of the Sevastopol Bay. Ecological Safety of Coastal and Shelf Zones and Comprehensive Use of Shelf Resources. Sevastopol: ECOSI-Gidrofizika 28:153–164

Vollenweider RA, Giovanardi F, Montanari G (1998) Characterization of the trophic conditions of marine coastal waters with special reference to the NW Adriatic Sea: proposal for a trophic scale, turbidity and generalized water quality index. Environmetrics. 9(3):329–357. https://doi.org/10.1002/(sici)1099-095x(199805/06)9:33.0.co;2-9

Geyser Eruption Mechanism as a Way of Extracting and Utilizing Underground Heat

A. M. Nechaev, A. A. Solovyev, and D. A. Solovyev

Abstract This article proposes a method for extracting hot water from a well drilled until the point where temperature exceeds the boiling point of water at a given depth. Core of this method lies on the physical mechanism of the eruption of a natural geyser. It bases on the cold water erupting from the well after entering inside from the surface of the earth due to the gas liquid imbalance (hereafter GLI). This column of water rises up due to the mutual movement between contacting liquid and gas by the pressure of the gas applied on it. The paper researches the physical conditions under which spontaneous outflow of hot water from the well to the surface begins. Described method of extraction of underground heat allows to eliminate the cost of energy required for the operation of circulating pumps of traditional geothermal power plants, since the heated water will be thrown out to the outside, obeying the GLI mechanism. Thus, this method can be considered as an environmentally friendly and significantly reduces the cost of extracting heated water from the well.

Keywords Lithosphere · Geysers · Volcanic eruptions · Underground heat · GLI-mechanism

1 Introduction

Geysers, in other words, natural fountains of hot water, are some of the most amazing and mysterious phenomena in nature. Geysers can be found in a variety of places around the world. For instance, in Kamchatka; in one of the regions of Tibet at the altitude of 4700 m above sea level, as well as in countries such as Iceland, New Zeeland and North America (Kiryukhin 2016; Rojstaczer et al. 2003; Barrick 2007).

A. M. Nechaev · A. A. Solovyev (✉)
Department of Geography, Lomonosov Moscow State University, Moscow, Russia
e-mail: solovev@ocean.ru

A. A. Solovyev
Moscow State Academy of Water Transport, Moscow, Russia

D. A. Solovyev
Shirshov Institute of Oceanology, Russian Academy of Sciences, Moscow, Russia

C. T. Olegovna (ed.), *Processes in GeoMedia—Volume I*,
Springer Geology, https://doi.org/10.1007/978-3-030-38177-6_13

Single geysers of a small size are found mostly in some other volcanic regions of the globe. There are different types of geysers exist. Some of them emit water in very low amounts or just merely spray it around the geyser. While other hot springs are more of the form of puddles, where water boils and bubbles. Typically, a regular geyser has a pool or a shallow crater of a few meters in diameter around it. The edges of such a pool and the adjacent area are covered with deposits of silica contained in boiling water. These deposits are called geyserite. Cones of geyserite are formed up to several meters high near some of the geysers. Before the eruption process starts, the water rises and slowly fills the pool, then bubbles and splashes, and only then the fountain of boiling water explodes high. After the eruption of the geyser, the pool is freed from water through the water-filled canal—a vent that goes deep into the earth (Rudolph et al. 2012).

Geysers and hot springs carry heat to the surface of the earth (Nechayev 2012a). There is yet no unequivocal answer to the question of what the precise mechanism of this process is (Nechayev 2012b). Usually, geysers are formed in areas where still hot magma is placed near to the earth's surface. The gases and vapors that are emerging from them rise up form depth and go a long way along the cracks. They dissolve in the underground waters and heat them. This highly heated water rises up to the surface of the Earth in the form of hot boiling springs, various mineral spring, and geysers. It is assumed that under the ground the geyser system consists of caves (chambers) and the connecting passages, cracks, and canals that can appear in cooled lava flows. These caves are filled with circulating hot underground water. This water under the temperature of superheated vapors, which rises from magmatic spots, is heated to temperatures above the boiling point of water. The power of geyser eruptions depends on different criteria. These are the size of underground chambers and canals; the location of cracks, through which heat from the magma spots rises; the amount and speed of the inflow of groundwater.

The pressure of the water column in the geyser canal increases the boiling point of water at the bottom of the canal. The heated lower layer of water becomes less dense and rises up to the surface, while the colder water from the surface descends, where it warms, in turn, then rises, etc. Overheated water that rises up through the canal will reach the level at which the pressure of the water column decreases so much that this water can boil. Moreover, the elasticity of the water vapor will be able to throw it with huge force upwards in the form of a boiling fountain.

If the geyser canal is wide enough and more or less in a regular shape, then water can mix, boil and periodically splash onto the surface in the form of a hot fountain. If the canal form is tortuous and narrow, water cannot be mixed and warmed evenly. Due to the pressure on top of the water column, the lower layers of water turn out to be overheated and do not turn into steam. Steam is emitted only through individual bubbles. The compressed vapor that is accumulated on the lower levels tends to expand. This expansion results in increased pressure on the upper layer of water in the canal. This causes upper water levels to lift up and spill out onto the surface of the Earth in a form of small fountains—precursors of geyser eruption.

Then the weight of the water column in the canal is reduced, consequently, the pressure at the depth is reduced as well and superheated water, while is still in a form of water heated above the boiling point instantaneously transforms into a steam. The steam pressure from below is so great that it pushes water out of the canal—thus, a huge fountain of boiling water and steam are thrown into the air. Cooled water partially falls into the geyser bowl and enters its canal. Some water rises from the depths, but most of it usually seeps into the canal from side bedrocks. In the channel it is heated until the superheated state in its lower parts is reached, then steam appears again, and steam mixture emissions occur, it means that the geyser is starting to erupt in a full power. Thus, the periodicity of geysers action depends on the channel sizes (but not of its shape), on the time necessary to fill geyser with water and on heating to a temperature somewhat above the boiling point of water at the location of a geyser, which depends on the altitude of the terrain.

The use of renewable geothermal energy of the Earth's bowels will be especially effective in areas of active volcanic activity and where the numerous geysers are present (the Kamchatka peninsula, the Kuril Islands, the islands of the Japanese archipelago, the island of Iceland, New Zealand) (Orlova 2014; Sevastopol'skiy 1975; Renner and Reed 2017; Trukhin et al. 2005). In Russia, geothermal energy, by the utilization of the geysers heat, can be effectively obtained in the Far East (Kamchatka, Sakhalin and the Kuril Islands). However, the idea and projects in implementation of geyser heat in human needs in those regions are not feasible due to the high degree of seismicity, as well as the difficult terrain, sparsely populated cities and poorly developed infrastructure. At the moment, the potential of geothermal power engineering in Russia is not fully utilized (only 5 stations have been built: Verkhne-Mutnovskaya GeoPP, Mendeleevskaya GeoTES, Mutnovskaya GeoPP, Ocean GeoTES and Pauzhetskaya geothermal power plant) (Svalova and Povarov 2015). Moreover, some of them are out of service and mothballed.

2 Study Objective

This article describes a relatively simple physical mechanism that helps to an ecologically flawless and renewable way of extraction to be realizable. This method extracts from the ground the natural heat contained therein in the form of hot water or another heated liquid and allows to eliminate the cost of energy required for the operation of circulating pumps of traditional geothermal power plants. This physical mechanism of instability in the structure "hydrostatic liquid"—"ideal gas", is designated as the mechanism of gas-liquid instability "GLI (Gas-Liquid-Imbalance) that allows explaining the phenomenon of volcanic eruptions and more. The essence of the GLI mechanism is that at certain parameters of the contacting liquid and gas their uncontrolled mutual displacement begins: the liquid column erupts under the action of a gas pressing on it. This mechanism, amenable to an easy experimental verification, and was described in (Belousov et al. 2013; Nechayev 2016, 2017). It is of a fundamental nature: the nature of liquid and gas, as well as their density, viscosity, and temperature, do not play a fundamental role in it. Theoretical description of this mechanism on the model "liquid-gas" structure is described below, with the definition of the instability criterion. And it will be shown how this mechanism can be applied to realize the method of extracting underground heat with minimal energy costs.

3 Theoretical Description of the Physical Eruption Mechanism of the Natural Geyser

Assuming that there is a vertical tank (rectangular or cylindrical) with solid walls (a canal or a crack) that is filled completely with some liquid. There is also a closed volume filled with gas. Initially, the liquid and gas are in equilibrium and have a region of direct contact in accordance with Fig. 1. The section of the tank is S, the volume of the gas cavity is V, the gas pressure p_g. In the contact area at depth $Z = 0$, the hydrostatic pressure of the liquid is equal to $p_0 + \rho g H$, where ρ—is the density of the liquid, and p_0 atmospheric pressure at $z = H$ (Fig. 1). The gas pressure p obeys the equation of state of an ideal gas, which in the case of an adiabatic process has the form:

$$p_g V^\gamma = A = \text{const} \tag{1}$$

where γ is the adiabatic coefficient for a given gas ($\gamma = 1,4$ for water vapor). Since the change in the volume of a gas is a process that goes much faster than all the processes of heat exchange, it can be regarded as adiabatic.

Assuming that the volume of the gas cavity has increased by ΔV a small amount because some part of the gas penetrated into the canal. In this case, the same volume of liquid ΔV, equal to $S\Delta z$ will be displaced from the canal, where Δz—liquid column height decrease. The corresponding decrease in the hydrostatic pressure of

Fig. 1 The diagram illustrating the GLI mechanism

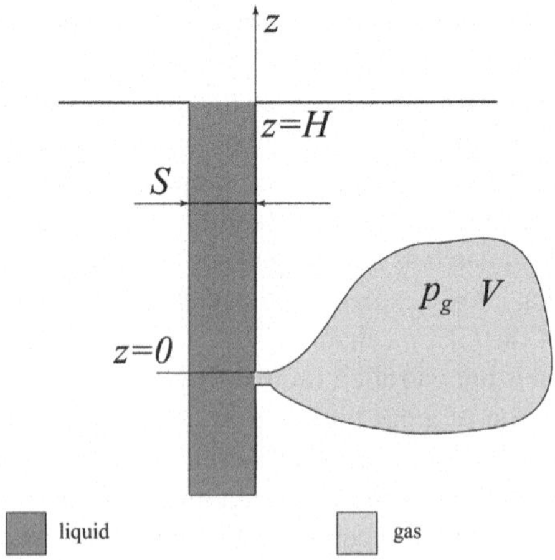

the liquid Δp_l in the contact area is:

$$\Delta p_l = \rho g \Delta z = -\rho g \Delta V / S \tag{2}$$

The pressure of the gas volume also fell by the value Δp_g in accordance with Eq. (1):

$$\Delta p_g = \frac{\partial p_g}{\partial V} \Delta V = -\frac{\gamma A}{V^{\gamma+1}} \Delta V \tag{3}$$

Thus, the larger the volume of the gas cavity, the smaller the pressure drop of the gas upon its expansion into the liquid canal. If the structure parameters (H, S, V) are such that $|\Delta p_g| < |\Delta p_l|$, then the pressure of the liquid column in the contact area is decreased faster than the pressure in the gas cavity, and the gas will begin to push the liquid out of the canal, like a piston. Let's define the critical parameters of the "gas piston" effect.

The value of the constant A in (3) is obtained from the condition that the pressures of the gas and the liquid column in the contact region at the beginning of the process are equal:

$$A / V^{\gamma} = p_0 + \rho g H \tag{4}$$

From (1), (3) and the instability conditions, we obtain a criterion for the eruption of liquid from the channel:

$$V > \gamma S(H + p_0/\rho g) \equiv V_{\text{cr}} \tag{5}$$

For example, in the case of a geyser, erupting liquid is water, gas then is water vapor that accumulates when boiling water in an underground cavity. Consequently, $\gamma = 1,4$; $p_0/\rho g = 10\mathcal{M}$. The value of the right-hand side of the formula (5) is a certain critical volume V_{cr}. If the volume of the gas cavity V is much higher that the critical volume V_{cr}, then the pressure drop between the gas and the liquid is increased. The eruption of the liquid has an accelerating pattern until, for example, all liquid amount is pushed out of the canal and the gas is pushed outwards as well. If $V < V_{\text{cr}}$ holds, then there is no instability in the geyser. Gas safely penetrates into the liquid without pushing it out and rises up to the surface in the form of bubbles popping up.

4 Method

Method for extracting geothermal heat can be proposed on the base of natural mechanism of geyser eruption for a well with a temperature gradient. In the well-known methods, the cooled heat carrier (water) enters the casing pipe, and then, heated by

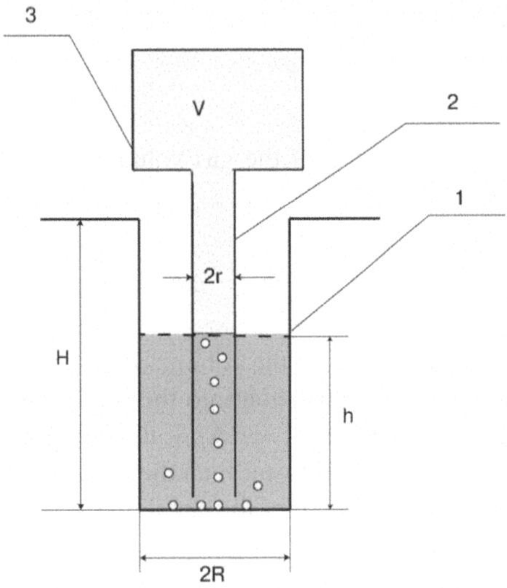

Fig. 2 The structure of the well for obtaining hot water. 1—casing, 2—internal pipe with leakproof tank 3. *H*—depth of the well; *h*—the water level in the pipe, *V*—the volume of the tank, *r*, *R*—the radii of the inner and outer pipes

underground heat, rises through an internal pipe concentrically arranged in the casing (Waring 1915). A disadvantage of those methods is that for extracting hot water it is necessary to spend large energy supplying the electric motor of the circulation pump. The method in this article allows the exclusion of these energy costs because the heated water is discharged outside itself, subject to the GLI mechanism.

The essence of the method is illustrated in Fig. 2. The device comprises a casing 1 and an inner tube 2 coaxially disposed in the borehole. At the lower end of the casing 1, a plug is made, and the casing 1 itself reaches a depth where the ground temperature is above the boiling point of water T_b corresponding to the depth *h*. There is a gap between the casing pipe plug and the bottom edge of the inner pipe. The upper part of the inner pipe 2 is covered by a sealed tank 3, whose volume *V* must exceeds the volume of the intertube space.

Work principle of the device is as follows. Water from the surface of the earth enters the intertube space until the level h is reached. At the bottom of the casing, the temperature should exceed the boiling point of the water under the hydrostatic pressure of the column of height h. Therefore, water vapor bubbles form on the bottom of the tube, and then must rise upwards both in the intertube space (and then out) and inside the pipe 2, fill the volume of the leakproof tank 3 and gradually push water from the pipe 2 into the intertube space. Further, the GLI mechanism described above, which determines the operation of natural geysers, should begin to function. The spontaneous outflow of water outside from the intertube begins when two conditions are met. First, water vapor fills the entire volume of the inner tube 2 along with the tank 3 and begins to penetrate the intertube space. Second: the water in the intertube space is squeezed by steam from the inner tube and after it reaches the upper edge of the casing it starts to flow outwards (Fig. 3).

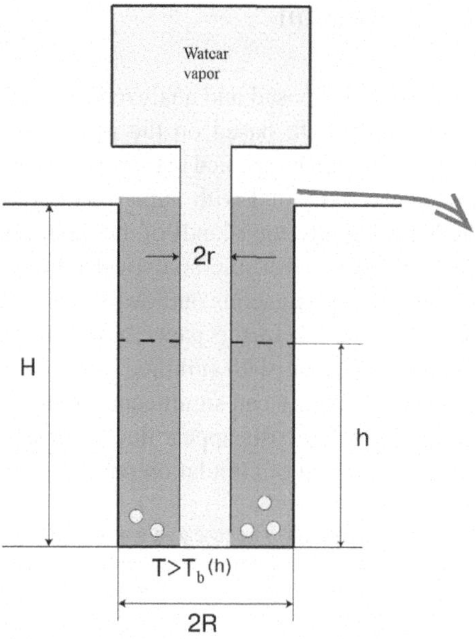

Fig. 3 The structure of the well before the "eruption" of hot water

In order to ensure the "eruption" of water, the volume of steam that is equal to the total volume of pipe 2 and tank 3 must exceed the critical volume that is equal to the volume of the intertube space with respect to condition (5):

$$V + \pi r^2 H > 1,4\pi H(R^2 - r^2) \tag{6}$$

where V—is the volume of the tank; R, r—are the radii of the casing and inner pipe, respectively.

The right-hand side of condition (6) is, in this case, the critical volume of steam that must be exceeded in order for the GLI mechanism to work so that the steam from the inner tube squeezes the hot water out of the intertube space outwards. With the help of condition (6), given the radii of the casing and inner pipe, a tank of volume V is produced for the operation of the GLI mechanism. The larger the volume V, the more intense the eruption of hot water outside.

Another condition must be imposed on the volume of water entering the casing and intended for heating. This volume, which is equal to $h\pi R^2$ in accordance with Fig. 2, should be enough for water that is squeezed by steam into the intertube space, to reach the edges of the casing. The corresponding condition for h is as follows:

$$h > H\left(1 - \frac{r^2}{R^2}\right) \tag{7}$$

Thus, the flow of water into the well, its heating, and extraction to the outside is done in portions that are equal to or greater than the volume of the intertube space.

5 Conclusion

This article proposed and analyzed the method for obtaining hot water from a high-temperature well, based on the physical mechanism of the eruption of a natural geyser. This can be applied in the development of geothermal energy in areas of active volcanic activity and with numerous geysers (the Kamchatka Peninsula, Sakhalin, the Kuril Islands, the islands of the Japanese archipelago, the island of Iceland, New Zealand). Moreover, the analytical relationship that determines the conditions for the onset of spontaneous outflow of hot water from the well to the surface is derived here. It is shown in the paper how this mechanism can be applied to extract the underground heat with minimal energy costs. Practical use of this mechanism in geothermal energy can significantly reduce the cost of extracting heated water from a well. Lowered costs appear due to the exclusion of the cost of electricity required for the operation of circulation pumps of geothermal power plants.

Acknowledgements The research was carried out within the state assignment of IO RAS, theme 0149-2019-0002 (methodical development of in situ data) and of M. V. Lomonosov MSU theme AAAA-A16-116032810088-8 (theoretical description).

References

Barrick KA (2007) Geyser decline and extinction in New Zealand-energy development impacts and implications for environmental management. Environ Manage 39:783–805. https://doi.org/10.1007/s00267-005-0195-1

Belousov A, Belousova M, Nechayev A (2013) Video observations inside conduits of erupting geysers in Kamchatka, Russia, and their geological framework: implications for the geyser mechanism. Geology 41:387–390. https://doi.org/10.1130/g33366.1

Kiryukhin A (2016) Modeling and observations of geyser activity in relation to catastrophic landslides-mudflows (Kronotsky nature reserve, Kamchatka, Russia). J Volcanol Geotherm Res 323:129–147. https://doi.org/10.1016/j.jvolgeores.2016.05.008

Nechayev AM (2012a) Kamchatka. Hot land at the cold sea, 3rd edn. LOGATA, Moscow

Nechayev AM (2012b) O mekhanizme izverzheniya vulkana. Fizicheskiye problemy ekologii (Ekologicheskaya fizika), pp 551–568

Nechayev A (2016) On the mechanism of catastrophic caldera-forming eruptions: yellowstone's approval. J Geogr Environ Earth Sci Int 6:1–9. https://doi.org/10.9734/jgeesi/2016/27330

Nechayev A (2017) Geyser eruption mechanism: natural and empirical verification. J Geogr Environ Earth Sci Int 9:1–9. https://doi.org/10.9734/jgeesi/2017/31316

Orlova YY (2014) Ispol'zovaniye geotermal'nogo tepla, kak al'ternativnogo istochnika energii (na primere respubliki Islandiya). In: Vvedeniye v energetiku. Sbornik materialov I Vserossiyskoy molodezhnoy nauchno-prakticheskoy konferentsii., p 18

Renner JL, Reed MJ (2017) Geothermal energy. In: Energy conversion, 2nd edn, p 12

Rojstaczer S, Galloway DL, Ingebritsen SE, Rubin DM (2003) Variability in geyser eruptive timing and its causes: yellowstone national park. Geophys Res Lett 30. https://doi.org/10.1029/2003gl017853

Rudolph ML, Manga M, Hurwitz S, Johnston M, Karlstrom L, Wang CY (2012) Mechanics of old faithful Geyser, Calistoga, California. Geophys Res Lett 39. https://doi.org/10.1029/2012gl054012

Sevastopol'skiy AY (1975) Geotermal'naya energiya: Resursy, razrabotka, ispol'zovaniye, Mir
Svalova V, Povarov K (2015) Geothermal energy use in Russia. Country update for 2010–2015. In: World geothermal congress
Trukhin VI, Pokazeev KV, Kunitsyn VE (2005) General and ecological geophysics. Fizmatlit, Moscow
Waring GA (1915) Springs of California. US Government Printing Office

Analytical Solutions of Internal Gravity Waves Equation for Model Buoyancy Frequency Distribution

V. V. Bulatov and Yu. V. Vladimirov

Abstract The problem of constructing analytical solutions for the equation of internal gravity waves in a stratified medium with variable Brunt-Vaisala frequency is considered. Four model buoyancy frequency distributions that describe the existence of water density jump layer in the ocean are proposed. For each model, analytical solutions describing the dispersion curves and eigenfunctions of the main vertical spectral problem are obtained. The results of numerical calculations for the parameters characteristic for the real ocean are presented.

Keywords Internal gravity waves · Variable buoyancy frequency · Eigenfunctions · Dispersion curves

An important mechanism for exciting fields of internal gravity waves (IGW) in the ocean is their generation by sources of perturbations of various physical natures, i.e., of the natural (moving typhoon, wind waves, flow over unevenness of the bottom topography, variations in the density and flow fields, leeward mountains) and anthropogenic (offshore technological structures, collapse of the turbulent mixing region, underwater explosions) origin (Miropol'skii and Shishkina 2001; Bulatov and Vladimirov 2012, 2019; Mei et al. 2017; Morozov 2018; Velarde et al. 2018). For example, moving atmospheric cyclones have a significant impact on ocean circulation, local sea surface temperature, and the internal gravity waves generation. Wave fields induced by this type of generation mechanism can play prominent part in variations of energy transport in the ocean depth. The drag force of the wind generated by a moving hurricane cause the formation of the wave plume or wake structures in the ocean. Experimental detection of these wave structures is one of the impressive achievements of modern Oceanology (Voelker et al. 2019). The propagation of dispersing IGW in stratified ocean has features associated directly with the dependence of the propagation velocity on the wavelength. If some source of perturbation moves in such a dispersing medium, it creates an encircling wave pattern with the pronounced lines of the constant phase as the main feature. The structure of wave patterns at large distances from a moving source (far field, distance well above

V. V. Bulatov (✉) · Yu. V. Vladimirov
Ishlinsky Institute for Problems in Mechanics RAS, Moscow, Russia
e-mail: internalwave@mail.ru

C. T. Olegovna (ed.), *Processes in GeoMedia—Volume I*,
Springer Geology, https://doi.org/10.1007/978-3-030-38177-6_14

size of source) practically does not depend on its shape and is determined mainly by the law of dispersion and the velocity of the source (Svirkunov and Kalashnik 2014).

The system of hydrodynamic equations describing wave perturbations is a rather complex mathematical problem and the principal solutions of the internal wave generation problems can be obtained only in the overall integral form or numerically. In numerical calculations, the Ocean is usually depicted by simplified system of hydrodynamic equations with a model density distribution. Integral representations of solutions require the development of asymptotic methods for their study, allowing qualitative analysis and quick evaluation of the obtained solutions during full-scale measurements of IGW characteristics in the Ocean. Therefore, simplified analytical models are widely used in contemporary scientific research and analysis of wave phenomena in the real stratified ocean. In the linear approximation, the existing approaches to the description of the wave pattern of the emergent IGW fields are based on the representation of wave fields by Fourier integrals and the analysis of the obtained spectral problems. The goal of this paper is construction of analytical solutions of the IGW equation for model distributions of the buoyancy frequency.

In this paper, we consider the problem of the free fields of internal gravity waves characterized by a harmonic time dependence in the form $\exp(i\omega t)$. Assuming the Boussinesq approximation, we have the following equation, for example, for the vertical displacement of isopycnic lines $\eta(x, y, z)$ (lines of equal density) (Miropol'skii and Shishkina 2001; Bulatov and Vladimirov 2012, 2019; Mei et al. 2017):

$$\omega^2\left(\frac{\partial^2}{\partial x^2}+\frac{\partial^2}{\partial y^2}+\frac{\partial^2}{\partial z^2}\right)\eta+N^2(z)\left(\frac{\partial^2}{\partial x^2}+\frac{\partial^2}{\partial y^2}\right)\eta=0$$

where $N^2(z) = -g/\rho_0(z)\, d\rho_0(z)/dz$ is the square of the Brunt-Vaisala frequency (buoyancy frequency), $\rho_0(z)$—the unperturbed density of the stratified medium considered as a function of the depth, g—the gravity acceleration. The solution is sought in the form of the Fourier integrals:

$$\eta(x, y, z)=\frac{1}{4\pi^2}\int_{-\infty}^{\infty}d\nu\int_{-\infty}^{\infty}\varphi(\mu, \nu, z)\exp(-i(\mu x+\nu y))d\mu$$

Then, for determining function $\varphi(\mu, \nu, z)$ it is necessary to solve the basic vertical boundary value problem of internal waves (Bulatov and Vladimirov 2012, 2018, 2019)

$$\frac{\partial^2\varphi(k, z)}{\partial z^2}+k^2\left(N^2(z)\omega^{-2}-1\right)\varphi(k, z)=0,\quad k^2=\mu^2+\nu^2$$

Further, for the analytical solution of this spectral problem, various model distributions of buoyancy frequency will be used. The distributions are applied in oceanologic calculations to study the internal gravity waves dynamics in the presence of a constant thermocline (a density jump layer). The presence of a constant layer of the density jump which is the area of rapid temperature variation is a remarkable feature of the World Ocean. The dependence of the Brunt-Vaisala frequency on the depth in model representations can differ from the empirical dependences, which, however, are characterized by the existence of maximum $N^2(z)$ in the layer of jump of density of the marine environment. The model distributions of buoyancy frequency allow solving the problem analytically, while using empirical dependencies requires application of only numerical methods. However, numerous researches confirm the fact that the main qualitative results on the description of the internal gravity waves dynamics depend, as a rule, not on the specific analytical form of the buoyancy frequency approximation, and on existence of maximum $N^2(z)$ in the layer of jump of oceanic water density (Mei et al. 2017; Morozov 2018; Velarde et al. 2018).

The first model is an infinite medium with the quadratic buoyancy frequency distribution: $N^2(z) = N_0^2 - 4\chi^2 z^2$ $(-\infty < z < \infty)$. In non-dimensional coordinates and variables

$$x^* = x/L, \quad y^* = y/L, \quad z^* = z/L, \quad k^* = kL,$$
$$\omega^* = \omega/N_0, L = N_0/2\chi$$

the main spectral problem is presented in the form (index * is omitted hereafter)

$$\frac{\partial^2\varphi(z,k)}{\partial z^2} + k^2\left((1-z^2)\omega^{-2} - 1\right)\varphi(z,k) = 0, \quad \varphi \to 0 \; at \; z \to \pm\infty \tag{1}$$

Functions $D_\lambda\left(\pm z\sqrt{2k/\omega}\right)$, where $\lambda = \left(k - \omega - k\omega^2\right)/2\omega$ and $D_\lambda(\tau)$ is the function of parabolic cylinder (Watson 1944), are the linearly independent solutions of this equation. These solutions satisfy the boundary conditions at $\lambda = n$ $(n = 0, 1, 2, \ldots)$. These conditions determine the dispersion relations: $\omega_n^1(k) = \left(-2n - 1 + \sqrt{(2n+1)^2 + 4k^2}\right)/2k$ and the eigenfunctions: $\varphi_n(z,k) = D_n\left(z\sqrt{2k/\omega}\right) D_n(-\tau) = (-1)^n D_n(\tau)$. The eigenfunctions can be expressed through Hermite polynoms, using the equality: $D_n(\tau) = 2^{-n/2} H_n(\tau)\exp\left(-\tau^2/2\right)$.

The second model is a three-layer medium of finite depth Z: $N^2(z) = N_0^2 - 4\chi^2 z^2$ $|z| \le N_0/2\chi$, $N^2(z) = 0$ at $N_0/2\chi < |z| < Z$. Then, at $|z| < 1$, in the non-dimensional variables, the main spectral problem can be written in the form (1). At $1 < |z| < h$, $h = Z/L$ we have:

$$\frac{\partial^2\varphi(z,k)}{\partial z^2} - k^2\varphi(z,k) = 0$$
$$\varphi(h,k) = \varphi(-h,k) = 0$$

Let us introduce designations: $F_\pm(z) = D_\lambda\left(\pm z\sqrt{2k/\omega}\right)$. In the interval $(-h, -1)$ the solution can be written as: $\varphi_-(z) = \operatorname{sh} k(z+h)$. In the interval $(-1, 1)$ the solution is sought in the form: $\varphi(z) = C_1 F_-(z) + C_2 F_+(z)$. The matching condition which is an equality of the solutions $\varphi_-(z)$, $\varphi(z)$ and their derivatives, at $z = -1$

gives

$$C_1 = \left(\operatorname{sh} k(h-1)F_+{}'(-1) - k\operatorname{ch} k(h-1)F_+(-1)\right)/W(k,\omega)$$
$$C_2 = \left(\operatorname{sh} k(h-1)F_-{}'(-1) - k\operatorname{ch} k(h-1)F_-(-1)\right)/W(k,\omega)$$
$$W(k,\omega) = 2\sqrt{\pi k/\omega}/\Gamma\left((-k+\omega+k\omega^2)/2\omega\right)$$

where $W(k,\omega)$ is Wronskian of functions $F_\pm(z)$, $\Gamma(\tau)$ is Gamma function. The matching condition at $z = 1$ gives the solution in the interval $(1, h)$ in the form: $\varphi_+(z) = \varphi'(1)\operatorname{sh} k(z-h)/\operatorname{sh} k(1-h)$. The respective dispersion relation is of the form: $\varphi'(1)/\varphi(1) = k\operatorname{cth} k(1-h)$. Solving this equation for variable k, we get a set of dispersion curves: $\omega_n^2(k)$, $n = 0, 1, 2, \ldots$.

The third model is an infinite three-layered medium: $N^2(z) = N_0^2 - 4\chi^2z^2$ at $|z| \le N_0/2\chi$, $N^2(z) = 0$ at $z > N_0/2\chi$. At $|z| \le 1$, in non-dimensional variables, the main spectral problem can be represented in the form (1). At $|z| > 1$ we have

$$\frac{\partial^2\varphi(z,k)}{\partial z^2} - k^2\varphi(z,k) = 0$$

Using matching conditions at $z = \pm 1$, for eigenfunctions in the interval $(-\infty, -1)$ one can obtain: $\varphi_-(z) = \exp(kz)$. In the interval $(-1, 1)$ we have

$$\varphi(z) = \exp(-k)F_-(z)\left(F_+{}'(-1) - kF_+(-1)\right)/W(k,\omega)$$
$$+\exp(-k)F_+(z)\left(-F_-{}'(-1) - kF_-(-1)\right)/W(k,\omega)$$

In the interval $(1, \infty)$, the eigen functions are of the form: $\varphi_+(z) = \varphi(1)\exp k(1-z)$. The equation $\varphi'(1)/\varphi(1) = -k$ defines dispersion curves $\omega_n^3(k)$, $n = 0, 1, 2, \ldots$ Obviously, at $h \to \infty$, dispersion curves $\omega_n^3(k)$ approach $\omega_n^2(k)$.

The fourth model is a three-layer infinite medium: $N(z) = N_0$ at $|z| \le L$, $N(z) = 0$ at $|z| > L$. In non-dimensional variables, the spectral problem can be represented in the form

$$\frac{\partial^2\varphi(z,k)}{\partial z^2} - k^2\left(\omega^{-2} - 1\right)\varphi(z,k) = 0 \text{ at } |z| \le 1$$
$$\frac{\partial^2\varphi(z,k)}{\partial z^2} - k^2\varphi(z,k) = 0 \text{ at } |z| > 1,$$
$$\varphi \to 0 \text{ at } z \to \pm\infty$$

Let us denote $\Omega = k\sqrt{1-\omega^2}/\omega$. Next, it is necessary to use the matching conditions. Then in the interval $z < -1$ one can get: $\varphi_-(z) = \exp(kz)$. In the interval $|z| \le 1$ we have:

$$\varphi(z) = \exp(-k)((k\cos\Omega - \Omega\sin\Omega)\sin\Omega z$$
$$+(k\sin\Omega + \Omega\cos\Omega)\cos\Omega)/\Omega$$

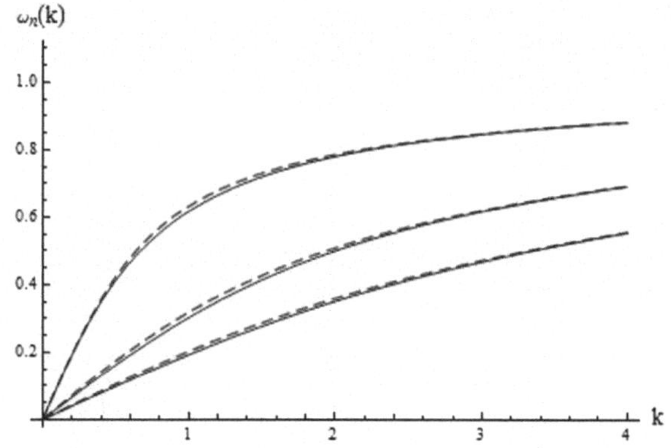

Fig. 1 Dispersion curves $\omega_n^1(k)$ (solid lines), $\omega_n^2(k)$ (dashed lines) for first three modes

In the interval $z > 1$ we obtain: $\varphi_+(z) = \exp(-kz)(k\sin 2\Omega + \Omega\cos 2\Omega)/\Omega$. Solving the dispersion relation that can be written in the form $\mathrm{tg}2\,\Omega = 2k\Omega/(\Omega^2 - k^2)$ with respect to ω numerically, we can get a countable set of dispersion curves: $\omega_n^4(k)$, $n = 0, 1, 2, \ldots$. Using the equality $\mathrm{tg}\,2\,\Omega = 2\,\mathrm{tg}\,\Omega/(1 - \mathrm{tg}^2\Omega)$ for determining dispersion dependencies, we obtain two equations: $\mathrm{tg}\,\Omega = k/\Omega$ and $\mathrm{tg}\,\Omega = -\Omega/k$. The solutions of the first equation define the dispersion curves with even mode numbers: $\omega_n^4(k)$, $n = 0, 2, 4, \ldots$. The eigenfunctions in this case are even and are of the form: $\varphi_-(z) = \exp(kz)$, $\varphi(z) = \exp(-kz)\cos\Omega z/\cos\Omega$, $\varphi_+(z) = \exp(-kz)$. Solving the second equation, we will obtain a dispersion relation with odd mode numbers: $\omega_n^4(k)$, $n = 1, 3, 5, \ldots$ and the respective eigenfunctions: $\varphi_-(z) = \exp(kz)$, $\varphi(z) = -\exp(-k)\sin\Omega z/\sin\Omega$, $\varphi_+(z) = -\exp(-kz)$.

Figure 1 shows the dispersion curves of the first three modes $\omega_n^1(k)$ (solid line), $\omega_n^2(k)$ at $h = 2$ (dashed line). It can be seen from the figures that the first model with negative values of the square of the buoyancy frequency well approximates the dispersion curves of the second model. Figure 2 shows the eigenfunctions of the first three modes at $k = 0.1$, the solid line represents the first model, the dashed line represents the second model. Dispersion curves of the first three modes $\omega_n^1(k)$ (solid line), $\omega_n^3(k)$ at $h = 2$ (dashed line) are shown in Fig. 3. From the results presented, it can be seen that the dispersion curves of the first mode differ in qualitative behavior at small k: the dispersion curve $\omega_0^1(k)$ has a limited phase velocity, and $\omega_0^3(k)$ has an unbounded phase velocity at small wave numbers $\left(\omega_0^3(k) \approx \sqrt{k}\right)$. Thus, the dispersion curve $\omega_0^3(k)$ combines both the properties of surface waves at small k and internal waves at large k. The wave modes $\omega_n^3(k)$, starting with the first one, have a limited phase velocity near zero. Figure 4 presents the results of calculating the dispersion curves $\omega_n^4(k), n = 0, 2, 4, \ldots$ (dashed line) and $\omega_n^4(k), n = 1, 3, 5, \ldots$ (solid line).

Thus, analytical solutions of the basic boundary problem obtained for model distributions of buoyancy frequency allow, further, efficient calculating amplitude-phase characteristics of fields of internal gravity waves in a stratified medium with a nonconstant Brunt-Vaisala frequency and, in addition, qualitative analyzing the

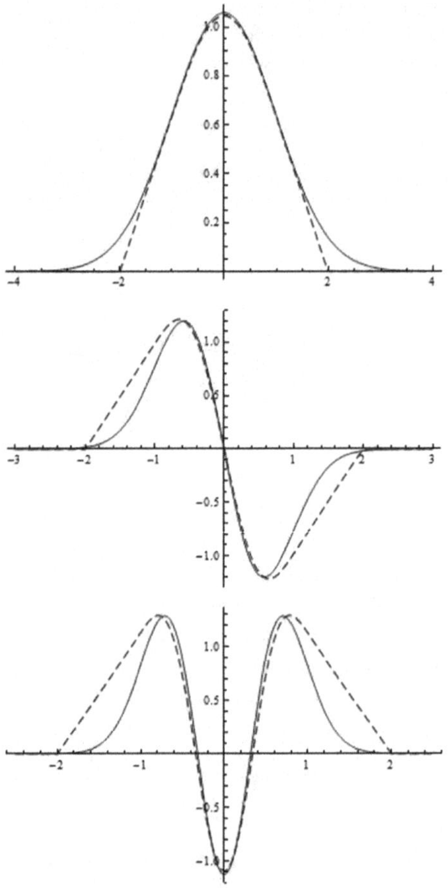

Fig. 2 Eigen functions of first three modes for first (solid lines) and second (dashed lines) models

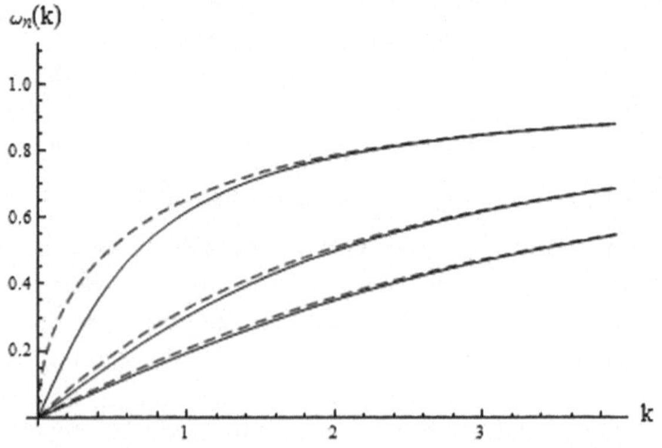

Fig. 3 Dispersion curves for first three modes $\omega_n^1(k)$ (solid lines), $\omega_n^3(k)$ (dashed lines)

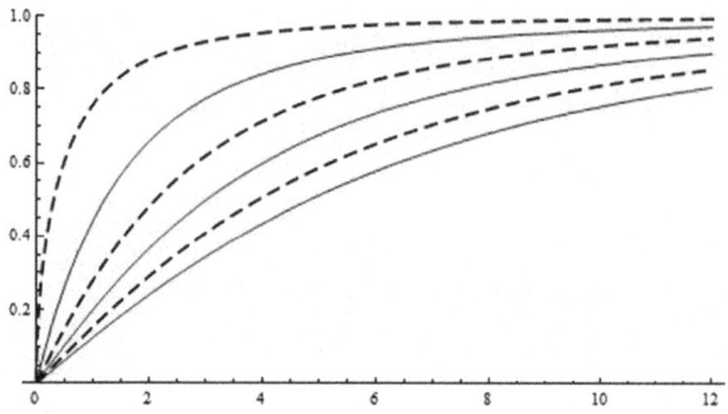

Fig. 4 Dispersion curves $\omega_n^4(k)$, $n = 0, 2, 4, \ldots$ (dashed lines) and $\omega_n^4(k)$, $n = 1, 3, 5, \ldots$ (solid lines)

solutions obtained. It is important for the correct formulation of mathematical models of wave dynamics of the real ocean.

Acknowledgements The work was carried out within the framework of RFBR project 20-01-00111A.

References

Bulatov VV, Vladimirov YuV (2012) Wave dynamics of stratified mediums. Nauka Publishers, Moscow, 584 p

Bulatov VV, Vladimirov YuV (2018) Unsteady regimes of internal gravity wave generation in the ocean. Russ J Earth Sci 18:ES2004

Bulatov VV, Vladimirov YuV (2019) A general approach to ocean wave dynamics research: modelling, asymptotics, measurements. OntoPrint Publishers, Moscow, 587 p

Mei CC, Stiassnie M, Yue DK-P (2017) Theory and applications of ocean surface waves. Advanced series of ocean engineering, vol 42. World Scientific Publishing, 1500 p

Miropol'skii YuZ, Shishkina OV (2001) Dynamics of internal gravity waves in the ocean. Kluwer Academic Publishers, Boston, p 406

Morozov TG (2018) Oceanic internal tides. Observations, analysis and modeling. Springer, 317 p

Svirkunov PN, Kalashnik MV (2014) Phase patterns of dispersive waves from moving localized sources. Phys Usp 57(1):80

Velarde G, Tarakanov RYu, Marchenko AV (eds) (2018) The ocean in motion. Springer Oceanography. Springer International Publishing AG, part of Springer Nature, 625 p

Voelker GS, Myers PG, Walter M, Sutherland BR (2019) Generation of oceanic internal gravity waves by a cyclonic surface stress disturbance. Dyn Atm Oceans 86:16

Watson GN (1944) A treatise on the theory of Bessel functions, 2nd edn. Cambridge University Press, Cambridge, 804 p. https://doi.org/10.2307/3609752

Dynamics of Riverine Water in the Black Sea Shelf Zone

M. V. Tsyganova and E. M. Lemeshko

Abstract The paper investigates the distribution of Danube riverine water and formation of the water hydrological structure, the plume and buoyancy current response to various types of wind forcing on the North-Western shelf of the Black Sea using numerical modeling and satellite data analysis. Numerical experiments were conducted for a 3-dimensional σ-coordinate model, adapted to the North-Western shelf conditions, taking into account the bottom topography of the computational area and the actual configuration of the coastline. Wind impact scenarios for numerical experiments were formed using the assessment of winds different directions recurrence at the North-Western shelf and in the area of the Danube delta according to the ERA-Interim reanalysis. As a result, patterns of formation and evolution of the discharge thermohaline front and buoyancy current depending on the wind forcing of different directions were revealed. Comparison of calculations with satellite maps for upward radiation at a wavelength of 551 nm confirms the simulation results.

Keywords Shelf · Coastal dynamics · Plume · Baroclinic current · River discharge · Thermohaline front · Numerical modelling · Wind mixing

1 Introduction

The fundamental scientific problem is the study of plume formation and the distribution of river water on the shelf as a result of the driving factors: shelf water stratification, wind forcing, river runoff in different seasons, the influence of bottom topography. The aim of the paper is to study the distribution of riverine water from the Danube, formation of the water hydrological structure, plume and alongshore buoyancy current under various types of wind forcing on the Black Sea North-Western shelf by using numerical simulation taking into account the actual bottom topography.

When the river water flows into the sea it forms mesoscale structures in the estuary, which have low salinity and temperature compared to the surrounding water, high

M. V. Tsyganova (✉) · E. M. Lemeshko
Marine Hydrophysical Institute, Russian Academy of Sciences, Sevastopol, Russia
e-mail: m.tsyganova@mhi-ras.ru

C. T. Olegovna (ed.), *Processes in GeoMedia—Volume I*,
Springer Geology, https://doi.org/10.1007/978-3-030-38177-6_15

content of suspended matter and dissolved organics. In modern papers, such structures are called "plumes" (Tsyganova and Lemeshko 2017; Garvine 1995; Yankovsky et al. 2004; Fong and Geyer 2002). Under free inertial movement, the plume takes the shape of a circle or bulge with an anticyclonic current. In the anticyclonic direction from the river mouth (in the Northern Hemisphere), the discharge current forms. In case the plume is in contact with the bottom, the Ekman transfer of riverine water in the bottom boundary layer leads to an additional flow of buoyancy from the coast. Thus, the plume does not form a quasi-circular anticyclonic structure, but takes the form of a strip that expands from the source along the coast south of the river mouth (Zhurbass et al. 2011; Vlasenko et al. 2013).

In accordance with the geostrophic balance of forces, riverine water spreads along the coast and forms alongshore current, which is directly related to the river water flowing into the sea (Fong and Geyer 2002; Zhurbass et al. 2011). The least studied aspects of the river plume dynamics, formation and evolution of the thermohaline front and coastal currents are their responses to the variability of river inflow, wind forcing, and interaction with dynamic structures (eddies, filaments, upwelling) on the shelf, the effects of various types of water stratification in the coastal zone, bottom topography and coastline configuration. To study these processes, it was used a three-dimensional sigma-coordinate numerical model of the joint dynamics of the shallow sea (Zhurbass et al. 2011), which gives a qualitatively correct position of the discharge thermohaline front near the Danube estuary and adequately describes its evolution in the coastal zone and the formation of alongshore buoyancy current under forcing of different wind directions.

2 Data and Methods

We used a three-dimensional sigma-coordinate numerical model of the joint dynamics of the shallow sea (Fong and Geyer 2002; Vlasenko et al. 2013; Ivanov and Fomin 2008). The model is based on three-dimensional equations of circulation in the σ-coordinate system $\sigma = (z - \eta)/H$, where x, y, z are Cartesian coordinates, $H = h + \eta$ is the sea depth, h (x, y) is the sea bottom relief, η (x, y, t) is the sea level. The initial system of equations is (Fong and Geyer 2002; Vlasenko et al. 2013):

$$\frac{\partial uH}{\partial t} + gH\frac{\partial \eta}{\partial x} - fvH + \frac{H}{\rho_0}\frac{\partial p}{\partial x} = \frac{\partial}{\partial \sigma}\left[\frac{K_M}{H}\left(\frac{\partial u}{\partial \sigma}\right)\right] + HF_X + G_X, \quad (1)$$

$$\frac{\partial vH}{\partial t} + gH\frac{\partial \eta}{\partial y} + fuH + \frac{H}{\rho_0}\frac{\partial p}{\partial y} = \frac{\partial}{\partial \sigma}\left[\frac{K_M}{H}\left(\frac{\partial v}{\partial \sigma}\right)\right] + HF_Y + G_Y, \quad (2)$$

$$\frac{\partial \eta}{\partial t} + \frac{\partial uH}{\partial x} + \frac{\partial vH}{\partial y} + \frac{\partial w}{\partial \sigma} = 0, \quad (3)$$

$$\frac{\partial TH}{\partial t} + \Lambda T = \frac{\partial}{\partial \sigma}\left[\frac{K_H}{H}\left(\frac{\partial T}{\partial \sigma}\right)\right] + HF_T, \quad (4)$$

$$\frac{\partial SH}{\partial t} + \Lambda S = \frac{\partial}{\partial \sigma}\left[\frac{K_H}{H}\left(\frac{\partial S}{\partial \sigma}\right)\right] + HF_S. \quad (5)$$

here u, v, w are the components of velocity in x, y and σ, respectively; ρ_0 is the average density, f is the Coriolis parameter, T, S are the water temperature and salinity. The terms HF_X, HF_Y, HF_T, HF_S parameterize the horizontal turbulent viscosity.

The coefficients of vertical turbulent viscosity and diffusion are determined by the formulas (Fong and Geyer 2002; Vlasenko et al. 2013): $K_M = \max(LqS_M, K_{Mf}), K_H = \max(LqS_H, K_{Hf})$, where $q^2/2$ is the turbulent kinetic energy; L is the macroscale of turbulence; S_M, S_H are the functions of the dynamic Richardson number; K_{Mf}, K_{Hf} constants are the background values. The q, L functions are found by solving the balance equations for turbulent energy, which are solved together with the main problem.

The following boundary conditions are set on the surface and bottom:

$$\begin{aligned}
&\sigma = 0,\ w = 0,\ \frac{K_M}{H}\left(\frac{\partial u}{\partial \sigma}, \frac{\partial v}{\partial \sigma}\right) = \frac{1}{\rho_0}\left(\tau_X^0, \tau_Y^0\right),\\
&\frac{K_H}{H}\left(\frac{\partial T}{\partial \sigma}, \frac{\partial S}{\partial \sigma}\right) = \frac{1}{\rho_0}(Q_T, Q_S),\\
&\sigma = -1,\ w = 0,\ \frac{K_M}{H}\left(\frac{\partial u}{\partial \sigma}, \frac{\partial v}{\partial \sigma}\right) = \frac{1}{\rho_0}\left(\tau_X^B, \tau_Y^B\right),\\
&\frac{K_H}{H}\left(\frac{\partial T}{\partial \sigma}, \frac{\partial S}{\partial \sigma}\right) = 0,
\end{aligned} \quad (6)$$

where Q_T, Q_S are the heat and salt fluxes. Equations for tangential stresses on the free surface and on the bottom, respectively are: $\left(\tau_X^0, \tau_Y^0\right) = \rho_a C_A |W| (w_X, w_Y)$, $(\tau_X^B, \tau_Y^B) = \rho_a C_D |U| (u, v)$, ρ_a is the air density, $C_{A,}$ C_D are the friction coefficients, $W = (w_x, w_y)$ is the wind speed at a height of 10 m, $U = (u, v)$ is the horizontal current velocity. On solid side boundaries, velocity, heat and salt fluxes are equal to zero.

The model was adapted to the conditions of the North-Western shelf of the Black Sea: coordinates of the computational area are: 28–31° E and 43–46° N, the number of grid nodes in X = 119, the number of grid nodes in Y = 167, grid spacing is 2 km, time step is 2 min, number of sigma horizons is 25. In the experiments, climatic fields of temperature and salinity were set for the conditions of April and August, the salinity of the incoming water off the Danube estuary, the water temperature at the mouth and the river flow rate water were set by climatic values.

3 Typification of Wind Fields

To select the wind forcing scenario for the numerical experiments, it was assessed the frequency of winds of various directions at the North-Western shelf off the Danube estuary. Six-hour values of surface atmospheric pressure and surface wind with a spatial resolution of 0.125° of the ERA-Interim re-analysis (http://data-portal.ecmwf.int/data/d/interim_full_moda) for 2010–2014 were used as input data. Using the Kohonen method of self-organizing maps (Lemeshko et al. 2013) it was obtained the typical synoptic situations over the North-Western shelf of the Black Sea (Fig. 1). Table 1 shows the repeatability in percent for each of the twelve composite self-organizing maps of 6-h data of atmospheric pressure and wind speed for 2010–2014.

According to the classification, six main types of the pressure field were identified, and it was found that the high wind speeds (over 20 m/s) of the North-West direction are mainly connected with the passage of deep cyclones through the central part of the sea. Accordingly, two main scenarios of wind forcing were chosen for the numerical experiments. The first is the North wind—maps No. 1, 3, 6, 9, 11, 12, the total repeatability of the North winds is 47% for the selected observation period (Table 1), typical wind situation is shown in Fig. 1a. The second scenario takes into

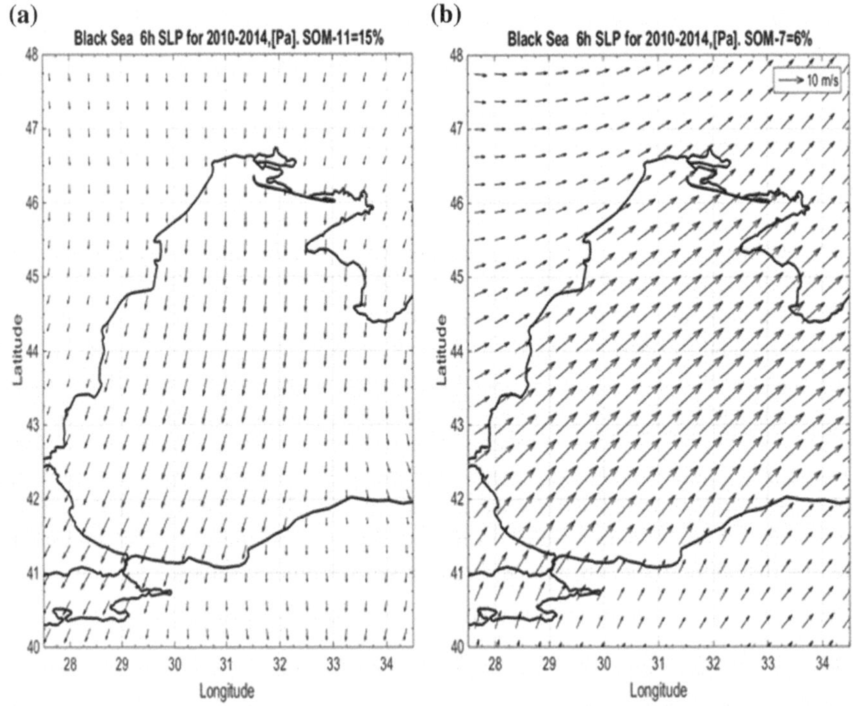

Fig. 1 Self-organizing wind speed map: **a** map No. 11, 15% repeatability and **b** map No. 7, 6% repeatability (Table 1)

Table 1 Repeatability of self-organizing surface atmospheric pressure maps according to re-analysis and wind speed in percent for 2010–2014

Map No.	1	3	4	5	6	7	8	9	10	11	12
Repeatability (%), (rounded tc integers)	7	4	5	7	9	6	10	13	7	15	10
Average wind direction over the Western part of the sea, rhumb	NW	NE	SW	S-SE	NE	SW	SE	NE	W	N-NE	NW

account the South wind—maps No. 4, 5, 7, 8, the total repeatability of the South wind is 28%, a typical field is shown in Fig. 1, b. East wind (map No. 2) and West wind (map No. 10) have the same frequency of 7% in this period and were also used in our work, however, this paper does not consider the results of these calculations.

4 Numerical Experiments

A number of numerical experiments were conducted to study the effect of wind and the actual configuration of the Black Sea western coastline on the formation of the coastal thermohaline front and density current in different seasons, taking into account the water stratification and the discharge rate of the Danube. Thus, the computational domain was prepared taking into account the coastline configuration (Fig. 2). When river water enters the coastal area, it interacts with more saline sea water thus forming thermohaline front off the Danube mouth. Under the influence of the South-West wind, the transfer of river water to the South along the coast is blocked, and the resulting front spreads North-East and East towards the inner shelf. At the same time, the riverine water moves in the anticyclonic direction and is carried to deeper to the western shelf. The North-East wind presses the plume to

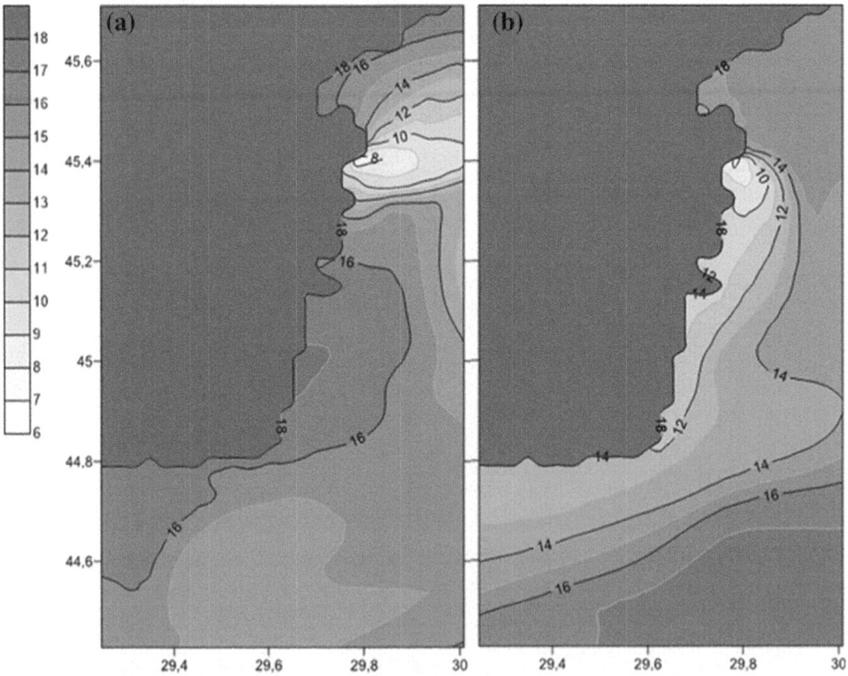

Fig. 2 Surface salinity according to simulation: **a** South wind; **b** North wind

the western shore, resulting in a coastal alongshore current that transfers the riverine water South. The North-West wind shifts the boundary of riverine water to the East. South wind (Fig. 2a) prevents the spread of riverine water downstream. Under the action of the Southern wind of 5–10 m/s, on average within 3–5 days, the resulting front shifts to the North of the mouth causing the phenomenon of blocking of the current which transfers the riverine water to the South (Fig. 2a). Accordingly, the North wind forms a coastal alongshore current, which transports the riverine water in a southerly direction (Fig. 2b).

5 Comparison with Satellite Data

In order to compare the position of the discharge thermohaline front and the alongshore current obtained as a result of numerical simulation with remote sensing data for situations with the North and South wind, its were selected MODIS satellite data (http://dvs.net.ru/mp/data) on upward radiation at a wavelength of 551 nm (the signal in green spectral band is suitable for identification of pigments, optical properties of water, suspended matter) with a spatial resolution of ~1 km. Propagation of the Danube water under the forcing of the South winds is traced in the image of the ascending radiation at a wavelength of 551 nm at 11:20 on 25.04.2018 (Fig. 3a). The wind situation developed as follows: from 23.04.2018 to 2 p.m. 25.04.2018 the South wind was blowing, with the average wind speed of ~5 m/s, the maximum

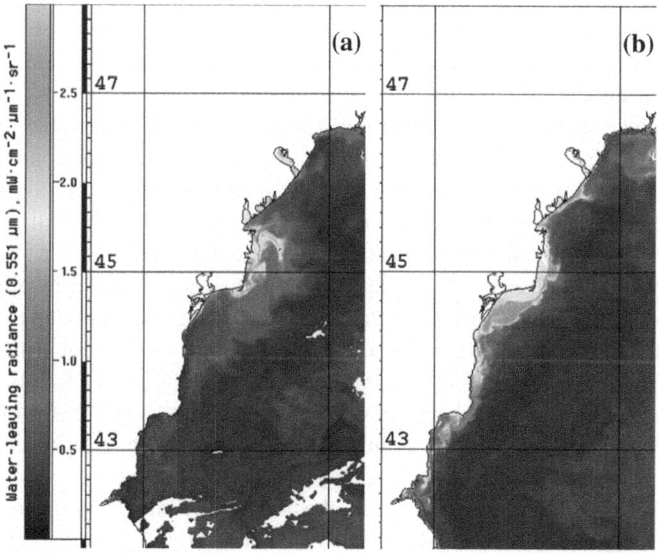

Fig. 3 Maps of upward radiation at a wavelength of 551 nm: **a** 24.04.2018, under the South wind; and **b** 10.04.2015, under the North wind

values reached 10 m/s according to the weather station in Sulina, Romania. This wind forcing blocked the Danube water propagation in a southerly direction, and even reversed the currents to the North at each of the three main branches of the delta (Fig. 3a). In addition, on 24.04.2018, the South wind caused the upwelling South of St. George Cape, which also contributed to blocking of the Danube water motion to the South.

Distribution of the current formed by the flow of riverine water from the Danube Delta to the South stands out well in the map dated 10.04.2015 (Fig. 3b). Note that during the five preceding days, April 6–10, 2015, the North wind was blowing steadily, with the average wind speed of 12 m/s and the maximum values reaching up to 21 m/s according to the weather station in Sulina, Romania (http://rp5.ru/). Such long action of a strong North wind, sometimes even stormy, led to the fact that the Danube plume shifted to the South, with the Danube water spreading South along the coast to 42.5° N.

6 Conclusions

Under the influence of the South-West wind, the transfer of river water to the South along the coast is blocked, the resulting front spreads to the North-East and East towards the deepwater area. At the same time, the riverine water moves in the anticyclonic direction and it is carried further to the western shelf. The North-East wind presses the plume to the Western shore, resulting in an intensification of coastal alongshore current that transfers the riverine water in a southerly direction. The North-West wind shifts the boundary of riverine water to the East. Comparison of the position of the outer boundary of the plume in satellite images is in good agreement with the calculated distances from the hydrological survey data.

The analysis of the model simulations output data made it possible to estimate the typical vertical scales and distances from the coast to the outer boundary of the discharge front of the Danube River. The analysis of hydrological archive CTD data on the shelf made it possible to obtain statistical characteristics of the distance to the outer boundary of the discharge front for typical conditions of water stratification on the shelf and discharge rate of the Danube River; stratification parameters and Rossby radius of deformation R_{di} were calculated using the formulas (Yankovsky et al. 2004). The minimum distance of the Danube water distribution from the coast into the sea without taking into account the wind effect was $y = 4.24\ R_{di}$, and the maximum $y = 5.46\ R_{di}$, which agrees well with both the results of hydrological data analysis (Yankovsky et al. 2004), simulation (Fig. 2), and with satellite data (Fig. 3). The obtained results give an idea of the typical spatial and temporal scale of plume formation—the vertical scale was 4–5 m, the horizontal scale (from the mouth to the outer boundary of the thermohaline front)—25–50 km, the characteristic time for establishing a stationary density flux without wind forcing was 5–10 days.

Thus, the model of joint dynamics of the shallow sea gives proper positions of the thermohaline front and the alongshore current depending on the wind direction, which is confirmed by comparing the simulation results with hydrological maps of surface salinity and satellite images of the upward radiation at a wavelength of 551 nm, which is consistent with the results (Zatsepin et al. 2017; Lavrova et al. 2016).

Acknowledgements This study was supported by the Russian Federation State Tasks No. 0827-2019-0004 and No. 075-00803-19-01.

References

Fong DA, Geyer WR (2002) The alongshore transport of freshwater in a surface-trapped river plume. J Phys Oceanogr 32:957–972
Garvine RW (1995) A dynamical system of classifying buoyant coastal discharges. Cont Shelf Res 15:1585–1596
http://data-portal.ecmwf.int/data/d/interim_full_moda
http://dvs.net.ru/mp/data
http://rp5.ru/
Ivanov VA, Fomin VV (2008) Mathematical modeling of dynamic processes in the sea-land zone. In: MHI UAS, Sevastopol, 363 p. (2008)
Lavrova OY, Mityagina MI, Kostyanoy AG (2016) Satellite methods for detecting and monitoring marine zones of ecological risk. Space Research Institute, Moscow, 334 p
Lemeshko EE, Repina IA, Lemeshko EM (2013) Identification of upwelling events by the self-organizing maps of the Black sea surface temperature. Syst Environ Control 20(19):135–139
Tsyganova MV, Lemeshko EM (2017) Interannual variability of the chlorophyll concentrations in the Black Sea and its relations with elements of the large-scale atmosphere circulation. Syst Anal Mod Econ Environ Syst Pap 2(1):98–104
Vlasenko V, Stashchuk N, McEwan R (2013) High-resolution modelling of a large-scale river plume. Ocean Dyn 63:1307–1320
Yankovsky AE, Lemeshko EM, Ilyin YP (2004) The influence of shelfbreak forcing on the alongshelf penetration of the Danube buoyant water Black Sea. Cont Shelf Res 24:1083–1098
Zatsepin AG, Zavialov PO, Kremenetskiy VV, Nedospasov AA, Poyarkov SG, Baranov VI, Ocherednik VV (2017) On the mechanism of wind-induced transformation of a river runoff water lens in the Kara Sea". Oceanology 1(57):5–12
Zhurbass VM, Zav'yalov PO, Sviridov AS, Lyzhkov DA, Andrulionis EE (2011) O perenose stoka malykh rek vdol'beregovym baroklinnym morskim techeniem. Okeanologiya 3(51):440–449

Optimization of Pollutant Emissions for Air Quality Modeling in Moscow

N. A. Ponomarev, N. F. Elansky, V. I. Zakharov, and Y. M. Verevkin

Abstract Application of chemical transport models in studying causes and mechanisms of variations in the composition of the atmosphere over megacities like Moscow requires the knowledge of the spatial distribution of pollutant emissions and their time variations. The emissions of anthropogenic pollutants have been calculated on the basis of multiyear data on the key pollutants measured at the Moscow network of atmospheric monitoring stations. Using the SILAM model, a few series of numerical experiments with calculated emissions and emissions from the TNO European inventory database have been carried out. Despite some differences between calculation and measurement results, the obtained optimum emission matrix, on the whole, adequately reflects atmospheric-composition variations that are characteristic of Moscow. This makes it possible to efficiently use the SILAM regional chemical transport model in forecasting the air quality in the Moscow megacity.

Keywords Megapolis · Atmospheric composition · Trace gases · Emissions · Numerical simulation · Air quality

1 Introduction

Determining the causes and mechanisms of variations in the composition of the lower atmosphere is one of the key problems to be solved for effective forecasting climate changes, extreme weather situations, or the presence of toxic air pollutants that are hazardous to human health and the entire ecological system. Large megacities are of great interest in studying the dynamics and spatial variability of atmospheric composition, because there are a lot of different pollutant sources, such as motor

N. A. Ponomarev · V. I. Zakharov (✉)
Faculty of Physics, M.V. Lomonosov Moscow State University, Moscow, Russia
e-mail: zvi_555@list.ru

N. A. Ponomarev · N. F. Elansky · V. I. Zakharov
A.M. Obukhov Institute of Atmospheric Physics, RAS, Moscow, Russia

Y. M. Verevkin
Dalhousie University, 6310 Coburg Road, Halifax, NS, Canada

C. T. Olegovna (ed.), *Processes in GeoMedia—Volume I*,
Springer Geology, https://doi.org/10.1007/978-3-030-38177-6_16

transport, chemical and petrochemical enterprises, non-ferrous and ferrous metallurgy enterprises, food industry enterprises, treatment facilities, and thermal power plants. Measurement data on atmospheric pollutants, such as CO, NO, NO_2, SO_2, and CH_4 are of great importance in testing and validating available methods of analyzing the air quality for densely populated territories and in adapting local and mesoscale chemical transport models.

After their averaging, systematized data on pollutant concentrations measured at the Moscow Ecological Monitoring (MEM) network of stations make it possible to determine the characteristic features of time variations in the composition and quality of the surface air layer over Moscow. Numerical simulation methods based on information about pollutant emissions from urban sources and both thermodynamical and photochemical properties of the atmospheric boundary layer (ABL) over the city are used to estimate the influence of anthropogenic emissions on concentrations of trace gases and aerosol particles. The earlier studies (Elansky et al. 2016, 2018, 2019) have made it possible to obtain necessary data on the state of the urban atmosphere and reveal its response to changes in the urban infrastructure. On the basis of these data, input parameters for the SILAM (System for Integrated modeLing of Atmospheric coMposition) dispersion model used for simulating the state of the atmosphere over the Moscow megacity were determined. This model allows one to calculate four-dimensional concentration fields with respect to spatial coordinates and time for every 10 min with steps of 0.1° in longitude and 0.05° in latitude. Continuous (in space and time) data series of pollutant concentrations are used to validate satellite data (Valin et al. 2014), characterize local pollutant sources, and provide the on-line forecasting of the urban-air quality (Pierce et al. 2010) in order to prevent situations hazardous to Moscow inhabitants.

In this work, a few series of numerical experiments have been carried out to correct the spatiotemporal distribution of urban sources and improve results of forecasting the composition of urban air in the Moscow megacity, when compared to results obtained using the TNO European inventory data (www.tno.nl/emissions).

2 Model Structure and Choosing Input Parameters

Emission data obtained in (Elansky et al. 2018) for the Moscow megacity were used as input data for the SILAM chemical transport model. The SILAM model (its fifth version) is a global-to-meso-scale dispersion model operating in both Eulerian and Lagrangian coordinates. The schemes of dry deposition of gases and dry and wet deposition of aerosols are used in this model. The latter scheme is used for a wide range of aerosol-particle sizes, including coarse aerosol particles that are removed mainly through their deposition. The parameterization of wet deposition includes the absorption of aerosol particles by water drops under clouds and their intracloud rainout due to rain or snow. This model operates with both three- and four-dimensional data (Sofiev et al. 2015a).

Both initial and boundary conditions were assigned for a region with coordinates 50.75–60.7° in latitude and 32.6–42.5° in longitude. The model grid has a resolution of 0.1° in longitude and 0.5° in latitude, 69 vertical levels to describe meteorological parameters, and 9 altitudinal levels to describe chemical transformations of assigned pollutants. Observational data on the CO, SO_2, NO_x, and NH_3 concentrations (Elansky et al. 2018) for the Moscow megacity and the TNO-2011 inventory data on CO, SO_2, NO_x, NH_3, $PM_{2.5}$, and PM_{10} emissions (www.tno.nl/emissions) were used as initial data. In addition, data on emissions of volatile organic compounds (VOCs) were taken from the MACCity inventory database (https://www.gmes-atmosphere.eu). To solve the problem stated, it is necessary to specify pollutant emissions with a time step of 10 min. To this end, the annually mean emissions for Moscow taken from (Elansky et al. 2019) are recalculated with consideration for weights determined from seasonal variations and weekly cycles of surface concentrations calculated from observational data obtained at the MEM stations and the Ecological Station of the Institute of Atmospheric Physics (IAP) and Moscow State University (MSU) (Elansky et al. 2019) and with consideration for weights determined from daily emission variations taken from the TNO-2011 inventory data.

Assigning emissions used in current models is a subject of wide discussion. In particular, the MEGAPOLI project [Megacities: Emissions, urban, regional and Global Atmospheric POLlution and climate effects, and Integrated tools for assessment and mitigation (www.megapoli.info)] was organized in Europe to solve this problem. The objective of this project was to develop and assess available methods of determining pollutant emissions in megacities and to estimate the influence of megacities and other large pollution sources on local, regional, and global air quality and climate (Baklanov and Mahura 2014). Official emission data presented by special urban institutions on request of municipal authority and state environmental protection departments and data from European inventories developed by international groups of experts are usually used in assigning emissions. The TNO-2011 inventory (the last version of the European inventory) was used to assign emission sources outside the Moscow agglomeration. This inventory is based on estimates of gases and aerosols emitted into the atmosphere from each of 40 sectors of pollution sources. Inventory data are interpolated to a spatial grid with consideration for the real distribution of sources.

Local data and European inventory data have different spatial resolutions and, sometimes, different sets of pollution sources. Since the areas covered by each inventory are exactly determined, emissions obtained from European inventory data and local data were compared in (Denier van der Gon et al. 2005). Table 1 gives the sampled data obtained from such comparisons in the form of ratios between emissions from European inventory data and local (municipal or regional) inventory data. These comparisons show that there is a significant disagreement between different inventory data. Meteorological parameters were assigned using the High Resolution Limited Area Model [HIRLAM (https://hirlam.org)] reanalysis data projected onto the SILAM model grid with a time step of three hours.

The gas-phase chemical mechanism in the model is given as a CBM-4 algorithm. This calculation scheme is an updated version of the DMAT model algorithm (Denier

Table 1 Ratios between emissions obtained from European inventory data recalculated for regional scales and from local data (Denier van der Gon et al. 2005)

Megacity	NOx	PM_{10}	$PM_{2.5}$	NH_3	SO_2	NMVOC	CO
London	1.6	3.9	–	–	9.4	1.5	3.6
Paris	1.1	3.0	3.2	2.2	1.9	1.9	4.1
Rhein-Ruhr	0.9	1.6	1.2	1.6	1.1	2.4	0.7
The Po river valley	0.9	0.9	–	1.0	1.2	0.8	1.4

van der Gon et al. 2005; Sofiev 2000). This algorithm includes the mechanisms of forming secondary organic aerosols. Two sets of compounds—biogenic VOCs and sea salt—are assigned in the model. Calculations of bio-VOCs follow the calculation algorithm described in (Kukkonen et al. 2012) and take into account both isoprene and monoterpene emissions. Determining the sea-salt parameters is an original development (by the authors of the model) based on (Kouznetsov and Sofiev 2012) with refinements to the description of the formation mechanism, which are added to the model (version 5.2). At present, dust transported by wind is taken from the boundary conditions of the C-IFS—Composition Integrated Forecasting System (https://www.ecmwf.int/en/research/ modelling-and-prediction/atmospheric-composition).

3 Description of Transport Processes

At present, there are a lot of advection schemes used in atmospheric dispersion models. Two basic types use Lagrangian or Eulerian description. The first type operates with physically infinitely small air volumes (Lagrangian particles) having a mass and the other type operates with concentration fields discretely assigned on the model grid. There are several types of the Eulerian scheme used in the SILAM: the scheme operating with finite-volume concentration fluxes, the semi-Lagrangian scheme, or the scheme using an extension function. One of the main problems of the available schemes is a significant numerical diffusion that occurs during discretization due to space-time finite steps. This phenomenon cannot apparently be avoided in using the Eulerian schemes; however, such a problem does not exist in the Lagrangian advection schemes that contain no explicit discretization of particle motions. The Lagrangian domain is a continuous space but not a set of predetermined grid cells, and the position of particles may be exactly calculated. As a result, numerical diffusion in purely Lagrangian schemes is always zero; however, in this case, the concentration fields are strongly nonmonotonic due to the limited spatial representativeness of one Lagrangian particle and limited number of particles. Using the semi-Lagrangian algorithm developed by M. Gal'perin, whose updated version is used in the fifth version of the SILAM model, may be the way out of this situation (Yanenko 1971; Marchuk 1995; Seinfeld and Pandis 2006).

The operation quality of this advection algorithm was estimated in carrying out a large set of tests by modelers. First, model results were compared with a few classical algorithms widely used in other dispersion models. Then, main tests were repeated for the updated scheme of the calculation algorithm and additional tests were performed using real data on two-dimensional wind distribution. This method has recently been proposed in literature, which made it possible to estimate the operation quality of the calculation scheme under consideration when compared to other up-to-date approaches. Finally, with sufficiently modest computational burden, the efficiency of the advection scheme was completely comparable to other algorithms (Sofiev et al. 2015b).

4 Horizontal and Vertical Diffusions, Dry Deposition

Dry deposition is usually taken into account as a boundary condition for the vertical-diffusion equation. Calculation of dry deposition for gases almost always follows from the electrical analogy whose parameterization was first proposed by Wesely (1989). This approach was updated by Sofiev (2002), who combined it with the solution of the vertical-diffusion equation and related resulting solutions to Gal'perin's advection scheme. In the algorithm, the centers of mass of cells are not modified due to diffusion; instead, the effective diffusion coefficient between neighboring layers is used. This coefficient is inversely proportional to aerodynamic resistance between the centers of mass of layers, into which the atmosphere is divided, and directly proportional to the effective layer depth determined from the value of pressure gradient between neighboring mass centers, which, in turn, is calculated with consideration for the distribution of mixing coefficient. Meteorological vertical profiles of high resolution are necessary to perform calculations. The atmosphere is divided into layers determined from available cell data obtained from calculations with the advection scheme, which noticeably increases the accuracy of Wesely's initial calculation scheme.

5 Parameterization of Chemical Interactions

Emission data are the only source of information describing a model cell, in addition to advection itself: the coordinates of the location of sources are transformed into the coordinates of the position of mass centers. For point sources, mass is added to mass corresponding to a grid cell, which results in changes in the coordinates of mass centers:

$$\begin{aligned} M'_{ijk} &= M_{ijk} + M_{ems} \\ X'_{ijk} &= \left[X_{ijk}M_{ijk} + M_{ems}X_{ems}\right]/M'_{ijk} \end{aligned}$$

$$Y'_{ijk} = \left[Y_{ijk}M_{ijk} + M_{ems}Y_{ems}\right]/M'_{ijk}$$
$$Z'_{ijk} = \left[Z_{ijk}M_{ijk} + M_{ems}Z_{ems}\right]/M'_{ijk} \quad (1)$$

where M_{ems} is the mass of emitted emissions in cell (i, j, k) over one time step, X_{ems} and Y_{ems} are the coordinates of emission source, Z_{ems} is the effective emission height assigned by default in the center of layer k.

For extended sources, the approach depends on initial emission parameters. If there is no information about the distribution of sources in grid cells, then the coordinates of the center of mass are considered identical to those of the center of cell. If, in addition to absolute values, there is information on the function of source distribution in the grid cells, then the coordinates of the center of mass are calculated with consideration for the probability density function for assigned sources.

A module responsible for chemical transformations is simple. This module calculates only variations in concentrations of substances during chemical reactions, but it does not determine variations in the position of the coordinates of the center of mass in cell. A new substance mass obtained during the operation of this block is added to initial mass and, in this case, the center of the corresponding cell is used as its coordinates. If another substance was obtained during chemical transformations, this substance is also placed in the cell center (Sofiev et al. 2015b).

6 Numerical Experiments

Emission Optimization

As was noted above, the main objective of this work is to find the most optimum spatiotemporal distribution of the intensity of sources in the Moscow megacity.

For the model to correctly operate it was necessary to assign annually mean emissions calculated in (Elansky et al. 2018) with a time step of 10 min and to distribute them over the megacity territory on the model grid. To this end, a data array obtained at the MEM network from 2005 to 2014 was analyzed. This made it possible to determine relative weights by time taking into account seasonal variations and weekly cycles of pollutant concentrations and daily emission variations taken from the TNO inventory for the Moscow megacity. When distributing emissions over the territory, the city was considered as a circle (with a radius of 25 km) divided into five sectors: central, northeastern, southeastern, southwestern, and northwestern. The total value of emissions from the megacity was assigned for each model cell with consideration for two weights: the first one was determined from the dependence of mean pollutant concentration on the distance from the city center, and the other was determined from concentration distribution over the chosen sectors (Elansky et al. 2018).

However, the spatial emission distribution obtained in such a way from an analysis of concentrations averaged over 10 years may differ from the real one. Therefore, after conducting tests and comparisons between model and observational results, it was decided to conduct a series of tests by simultaneously varying the assignment of emission distributions horizontally and vertically. The Student criterion and correlation function were used as criteria of comparing observational data with calculation results for space and time, respectively. In this case, the features of spatiotemporal variations in sources of each pollutant (Elansky et al. 2018, 2019) were taken into account. Thus, for example, in assigning nitrogen oxides, the influence of emissions from heat power stations was taken into account, and additional sources within a layer of 45–270 m, whose distribution varied in height, were assigned for the cells with these stations. Finally, four types of vertical distribution of sources, four types of their horizontal distribution, and additionally four types of vertical distribution in cells with heat power stations were chosen. All in all, 64 model runs were conducted simultaneously varying these parameters. The correlation coefficient (Cor) and the Student criterion parameter $Z = (N_{av}(SILAM) - N_{av}(MEM))/(\sigma^2(SILAM) + \sigma^2(MEM))^{0.5}$ (where N_{av} is the average concentration in cell and σ is the error of average concentration in cell; SILAM—calculation results and MEM—measurement data) were determined for each run. Then, parameter $D = (Cor + 1/Z)/2$ was calculated. The highest value of D was taken as an optimum emission distribution.

Comparison Between Calculated and Measured Data

For comparison with the most optimum version of emission assignment, an additional run of the SILAM model was carried out for the same calculation period with CO and NO_x emissions taken from the TNO-2011 inventory for the Moscow megacity. When comparing concentration fields obtained in both the runs, time variations in pollutant concentrations (f(t)) averaged over cells, in which MEM stations were located, were determined. Figure 1 gives an example of f(t) for CO, which was calculated using SILAM and measurement data obtained at the MEM stations for January 15–31, 2014. For emission data obtained in this work, there are no concentration peaks in the time series of mean nitric oxide concentrations, which do not correspond to measured data, and the coefficient of correlation with observational data has increased for carbon monoxide, as opposed to scenario with TNO emissions.

Thus, one can conclude that our optimally assigned values for both carbon monoxide and nitrogen oxide emissions, which are used in SILAM calculations, yield results that are in good agreement with observational data.

In order to estimate the air quality in Moscow during the time period under consideration, we used the multipollutant index (MPI) determined from the formula:

$$MPI = \frac{1}{N}\sum_{i=1}^{N}\frac{\left(n_i^{mean} - n_i^{mpc}\right)}{n_i^{mpc}}, \qquad (2)$$

where n_i^{mean} is the annually mean concentration averaged over all stations for the i-th pollutant, n_i^{mpc} is the daily mean of the maximum permissible concentration (MPC) for the i-th pollutant, and N is the total amount of pollutants, which is used in calculating the MPI.

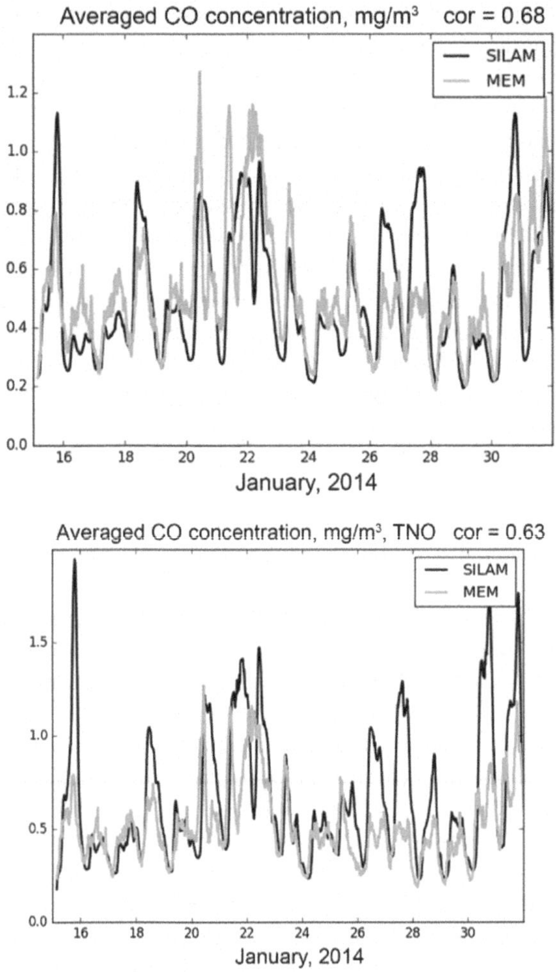

Fig. 1 Time variations in CO concentrations (for January 15–31, 2014) averaged over cells, in which the MEM stations are located, and calculated using SILAM model and measurement data for the optimum version of emissions (top) and TNO inventory emissions (bottom), SILAM—calculation data and MEM—measurement data

This index, which is an integral characteristic of the atmospheric-pollution level in a city, was first used in (Butler et al. 2008). This index varies from −1, when the atmosphere is free of pollutants, to plus infinity. If the MPI reaches zero or higher values, this implies that the concentration of, at least, one pollutant has reached maximum permissible values, which is hazardous to human health. Figure 2 gives the MPI time-dependent values for January 15–31, 2014, for the four main pollutants NO_2, SO_2, CO, and PM_{10}. A sufficiently high correlation (0.59) of the MPI time variability and (with consideration for determination error) agreement between mean (over space and time) MPI values (−0.75 for SILAM and −0.52 for MEM stations) allow one to make an integral assessment of the operation of both chemical and dynamical blocks for CO, NO_2, SO_2, and PM_{10} pollutants and conclude that the

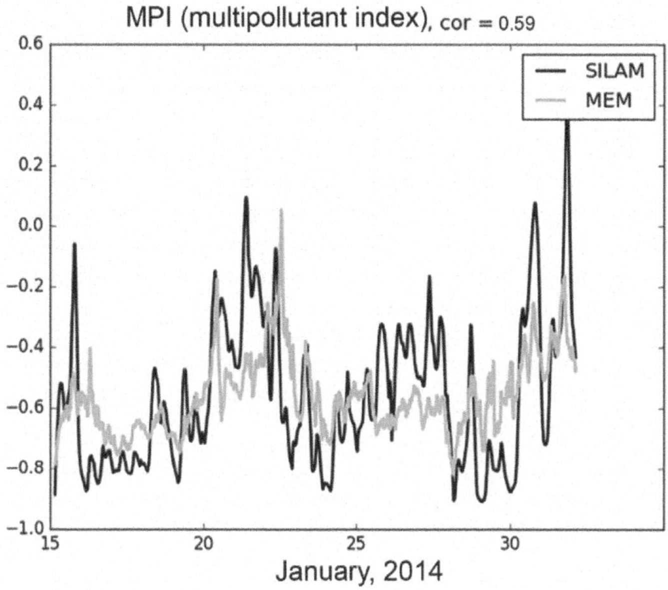

Fig. 2 MPI time variations within a period of January 15–31, 2014, for the Moscow megacity, SILAM—calculation results and MEM—measurement results

model has adequately reproduced the spatiotemporal variability of CO, NO_2, SO_2, and PM_{10} surface concentration fields in Moscow on days with and without stable air masses.

7 Conclusions

In this work, the SILAM regional chemical transport model has been adapted to the conditions of the Moscow megacity. Necessary changes were introduced into both photochemical and dynamical blocks of the model. Series of data on CO, NO, NO_2, SO_2, and PM_{10} surface concentrations measured at the MEM stations and the IAP RAS from 2005 to 2014, their annually mean emissions calculated on the basis of these data, and the TNO inventory data were used for the correct selection of spatial distribution and time variations of pollutant emissions from urban sources. A few series of numerical experiments were carried out to construct a four-dimensional matrix with values at the points of the chosen grid. The optimum emission matrix was obtained through minimizing deviations of simulation results from observational data obtained at Moscow stations. Despite the presence of some disagreements between calculation and measurement results, on the whole, the obtained optimum spatiotemporal distribution of emissions adequately reflects the state of urban infrastructure and its changes with time. We think that a more detailed consideration of the operation of the chemical block of the model (in particular, improved description of photochemical processes and corrected assignment of non-methane hydrocarbons) is necessary for further correction of assigned sources. A more precise parameterization of mixing processes in the ABL (in particular, the turbulent diffusion coefficient determined with an essential error) is also necessary. On the whole, the SILAM chemical transport model adapted to the conditions of the Moscow megacity and verified makes it possible to study the causes and mechanisms of variations in the composition of the atmosphere over the Moscow megacity and to forecast unfavorable meteorological conditions and situations that are potentially hazardous to human health, for example, on the basis of calculations of the MPI.

Acknowledgements The authors thank R. Kouznetsov and M. Sofiev for providing the initial code of the model, data on both boundary and initial conditions, helpful remarks, and assistance in conducting numerical experiments.

This work was performed as the part of the State assignment of Moscow State University (N 01200408543) and partially supported by the Russian Foundation for Basic Research (project nos. 17-29-05102 and 19-35-90073). The estimation of emission matrices was performed with the support of the Russian Scientific Foundation (project № 16-17-10275).

References

Baklanov A, Mahura A (2014) Megacities: Emissions, urban, regional and Global Atmospheric POLlution and climate effects, and Integrated tools for assessment and mitigation-MEGAPOLI. Sci Rep, 11–24

Butler TM, Lowrence MG, Gurjar BR, van Aardenne J, Schultz M, Lelieveld J (2008) The representation of emission from megacities in global emission inventories. Atmos Environ 42:703–719

Denier van der Gon HAC, Kuenen J, Butler T (2010) A base year (2005) MEGAPOLI Global Gridded Emission Inventory (1st Version), 20 p

Elansky NF, Lavrova OV, Skorokhod AI, Belikov IB (2016) Trace gases in the atmosphere over Russian cities. Atmos Envion 143:108–119

Elansky NF, Ponomarev NA, Verevkin YM (2018) Air quality and pollutant emissions in the Moscow megacity in 2005–2014. Atmos Environ 175:54–64

Elansky NF, Shilkin AV, Semutnikova EG, Zaharova PV, Rakitin VS, Ponomarev NA, Verevkin YM (2019) Weekly cycle of pollutant concentrations in near-surface air over Moscow. Atmos Oceanic Optics 32:85–93

Kouznetsov R, Sofiev M (2012) A methodology for evaluation of vertical dispersion and dry deposition of atmospheric aerosols. J Geophys Res 117:D01202

Kukkonen J, Olsson T, Schultz DM et al (2012) A review of operational, regional-scale, chemical weather forecasting models in Europe. Atmos Chem Phys 12:1–87

Marchuk G (1995) Adjoint equations and analysis of complex systems. Kluwer Academic Publishers, Dordrecht, The Netherlands

Pierce T, Hogrefe C, Rao ST, Porter PS, Ku JY (2010) Dynamic evaluation of a regional air quality model: assessing the emissions-induced weekly ozone cycle. Atmos Environ 44:3583–3596

Seinfeld JH, Pandis SN (2006) Atmospheric chemistry and physics. From air pollution to climate change, 2nd edn. Wiley, Hoboken, New Jersey

Sofiev M (2000) A model for the evaluation of long-term airborne pollution transport at regional and continental scales. Atmos Environ 34(15):2481–2493

Sofiev M (2002) Extended resistance analogy for construction of the vertical diffusion scheme for dispersion models. J Geophys Res Atmos 107:ACH 10-1–ACH 10-8

Sofiev M, Berger U, Prank M et al (2015a) MACC regional multi-model ensemble simulations of birch pollen dispersion in Europe. Atmos Chem Phys 15:8115–8130

Sofiev M, Vira J, Kouznetsov R et al (2015b) Construction of the SILAM Eulerian atmospheric dispersion model based on the advection algorithm of Michael Galperin. Geosci Model Dev 8:3497–3522

Valin LC, Russell AR, Cohen RC (2014) Chemical feedback effects on the spatial patterns of the NO_x weekend effect: a sensitivity analysis. Atmos Chem Phys 14:1–9

Wesely M (1989) Parameterization of surface resistances to gaseous dry deposition in regional-scale numerical models. Atmos Environ 23:1293–1304

Yanenko NN (1971) The method of fractional steps: solution of problems of mathematical physics in several variables. Springer-Verlag, Berlin, Heidelberg, Berlin

On the Ratio of Biological and Aquatic Migration of Chemical Elements in the Continental Biosphere Sector

V. S. Savenko

Abstract The intensity of biological and aquatic migration of chemical elements within continental sector of the biosphere was quantified. Ratio of the amount of continental runoff of element *i* to the mass of this element, which during the year is involved on land in the biotic cycle, (β_i) characterizes relation of aquatic and biological migration and is as a quantitative measure of the biotic cycle openness degree. The β_i values were determined for various chemical elements. It was established that high degree of the biotic cycle closeness is appropriate for elements that both perform and do not perform certain physiological functions in organisms. It was shown that a relatively low selectivity of biological consumption by plants is observed for most chemical elements. Clearly expressed selective concentration is inherent to P, Mn, Rb, Co, K, Cu, Ba, Zn. Discrimination is peculiar to F, Cl, Br, I, Li, Na, U, Se, Sb.

Keywords Aquatic migration · Continental runoff · Biological migration · Biotic cycle · Selectivity of biological consumption

1 Introduction

At present, ideas about the decisive role of biota in regulating state of the biosphere and possibility of overcoming the negative effects of economic activity by restoring natural ecosystems to at least 90% of their pre-industrial prevalence are widespread. It is believed that the mechanism of biotic regulation consists in changing the degree of biotic cycle closure, the total output of which (200–300 billion tons of organic matter per year) is sufficient to compensate for the changes caused by abiotic processes (Gorshkov et al. 1994; Gorshkov 1995; Arskii et al. 1997; Kondrat'ev et al. 2003). The latter statement is not strictly proven, since it is based mainly on a comparison of biogenic and abiogenic carbon fluxes. Some data indicate that in the land biogeocenoses the intensity of transit fluxes of "ash" components associated with water migration is comparable to the intensity of their fluxes in the biotic cycle (Bazilevich 1976).

V. S. Savenko (✉)
Faculty of Geography, M. V. Lomonosov Moscow State University, Moscow, Russia
e-mail: alla_savenko@rambler.ru

C. T. Olegovna (ed.), *Processes in GeoMedia—Volume I*,
Springer Geology, https://doi.org/10.1007/978-3-030-38177-6_17

From geochemical point of view, the biotic cycle can control state of the biosphere only if the involvement of not only carbon, but also other chemical elements in it significantly exceeds their abiogenic fluxes. The purpose of this article is to compare the output of biological and aquatic migration of various chemical elements within the continental sector of the biosphere.

Biological migration of chemical elements is carried out in the form of production–destruction processes associated with the biotic cycle. Since the amounts of formed and decomposing organic matter differ by a small value, the output of the biotic cycle of element i on land (P_i) can be estimated by the value of primary land production of organic matter (P) and content of this element in it (B_i):

$$P_i = B_i P.$$

The value of primary land production is 141 billion tons of dry organic matter per year (Efimova 1977). The average chemical composition of land plants was calculated as a geometric mean for three most representative reports (Bowen 1966; Romankevich 1988; Savenko 1991).

On land, water migration is generally organized in the form of continental runoff, which is subdivided into external and internal runoffs, flowing into the World Ocean and inland water bodies or areas of accumulation of dry salts, respectively. Both external and internal runoffs consist of river, ice, and groundwater runoffs, which differ in discharge of water and its chemical composition. In external drain areas, river, ice, and groundwater runoffs amount to 41.7, 3.0, and 2.2 thousand km^3/year (World water balance and water resources of the earth 1974). For internal drain areas, there are only estimates of river runoff, which is equal about 1 thousand km^3/year (Area of internal runoff 2013). Due to the lack of information on the average composition of soluble substances of the ice runoff, we are forced to ignore this component of the continental runoff of dissolved chemical elements. However, due to the relatively low discharge of ice runoff (almost 14 times less than the river runoff) and lower mineralization of ices compared to river and ground waters, this should not have a significant impact on the final assessment of average composition of the continental runoff.

Groundwater runoff of internal drain areas is not precisely defined. Approximately it can be estimated based on the assumption that the ratio of volumes of groundwater (Q_{GW}) and river (Q_{RW}) runoffs in the external and internal drain areas are the same: $Q_{GW}/Q_{RW} = 0.053$. With a volume of internal river runoff of 1 thousand km^3/year, this assumption leads to the value of groundwater runoff of 0.053 thousand km^3/year. In this case, the total groundwater runoff (Q_{GW}) will be equal to 2.25 thousand km^3/year.

The composition of river runoff to the World Ocean is currently quite reliably established for most chemical elements (Meybeck 2003; Gordeev 2012; Gaillardet et al. 2013). The average composition of river waters of the internal drain areas is not determined, but the available information for large and medium rivers located here (Volga, Ural, Amu-Darya, Syr-Darya, Chari, etc.) makes it possible to consider this composition as a first approximation similar to the composition of river runoff into the World Ocean. The average content of many chemical elements in groundwater is given in a summary report of Shvartsev (Shvartsev 1998).

Taking into account these assumptions, the equation for the intensity of the lateral aquatic migration of chemical elements (L_i) can be written as

$$L_i = C_{i(RW)}(Q_{RW(ex)} + Q_{RW(in)}) + C_{i(GW)}Q_{GW},$$

where $C_{i(RW)}$ and $C_{i(GW)}$ are the average concentrations of element i in river and ground waters, respectively; $Q_{RW(ex)}$, $Q_{RW(in)}$, and Q_{GW} are discharges of the river external, river internal, and groundwater runoffs.

Lateral water migration, or continental runoff, permeates the entire biogeocenotic land cover and is the main factor of incomplete closeness of the biotic cycle. It is obvious that the ratio of continental runoff of element i to mass of this element, which during the year is involved in the biotic cycle on land, characterizes the ratio of aquatic and biological migration and at the same time serves as a quantitative measure of the openness degree (β_i) of the biotic cycle in the continental biosphere sector:

$$\beta_i = \frac{L_i}{P_i} = \frac{C_{i(RW)}(Q_{RW(ex)} + Q_{RW(in)}) + C_{i(GW)}Q_{GW}}{B_i P}.$$

Table 1 shows the average concentrations of chemical elements in the river runoff, groundwater of the hypergenesis zone, and in the land vegetation cover. It also contains the β_i values calculated for $P = 141 \times 10^9$ t/year, $Q_{RW(ex)} = 41{,}700$ km^3/year, $Q_{RW(in)} = 1000$ km^3/year, and $Q_{GW} = 2250$ km^3/year.

The data of Tables 1 and 2 indicate that high degree of the biotic cycle closure, which corresponds to low β_i values, is not typical for all elements. More than 99% of the biotic cycle is closed only for carbon. For Zn, P, Pb, Mn, Co, Cu, and K, the degree of closure is 99–95%. All these elements perform certain physiological functions in organisms. The exception is lead, for which the calculated β_i value may differ significantly from the natural value due to strong anthropogenic pollution of the environment. However, for many other biogenic and biologically active elements (B, N, Ni, Fe, V, Mo, Ca, Mg, S, and Se), the biotic cycle is closed by only 10–50%, and for iodine, lateral transport in the continental runoff significantly exceeds intensity of the biotic cycle. Along with this, high degree of the biotic cycle closure ($\beta_i = 0.05$–0.1) is observed for a number of elements that are not included in the list of substances necessary for the normal functioning of plants (Zr, Rb, Ra, Cs, Th, Be, Ti, Al, and Nb). Thus, the intensity of chemical elements involvement in the biotic cycle cannot be considered an exceptional factor determining state of the biosphere. Continental runoff, for example, plays an equally important role.

It is interesting to find out the mechanisms linking the biological and aquatic migration of chemical elements.

From the middle of the last century, there are two opposing views on the causes of formation of the chemical composition of surface land waters. According to B. B. Polynov, "the composition of natural river waters is determined not by simple abiotic

Table 1 Characteristics of the continental runoff and biotic cycle of chemical elements on land

Element	$C_{i(RW)}$ (μg/l)	$C_{i(GW)}$ (μg/l)	B_i (μg/kg)	L_i (10^3 t/year)	P_i (10^3 t/year)	β_i	$K_{BC(i)} = \frac{B_i}{C_{i(GW)}}$
Li	1.84	13.0	600	108	84.6	1.3	46.2
Be	0.009	0.19	70	0.812	9.87	0.082	368
B	10.2	77.9	4.0×10^4	611	5640	0.11	513
C	9570	52,300	4.5×10^8	5.26×10^5	6.35×10^7	0.0083	8600
N	14,500	2040	3.0×10^7	6.24×10^5	4.23×10^6	0.15	14,700
F	100	480	2000	5350	282	19.0	4.17
Na	5520	67,600	7.5×10^5	3.88×10^5	1.06×10^5	3.7	11.1
Mg	2980	18,200	3.4×10^6	1.68×10^5	4.79×10^5	0.35	187
Al	32	226	1.5×10^5	1.87×10^3	2.12×10^4	0.089	664
Si	4060	8350	1.0×10^6	1.92×10^5	1.41×10^5	1.4	120
P	140[a]	58.0	1.9×10^6	6.11×10^3	2.68×10^5	0.023	32,800
S	2800	25,600	3.4×10^6	1.77×10^5	4.79×10^5	0.37	133
Cl	5920	59,700	2.0×10^6	3.87×10^5	2.82×10^5	1.4	33.5
K	1720	5150	1.2×10^7	8.50×10^4	1.69×10^6	0.050	2330
Ca	11,900	39,200	1.3×10^7	5.96×10^5	1.83×10^6	0.33	332
Ti	0.49	17.4	5000	60.1	705	0.085	287
V	0.71	1.34	1200	33.3	169	0.20	896
Cr	0.7	3.0	500	36.6	70.5	0.52	167
Mn	34	54.5	3.1×10^5	1.57×10^3	4.37×10^4	0.036	5690
Fe	66	481	1.5×10^5	3.90×10^3	2.12×10^4	0.18	312
Co	0.15	0.39	1100	7.28	155	0.047	2820
Ni	0.80	3.6	2500	42.3	353	0.12	694
Cu	1.48	5.6	1.1×10^4	75.8	1551	0.049	1960

(continued)

Table 1 (continued)

Element	$C_{i(RW)}$ (μg/l)	$C_{i(GW)}$ (μg/l)	B_i (μg/kg)	L_i (10^3 t/year)	P_i (10^3 t/year)	β_i	$K_{BC(i)} = \frac{B_i}{C_{i(GW)}}$
Zn	0.60	41.4	6.0×10^4	119	8460	0.014	1450
Ga	0.03	0.37	40	2.11	5.64	0.37	108
As	0.62	1.46	300	29.8	42.3	0.70	205
Se	0.07	0.72	70	4.61	9.87	0.47	97.2
Br	20	85.2	8000	1.05×10^3	1.13×10^3	0.93	93.9
Rb	1.63	1.86	8000	73.8	1.13×10^3	0.065	4300
Sr	60	183	3.5×10^4	2.97×10^3	4.94×10^3	0.60	191
Zr	0.039	1.20	500	4.37	70.5	0.062	417
Nb	0.0017	0.45	80	1.09	11.3	0.096	178
Mo	0.42	1.75	600	21.9	84.6	0.26	343
Ag	0.1	0.26	40	4.86	5.64	0.86	154
Cd	0.08	0.24	90	3.96	12.7	0.31	375
Sn	0.4	0.39	300	18.0	42.3	0.42	769
Sb	0.07	0.68	15	4.52	2.12	2.1	22.1
I	2.0[b]	8.0	600	103	84.6	1.2	75.0
Cs	0.011	0.26	100	1.05	14.1	0.075	385
Ba	23	18.3	3.4×10^4	1023	4794	0.21	1860
La	0.12	0.67	200	6.63	28.2	0.24	299
Ce	0.26	0.26[c]	300	11.7	42.3	0.28	1150
Au	0.002	0.005	1.5	0.097	0.212	0.46	300
Hg	0.007	0.041	20	0.391	2.82	0.14	488
Pb	0.079	3.0	3000	10.1	423	0.024	1000
Ra	2.4×10^{-8}	4.6×10^{-7}	2.0×10^{-4}	2.06×10^{-6}	2.82×10^{-5}	0.073	435
Th	0.041	0.24	200	2.29	28.2	0.081	833
U	0.37	1.31	20	18.7	2.82	6.6	15.3

[a]Accepted according to (Savenko and Savenko 2007); [b]accepted according to (Vinogradov 1967); [c]accepted as for river runoff

Table 2 A number of elements ascending β_i value

β_i	Elements
<0.01	C (0.0083)
0.01–0.05	Zn (0.014), P (0.023), Pb (0.024), Mn (0.036), Co (0.047), Cu (0.049), K (0.050)
0.05–0.1	Zr (0.062), Rb (0.065), Ra (0.073), Cs (0.075), Th (0.081), Be (0.082), Ti (0.085), Al (0.089), Nb (0.096)
0.1–0.5	B (0.11), N (0.15), Ni (0.12), Hg (0.14), Fe (0.18), V (0.20), Ba (0.21), La (0.24), Mo (0.26), Ce (0.28), Cd (0.31), Ca (0.33), Mg (0.35), S (0.37), Ga (0.37), Sn (0.42), Au (0.46), Se (0.47)
0.5–1	Cr (0.52), Sr (0.60), As (0.70), Ag (0.86), Br (0.93)
1–5	I (1.2), Li (1.3), Cl (1.4), Si (1.4), Sb (2.1), Na (3.7)
5–10	U (6.6)
>10	F (19)

reactions of the action of water on igneous minerals, i.e. hydrolysis and dissolution, as invented in textbooks, but by more complex, though more quickly current process of elements extracting from minerals by organisms and dissolving the ash part of organisms in water during their mineralization" (Polynov 1956, p. 474). M. G. Valyashko, on the contrary, believed that "the abundance of elements in the earth's crust determines major components of natural waters, while solubility of compounds formed by the major components plays a decisive role in the formation of chemical composition of natural waters" (Valyashko 1954, p. 581). By now, it has been proven that litter mineralization, as well as leaching effect of free carbon dioxide and exometabolites plays an important role in the formation of surface waters composition. At the same time, the mobility numbers established for water migration of chemical elements definitely correlates with the solubility of minerals (Savenko 2000). Apparently, biological and chemical factors are equally involved in the formation of chemical composition of surface waters, and it is important to know how observed chemical composition of the land vegetation cover is related to composition of the waters.

For quantitative estimation of the biological fractionation of chemical elements in the continental biosphere sector, the coefficient of biological accumulation $K_{\boldsymbol{BA}(i)}$ equal to the ratio of content of element i in plant ash ($C_{i(A)}$) to its content in rocks ($C_{i(CR)}$) that are the primary source of biota substance is widely used:

$$K_{\boldsymbol{BA}(i)} = \frac{C_{i(A)}}{C_{i(CR)}}. \tag{1}$$

The exceptions are C, H, O, and N, which form the group of "non-ash" elements.

Differences in $K_{\boldsymbol{BA}(i)}$ values are usually considered as an occurrence of the selectivity of plants extraction of various chemical elements from mother rocks with which living organisms interact, forming thus bio-inert land systems. The range of $K_{\boldsymbol{BA}(i)}$ values reaches 10^6–10^7 times (Perel'man 1989; Dobrovolsky 1998), which creates the impression of exceptionally high selectivity of chemical elements accumulation in plants. It is known that chemical elements enter plants not directly from rocks, but mainly from groundwater, macro- and microelement composition of which is controlled not only by the abundance of chemical elements in rocks, but also by the solubility of mineral phases and the parameters of sorption–desorption equilibria. Living organisms influence in the content of dissolved forms of chemical elements in groundwater through chemically active exometabolites and products of destruction of dead organic matter, but these processes are involved only in the formation of aquatic environment from which plants extract the substances they need. Therefore, to quantify the selectivity of chemical elements accumulation in land plants, it would be more appropriate to use the coefficient of biological consumption $K_{BC(i)}$ instead of (1):

$$K_{BC(i)} = \frac{B_i}{C_{i(GW)}},$$

where B_i is the content of chemical element i in plants; $C_{i(GW)}$ is the concentration of element i in groundwater, from which the substances enter plants through the root system.

The concentrations of elements in the aquatic environment were taken in accordance with the estimate (Shvartsev 1998) for the average composition of waters of the hypergenesis zone. Comparison of B_i and $C_{i(GW)}$ values shows a close correlation between them (Fig. 1):

$$\lg B_i = 0.984 \lg C_{i(GW)} + 2.59,\ r = 0.94.$$

Figure 1 the relationship between the content of chemical elements in land plants (B_i) and groundwater ($C_{i(GW)}$).

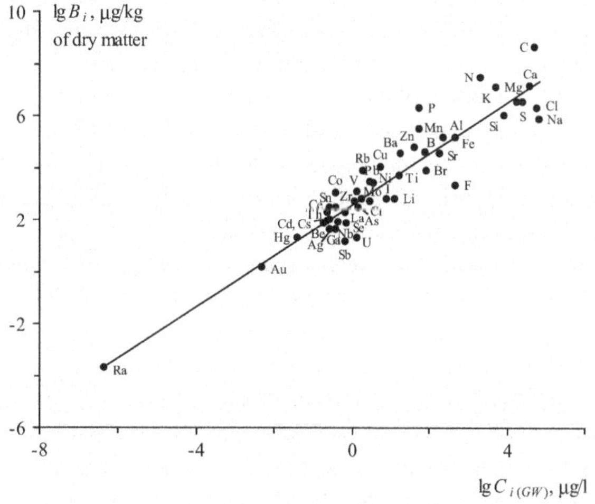

Fig. 1 The relationship between the content of chemical elements in land plants (B_i) and groundwater ($C_{i(GW)}$)

The range of $K_{BC(i)}$ values reaches four orders of magnitude (Table 1), but in most cases $\lg K_{BC(i)} = 2.5 \pm 0.5$, i.e. deviations from the average do not exceed 0.5 order of magnitude. These elements (Ga, Si, S, Ag, Nb, Cr, Mg, Sr, As, Au, Ti, La, Fe, Mo, Ca, Cd, Be, Cs, Ra, B, Al, Ni, Sn, Th, V, and Pb) enter plants without strong fractionation and approximately in the same proportions that are typical of groundwater.

Supposedly, two groups of elements with significant fractionation can be distinguished: elements-discriminant ($\lg K_{BC(i)} < 2.0$) and elements-concentrator ($\lg K_{BC(i)} > 3.0$).

The first group includes halogens, the degree of accumulation of which increases with an increase in atomic weight (F < Cl < Br ≈ I), light alkaline elements (Li, Na), and also U, Se, and Sb, for which the available data on B_i and $C_{i(GW)}$ contain great uncertainty.

If we exclude from consideration C and N, which enter plants from the atmosphere, then a clearly expressed selective concentration is inherent to 8 chemical elements, which, by decreasing the $K_{BC(i)}$ value, forms the following number:

$$P > Mn > Rb > Co > K > Cu > Ba > Zn.$$

All these elements perform certain physiological functions and their position in the group of elements-concentrator is not surprisingly (Rb is concentrated in plants, being a close chemical analogue of K). This group did not include Ce ($\lg K_{BC} =$ 1150), for which, in the report (Shvartsev 1998), the average content in groundwater was not determined and was equal to the content in river water.

Another biogeochemical aspect of $K_{BC(i)}$ is associated with revealing the factors that limit the primary production of vegetation. Value of the primary production of terrestrial ecosystems depends on the intensity of solar radiation and the availability of moisture, which is reflected in the close correlation of the primary production with value of the hydrothermic potential (Ryabchikov 1972). It was found that for implementation of the photosynthesis reaction on the sheet surface, transpiration of large volumes of water is required, 200–1000 times higher than the biomass increase (Budyko 1977). Along with this, transpiration is also needed for the supply of chemical elements to plants necessary for the formation of new biomass (Savenko 2004), and if the $K_{BC(i)}$ is small, then factor limiting the primary production of terrestrial ecosystems may be not hydrothermic potential, but insufficient concentration of physiologically necessary elements in the groundwater.

It is noteworthy that the average value of $K_{BC(i)} = 10^{2.5} = 316$ almost exactly coincides with the average value of the ratio of mass of transpirated moisture to biomass produced by plants, equal to 300 (Budyko 1977). This means that during the transpiration of moisture, which expended for ensure normal course of the photosynthesis reaction, plants receive sufficient quantities of almost all chemical elements. Limitation of the primary production can be expected only from the side of a small number of elements, showing the tendency of selective concentration in plants (P, Mn, Rb, Co, K, Cu, Ba, and Zn), and even then, if the reserve of mobile forms in soils and subsoils is less than 10 times increase the mass of dissolved forms of corresponding elements. As a rule, the latter condition in nature is not fulfilled and therefore the main factor determining value of the primary production of land vegetation cover is the hydrothermic potential.

2 Conclusions

1. Ratio of the continental runoff of element *i* to the mass of this element, which during the year is involved on land in the biotic cycle, characterizes relation of aquatic and biological migration and is as a quantitative measure of the biotic cycle openness degree in the continental biosphere sector.
2. High degree of the biotic cycle closeness is appropriate for elements that both perform and do not perform certain physiological functions in organisms. The intensity of chemical elements involvement in the biotic cycle cannot be considered an exceptional factor determining state of the biosphere. In particular, continental runoff of dissolved matter plays an equally important role.
3. Chemical compositions of the land plants and the lithogenic basis of landscapes are indirectly depended through groundwater. A relatively low selectivity of biological consumption by plants is observed for most chemical elements. Clearly expressed selective concentration is inherent to 8 chemical elements for which the coefficient of biological consumption $K_{BC(i)} > 1000$. By decreasing the $K_{BC(i)}$ value, the elements-concentrator forms a number: P > Mn > Rb > Co > K > Cu > Ba > Zn. Lower $K_{BC(i)}$ values are observed for F, Cl, Br, I, Li, Na, U, Se, and Sb.
4. The main factor controlling value of the primary production of terrestrial ecosystems on a global scale is apparently hydrothermic conditions, while variations in the geochemical background of physiologically necessary elements are of subordinate importance.

Acknowledgements This study was financially supported by the Russian Foundation for Basic Research (project no. 18-05-01133).

References

Area of internal runoff. In: Great Russian encyclopedia, vol 23. Great Russian Encyclopedia, Moscow, p 481 (2013)

Arskii YuM, Danilov-Danil'yan VI, Zalikhanov MCh et al (1997) Ecological problems: what happens, who is guilty, and what to do? MNEPU, Moscow, 330 p

Bazilevich NI (1976) Biogenic and abiogenic processes in forest, steppe and desert ecosystems. International Geography, vol 76. Section 4. Biogeography and Soil Geography. Moscow, pp 58–62

Bowen HJM (1966) Trace elements in biochemistry. Acad. Press, London, 241 p

Budyko MI (1977) Global ecology. Mysl', Moscow, 327 p
Dobrovolsky VV (1998) Basics of biogeochemistry. Vysshaya Shkola, Moscow, 413 p
Efimova NA (1977) Radiation factors of vegetation cover productivity. Gidrometeoizdat, Leningrad, 216 p
Gaillardet J, Viers J, Dupre B (2013) Trace elements in river waters. Treatise on geochemistry, vol 7, 2nd edn. Elsevier, Amsterdam, pp 195–236
Gorshkov VG (1995) Physical and biological bases of life sustainability. VINITI, Moscow, 470 p
Gordeev VV (2012) Geochemistry of the Rive–Sea System. Moscow, 452 p
Gorshkov VG, Kondrat'ev KYa, Losev KS (1994) Natural biological regulation of the environment. Proc Russ Geogr Soc 6:17–23
Kondrat'ev KYa, Losev KS, Ananicheva MD, Chesnokova IV (2003) Natural science bases of life sustainability. TzS AGO, Moscow, 240 p
Meybeck M (2003) Global occurrence of major elements in rivers. In: Treatise on geochemistry, vol 5. Elsevier–Pergamon, Oxford, pp 207–223
Perel'man AI (1989) Geochemistry. Vysshaya Shkola, Moscow, 528 p
Polynov BB (1956) On the geological role of organisms. In: Selected works. Publ. House of the Academy of Sciences of the USSR, Moscow, pp 466–476
Romankevich EA (1988) Living matter of the Earth (biogeochemical aspects of the problem). Geochem Int (2):292–306
Ryabchikov AM (1972) Structure and dynamics of the geosphere. Mysl', Moscow, 223 p
Savenko VS (1991) Natural and anthropogenic sources of air pollution. In: Results of Science and Technology. Ser. Nature Protection and Reproduction of Natural Resources, vol 31. VINITI, Moscow, 212 p
Savenko VS (2000) Water migration coefficients of chemical elements in the hypergenesis zone. Lithol Min Resour 35(4):345–350
Savenko VS (2004) What is the life? Geochemical approach to the problem. GEOS, Moscow, 203 p
Savenko VS, Savenko AV (2007) Geochemistry of phosphorus in the global hydrological cycle. GEOS, Moscow, 255 p
Shvartsev SL (1998) Hydrogeochemistry of the hypergenesis zone. Nedra, Moscow, 366 p
Valyashko MG (1954) The role of solubility in the formation of chemical composition of natural waters. Doklady Acad Sci USSR 49(4):581–584
Vinogradov AP (1967) Introduction to ocean geochemistry. Nauka, Moscow, 215 p
World water balance and water resources of the earth. Gidrometeoizdat, Leningrad, 636 p (1974)

Calculation and Analysis of the Average for Seasons Fields of the Currents Velocities of the Marmara Sea Waters

S. V. Dovgaya

Abstract Using a numerical three-dimensional non-linear thermohydrodynamic model of MHI, an experiment by simulate of the circulation of the waters of the Sea of Marmara on the example of 2008 was conducted. As a result of the calculation, three-dimensional fields of hydrodynamic characteristics were obtained for every day of the year for the entire basin. The analysis of the average seasonal distributions of the velocity fields of the currents of the sea water at different horizons is carried out. It was found that in the surface layer in winter, when the waters move from the Bosphorus to the Dardanelles, in the northwestern part of the basin there is an anticyclone, while in the south the movement with cyclonic vorticity prevails. In the spring hydrological season the extensive anticyclonic vortex is formed in the center of the basin, which is intensified in summer. In the autumn, the central anticyclone weakens and its area decreases, and the cyclonic vortex forms in the southeastern region of the sea. The such location of dynamic features in the surface layer of the sea mainly depends on the distribution of the average seasonal vorticities of the wind field over the region and the orography of the coastline. The regularities of water behavior below the density jump are mainly determined by atmospheric influence, the impulse flow through the straits and the bottom topography. At depths below 500 m, the seasonal signal in the velocity fields is traced weakly. The movement of waters at these depths is concentrated in three deep-water basins. In the western and central deep-water regions there are cyclonic eddies, in the east—the system of anticyclone and cyclone.

Keywords Marmara Sea · Numerical hydrodynamic model · Vorticity of wind field · Cyclone · Anticyclone · Current

S. V. Dovgaya (✉)
Marine Hydrophysical Institute of the Russian Academy of Sciences, Sevastopol, Russia
e-mail: dovgayasvetlana0309@yandex.ru

C. T. Olegovna (ed.), *Processes in GeoMedia—Volume I*,
Springer Geology, https://doi.org/10.1007/978-3-030-38177-6_18

1 Introduction

The Turkish Straits System, which unites the straits Bosphorus, Dardanelles and the Marmara Sea, is an important component of the formation of the hydrology like Black and so Aegean seas. There is an interest in numerical modeling of processes in this system and, in particular, in the Marmara Sea region. The surface area is 11,500 km^2, volume—3378 km^3. From west to east the size of the basin is ~250 km, from north to south—about 70 km. Three depressions occupy the northern part of the basin with maximum depths of 1097 m, 1389 m and 1238 m, respectively, from west to east. The sills between the western, central and eastern basins are about 10 km and 30 km wide, respectively. The southern part of the basin is occupied by a relatively shallow shelf area with an average depth of 100 m. The Marmara Sea is connected with the Black Sea through the Bosphorus (average depth $\approx$ 35 m) and with the Mediterranean through the Dardanelles (average depth $\approx$ 55 m). The characteristics water masses of the sea are mainly determined by the water exchange between the Black and Aegean seas through straits. Two different water masses of the Marmara Sea are separated by a strong pycnocline at depths of 20–30 m. The moderate wind climate over the basin is strongly determined by the topography of the adjacent territory. The whole region is a passage for cold wind systems from the north and for cyclones moving from the Mediterranean to the Black Sea. Northeast winds are prevalent throughout the year, southwest winds are secondary in importance.

Some of the first studies of this basin were conducted at the end of the nineteenth century on expeditions under the guidance of Russian scientists Makarov and Shpindler (Makarov 1885; Shpindler 1896). In Besiktepe et al. (1994) a simplified box model of circulation and hydrographic data of the waters of the sea, collected in the period 1986–1992 are presented. Numerical solutions based on the *ROMS* model, which reproduce the general circulation of the Marmara Sea for the period of September–December 2008 and February–March 2009 are given in Chiggiato et al. (2012). It shows that the displacement of the pycnocline in the western part of the basin is due to the northeast wind, and in the eastern part of the basin the pycnocline response is determined by the total effect of the wind and the flow of Bosphorus water. Using an unstructured grid and specifying real atmospheric disturbance from 2009 to 2013, numerical modeling of the dynamics of the water system, including the Bosporus, the Marmara Sea and the Dardanelles, is presented in Aydogdu et al. (2018). It shows the effect of maximums of the stress of the wind friction on the currents in the surface layer of the Marmara Sea and possible mechanisms to block water flows from the Bosphorus Strait.

This work is devoted to the study of the regularities of variability in the dynamics of the waters of the basin of the Marmara Sea in different hydrological seasons of the year and at different horizons.

2 Formulation of the Problem

For simulate hydrophysical processes in the Marmara Sea we used the eddy-resolving nonlinear thermo-hydrodynamic model of the MHI RAS, the adapted to the conditions of the Black Sea version of which was presented in Markova and Dymova (2019). The model equations are detailed in Demyshev et al. (2012). The Boussinesq approximations, hydrostatics and incompressible seawater were taken. The numerical experiment was carried out with a horizontal resolution of 1.22 km × 0.83 km. It was used 18 horizons at the version, time step—0.5 min. The velocities in the Bosphorus and Dardanelles straits were set constant over time and calculated taking into account the known flow through these straits (Besiktepe et al. 1994). The temperature of the waters of the upper layer of the Bosporus was determined taking into account its variability with time and with depth, the values of salinity—variable in depth (Zapevalov and Dovgaya 2007; Zapevalov 2005). In the lower current of the Dardanelles the salinity varied in depth, the temperature was constant (Besiktepe 2003).

The fields of temperature, salinity, horizontal velocities of the currents and level fields, corresponding to 6620 days of calculation by experiment with the same parameters but excluding atmospheric disturbance were used as initial fields. The duration of the integration period was determined by the establishment of currents in the deep layers. At the sea surface the fields of tangential stresses of the friction of the wind, heat flow, precipitation and evaporation in 2008 were set daily averaged per day. These fields were obtained in accordance with the calculation data for the regional atmospheric model *MM5* (LNCS Homepage 2014).

3 Results of the Numerical Experiment

As a result of the numerical experiment for the whole area of the Marmara Sea three-dimensional fields of hydrodynamic characteristics were obtained for each day of 2008. The analyze of the distributions of velocity fields averaged in different hydrological seasons of the year was conducted. The winter, spring, summer and autumn hydrological seasons are three-month intervals, which begin, respectively, on January 1, April 1, July 1, and October 1. So in Fig. 1 shows averaged over the hydrological seasons of the velocities fields of the Marmara Sea waters at the depth of 3 m. It's clear that in the northwestern part of the basin the anticyclone forms in the surface layer in the winter as the waters move from the Bosphorus Strait to the Dardanella Strait; in the southern part the movement of waters with cyclonic vorticity is dominated (Fig. 1a). In the spring season (Fig. 1b) the incoming Black Sea waters move to the south-western region of the sea (towards the Dardanelles Strait), wherein the extensive anticyclone is generated in the central part of the basin with maximum velocities of up to 18.21 cm/s. Cyclonic eddies are formed in Erdek Bay (to the south of the Marmara Island) and in the area of Gemlik Bay (south-eastern part of the sea).

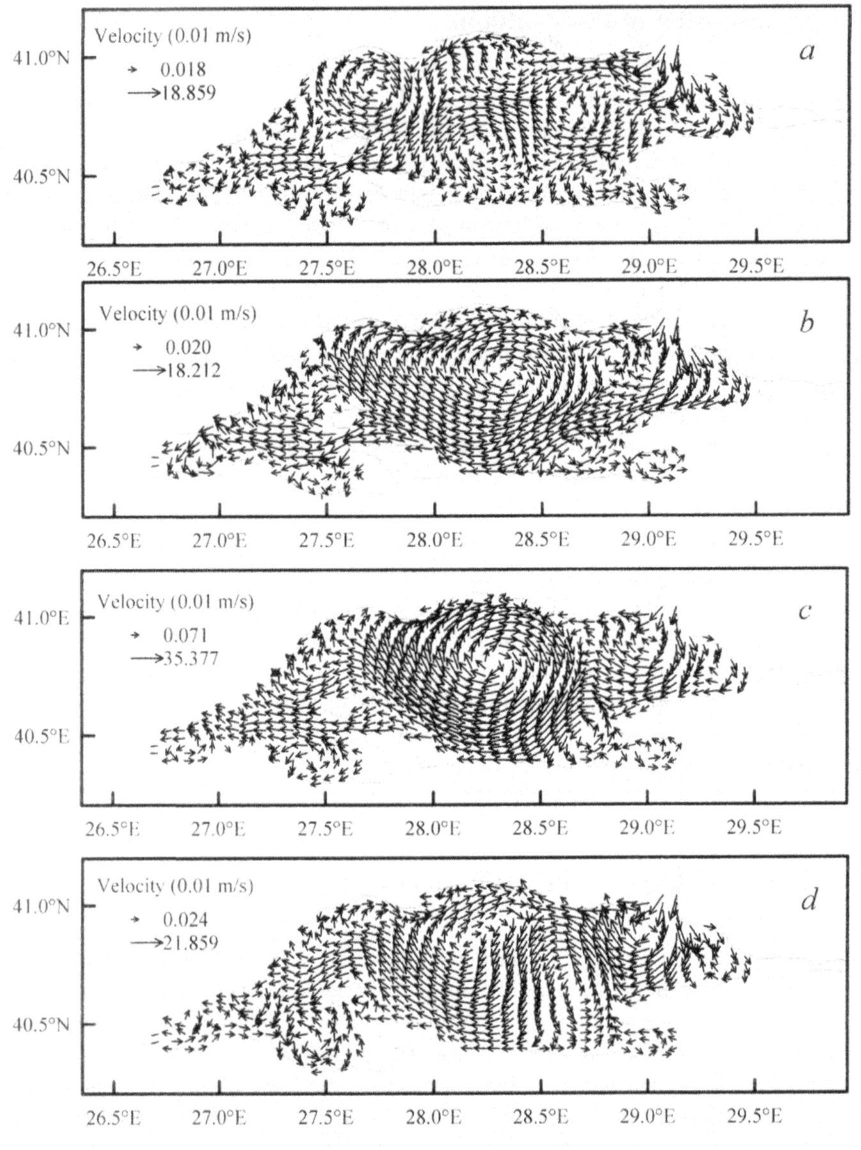

Fig. 1 Average for the winter (**a**), spring (**b**), summer (**c**) and autumn (**d**) hydrological seasons the fields of the velocities of the currents on the horizon of 3 m

In the summer period (Fig. 1c), the circulation pattern is similar to the spring one (Fig. 1b), the differences lie in the fact that the central anticyclone intensifies and the maximum velocities in this season reach 35.38 cm/s. In the autumn time (Fig. 1d) the area occupied by the anticyclone in the center of the basin decreases, and the maximum current velocities are 21.86 cm/s. Such the location of dynamic features in the surface layer of the sea mainly depends on the orography of the coastline and the distribution of the average seasonal vorticities of the wind field over the region, which are shown in Fig. 2. In this figure, the anticyclonic vorticity of wind corresponds to negative values, the cyclonic vorticity—to positive values. It is noteworthy that in the winter period the cyclonic vorticity of the wind field prevails over the region (Fig. 2a), in the spring and summer—the anticyclonic vorticity (Fig. 2b, c).

More dense waters from the neighboring Aegean Sea enters at the depths of 20–75 m what significantly influences on the dynamic processes in the lower layers. In

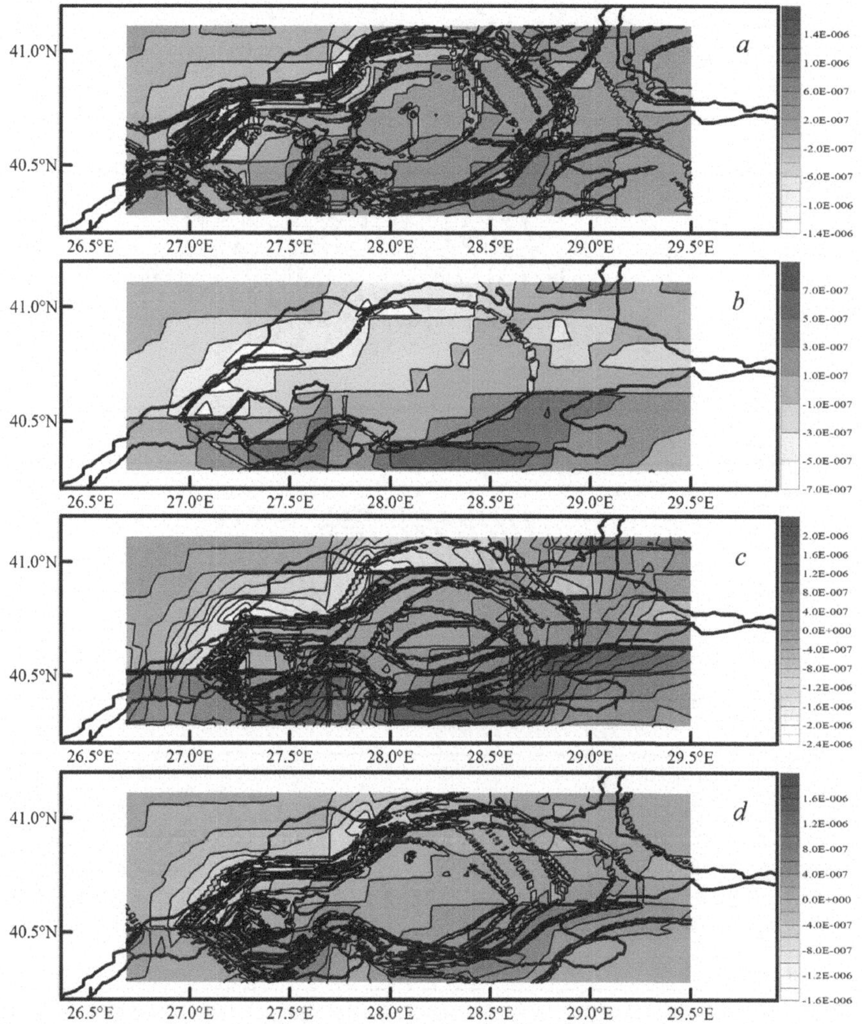

Fig. 2 Average for the winter (**a**), spring (**b**), summer (**c**) and autumn (**d**) hydrological seasons the wind stress curls (Nm^{-3}, shades) over region of the Marmara Sea

Fig. 3 shows the map of the velocities of the currents on the horizon of 30 m. Thus in winter at this depth (Fig. 3a) the waters flowing from the Dardanelles, displacing less dense waters into the southern part of the basin, are drawn into the anticyclonic eddy system. Herewith the maximum velocity of the currents is decreased twice when

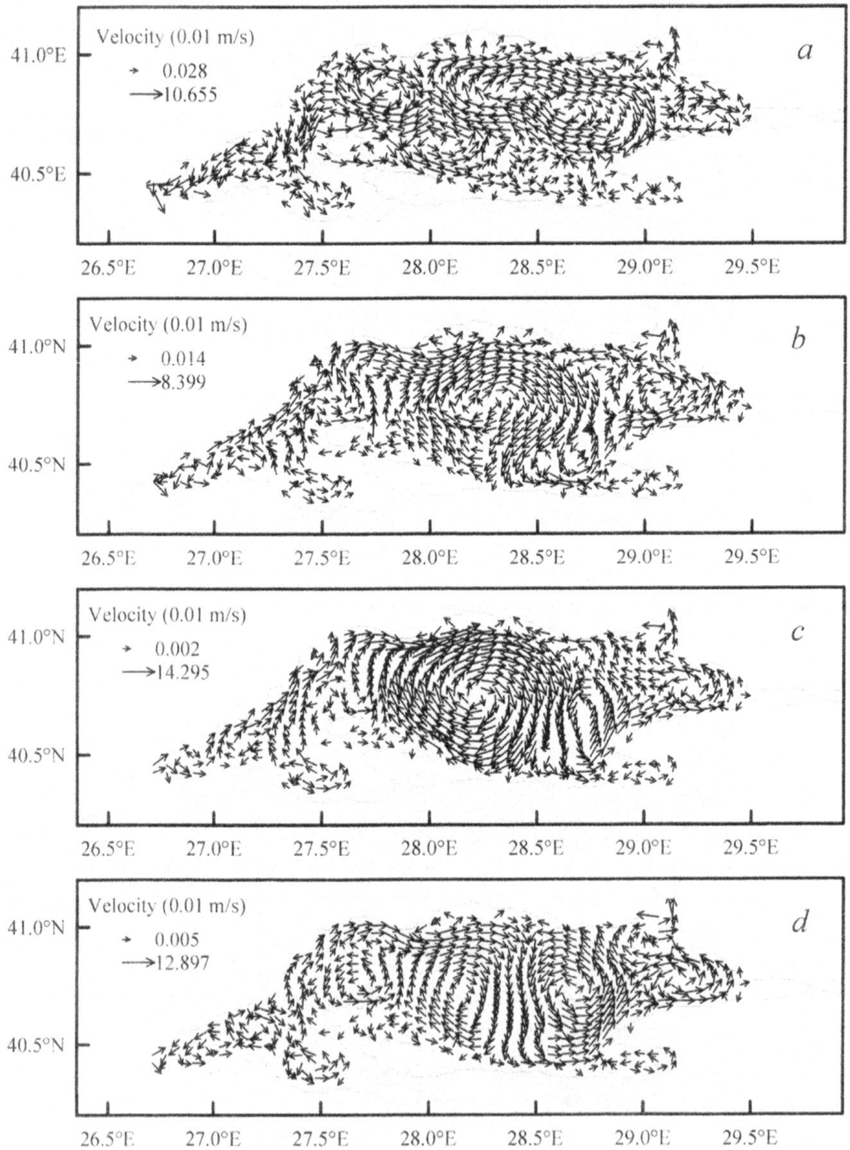

Fig. 3 Average for the winter (**a**), spring (**b**), summer (**c**) and autumn (**d**) hydrological seasons the fields of the velocities of the currents on the horizon of 30 m

comparing with the maximum velocity on the surface in the same season (Fig. 1a). During the spring period (Fig. 3b) the incoming Mediterranean waters, running with the reflected from the Marmara island waters, move along the western coast of the sea and flow into the extensive central anticyclonic circle. In the eastern part of the sea the portion of the waters goes to the northeastern and southeastern cyclonic eddies and some—to the Bosphorus. Wherein the maximum velocity is 8.40 cm/s.

In the summer (Fig. 3c) the map of the field of the currents velocities is qualitatively similar to the spring one, but the differences consist in the values of the achievable velocities. So in this season there is the intensification of dynamic processes and the maximum velocity of the flows reach 14.29 cm/s. In autumn, the central anticyclone shifts to the west, and in the eastern part of the sea there is one extensive cyclonic vortex (Fig. 3d). The submitted regularities of the behavior of the water below the density jump are mainly determined by atmospheric impact, the flow of impulse through the straits and the bottom topography.

Below 75 m the inflow and outflow of water through the straits are absent and the dynamics in the lower horizons is determined mainly by internal processes and the bottom topography. In Fig. 4 shows averaged for the hydrological seasons the fields of the currents velocities on the horizon of 100 m. In winter at this depth in the central and eastern parts of the sea there is a dynamic system consisting of two anticyclones with maximum velocities up to 8 cm/s (Fig. 4a).

In the western part of the basin, as a result of the reflection from the Marmara Island of the waters entering from the Dardanelles, there is the cyclonic circulation. In the spring (Fig. 4b) anticyclones in the central and eastern parts of the basin become less intense and the maximum velocities decrease to 4 cm/s. With the orbital velocities of up to 6 cm/s cyclonic eddies are located in the northwestern and eastern regions of the sea. In the summer season at the depth of 100 m in the center of the sea the intensification of the anticyclonic cycle (Fig. 4c) is tracked as a result of the intensification of the central anticyclone in the surface layer this season.

In the autumn the waters of the Marmara Sea are involved in extensive cyclonic movement practically along the entire perimeter of the basin (Fig. 4d). The exceptions are local anticyclonic cycles in the outside western and central regions of the sea.

Below 500 m the seasonal signal in the field of the velocities of the currents is poorly traced (Fig. 5). The movement of waters at these depths is concentrated in three deep-water basins. In the western and central deep-water regions there are cyclonic eddies, in the east one—the system of anticyclone and cyclone.

4 Conclusion

Thus using the numerical three-dimensional nonlinear thermos-hydrodynamic model of MHI the experiment on modeling the circulation of the Marmara Sea for example of 2008 was conducted. As a result of the calculation the three-dimensional fields of hydrodynamic characteristics of the circulation for each day were obtained. The analysis of the average seasonal distributions of the fields of the currents velocities

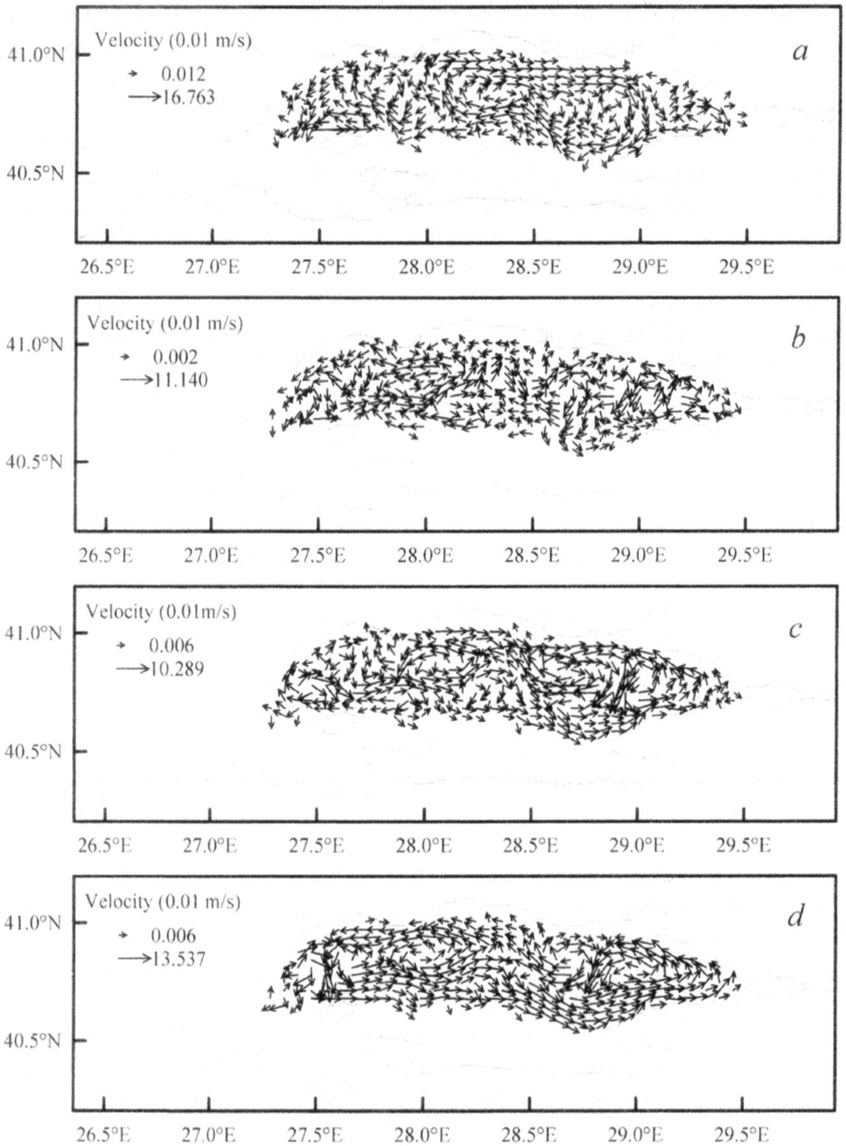

Fig. 4 Average for the winter (**a**), spring (**b**), summer (**c**) and autumn (**d**) hydrological seasons the fields of the velocities of the currents on the horizon of 100 m

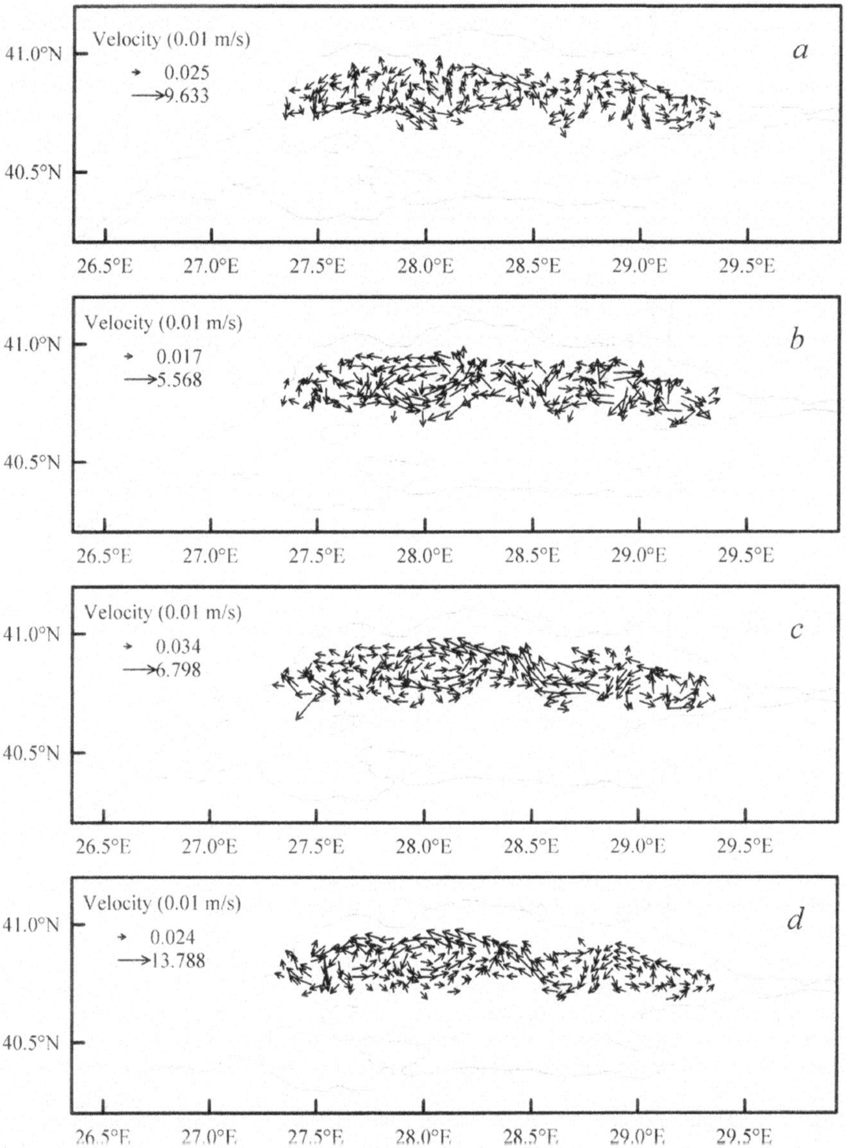

Fig. 5 Average for the winter (**a**), spring (**b**), summer (**c**) and autumn (**d**) hydrological seasons the fields of the velocities of the currents on the horizon of 700 m

of the sea water at different horizons was carried out. It was found that in the surface layer in the winter when the waters move from the Bosphorus to the Dardanelles in the northwestern part of the basin there is the anticyclone, and in the south part the movement of waters with cyclonic vorticity prevails. In the spring hydrological season in the center of the basin the extensive anticyclonic vortex is generated, which is intensified in the summer period. In the autumn the central anticyclone weakens and its area decreases and the cyclonic vortex forms in the southeastern region of the sea. The such location of dynamic features in the surface layer of the sea mainly depends on the distribution of the average seasonal vorticities of the wind field over the region and the orography of the coastline. The regularities of water behavior below the density jump are mainly determined by atmospheric influence, the impulse flow through the straits and the bottom topography. At the depths below 500 m the seasonal signal in the velocity field is traced weakly.

Acknowledgements The numerical experiment was carried out within the framework of the State assignment theme No. 0827-2019-0003.

References

Aydogdu A, Pinardi N, Ozsoy E et al (2018) Circulation of the Turkish straits system between 2008–2013 under complete atmospheric forcing. Ocean Sci. Discuss. https://doi.org/10.5194/os-2018-7

Besiktepe TS (2003) Density currents in the two-layer flow: an example of Dardanelles outflow. Oceanol Acta 26:243–253

Besiktepe ST, Sur HI, Ozsoy E et al (1994) The circulation and hydrography of the Marmara Sea. Prog Oceanogr 34:285–334

Chiggiato J, Jarosz E, Book JW et al (2012) Dynamics of the circulation in the Sea of Marmara: numerical modeling experiments and observations from the Turkish straits system experiment. Ocean Dyn 62:139–159

Demyshev SG, Dovgaya SV, Ivanov VA (2012) Numerical modeling of the influence of exchange through the Bosporus and Dardanelles Straits on the hydrophysical fields of the Marmara Sea. Izv Atmos Oceanic Phys 48(4):418–426

LNCS Homepage. http://www.ucar.edu/mm5/mm4/home.html. Last accessed 10 Dec 2014

Makarov SO (1885) Ob obmene vod Chernogo i Sredizemnogo morey. Prilozhenie k t. II "Zapisok AN". 6:1 - 147

Markova NV, Dymova OA (2019) Validatsiya rezuljtatov chislennogo modelirovaniya gidrofisicheskih poley Chernogo morya pod osnovnym piknoklinom s ispoljzovaniem dannih ARGO. Processes in GeoMedia, no 1, pp 45–50

Shpindler IB (1896) Materialy po gidrologii Mramornogo morya. Zapiski Imperatorskogo Russkogo Geograficheskogo Obschestva 33(2):1–70

Zapevalov AS (2005) Seasonal variability of vertical distributions of temperature and salinity in the Sea of Marmara. Russ Meteorol Hydrol 2:78–84

Zapevalov AS, Dovgaya SV (2007) Transformation of black-sea waters in the Sea of Marmara. Phys Oceanogr 17(2):106–112

Strain Properties of Materials with Gas-Filled Cracks

V. I. Karev

Abstract The paper studies the process of crack growth in a material with gas-filled pores during unloading from external compressive stresses. The variant of directional unloading is considered when a system of parallel cracks oriented perpendicular to the direction of the least compression is formed, and uniform all-round unloading when a system of randomly directed cracks is formed. The effective elastic characteristics of the material for these two cases are estimated. Regularities of crack growth in both cases are revealed depending on the porosity of the material.

Keywords Stress state · Gas pressure · Gas filled cracks · Unloading pattern · Effective elastic characteristics

1 Introduction

The task considered in the article arose in connection with the study of filtration processes in gas-saturated coal bed. The permeability of the coal bed in an intact massif is practically absent; the gas-methane is contained in isolated pores and cracks under pressure close to the mountain pressure (Khristianovich and Salganik 1980). The filtration capacity in the coal bed may appear due to the disturbance of its structure when the stress-strain state changes (Karev and Kovalenko 1988). It is caused by the formation of a filtration channel cohesive system in the bed. Such a system is formed as a result of either the plastic deformation of coal or the growth of cracks from the numerous gas-filled pores contained in the coal. Filtration characteristics of the bed are highly dependent on the size of cracks and the pressure of gas in them, so it is important to know how the cracks grow with decreasing compressive stress, and what the degree of their growth depend on.

Different systems of cracks may arise in dependence on the sequence of material unloading. So, if one of the main stresses is significantly reduced and the other two do not change significantly, a system of parallel cracks oriented perpendicular to the

V. I. Karev (✉)
Ishlinsky Institute for Problems in Mechanics of Russian Academy of Sciences,
Ave. Vernadskogo 101, Bldg 1, Moscow, Russia 119526
e-mail: wikarev@ipmnet.ru

C. T. Olegovna (ed.), *Processes in GeoMedia—Volume I*,
Springer Geology, https://doi.org/10.1007/978-3-030-38177-6_19

direction of least compression is formed in the material. If the material is unloaded in all directions equally, a system of randomly oriented cracks is formed. For example, in order to prevent coal and gas outbursts, the coal seam is unloaded in the direction perpendicular to the bedding plane and a system of cracks parallel to this plane is formed. If as a result of an explosion, an rock outburst, or a rock bump, a piece of rock or coal is came off from the massif, then it goes from a state of all-round compression to a state of unloading in all directions, and a chaotic system of cracks is formed.

As cracks grow, the effective strain characteristics of the material change, and the degree of the growth of cracks in turn depends on strain properties of material. It is interesting to see how the cracks grow and, accordingly, the effective characteristics of the material change in both cases, to find out whether there is a difference and, if so, what is the reason for it.

The paper is devoted to this issue. It was found out that the cracks grow differently when the reduction of compressive stresses in these two cases. In the case of a system of parallel cracks, they grow when compressive load decreasing more than an isolated crack in an isotropic elastic material without cracks, and in the case of a system of chaotically distributed cracks, they grow to a lesser extent. An explanation is offered for this fact.

The work (Kovalenko 1980) studied how effective elastic characteristics of the material with the system of parallel gas-filled cracks change taking into account their growth when the external compressive load changes by the method developed in Salganik (1973). Determination of effective characteristics in this work is based on the solution of the problem of one crack in the material with some average characteristics due to the presence of other cracks. Using this solution as well as knowing the type of crack distribution function in space and size $a(N)$, the author solved a system of differential equations expressing the relationship between the change of components of the pliability tensor and an increment of the function $a(N)$, and found the effective characteristics of the material. The effective characteristics of the material surrounding the crack will depend on the stress state achieved during crack growth. To solve the problem of determining the effective elastic characteristics of the material with a system of parallel gas-filled cracks the iterative procedure was used, it is briefly described below.

A similar problem is solved in the present paper but for the material with gas-filled cracks distributed chaotically in all directions in space. In such a material, the averaged characteristics in any direction are the same; therefore, it can effectively be considered isotropic.

To determine the effective characteristics of such a material, it can be use the solution of the problem of one crack in an elastically isotropic medium (Kovalenko 1980) and the system of differential equations, which establishes a relationship between the components of the pliability tensor a_{ij} and the function $a(N)$ for an isotropic case obtained in Salganik (1973). However, the ratios found in Salganik (1973) do not take into account the influence of gas contained in the cracks, they are valid for empty cracks. At high gas pressures, small crack openings and low modulus of elasticity of the surrounding material, this effect can be significant.

Let us get the specified ratios for gas-filled cracks. So from (Salganik 1973) we have

$$\delta\varepsilon_{ik} = \delta\varepsilon^0_{ik} + \frac{1}{2}\int (n_i\delta V_k + n_k\delta V_i) F dY = S_{iklm}\delta\sigma_{lm} \quad (1)$$

$$S_{iklm} = S^0_{iklm} + \frac{1}{2}\int (n_i\frac{\delta V_k}{\delta\sigma_{lm}} + n_k\frac{\delta V_k}{\delta\sigma_{lm}})\ FdY$$

In (1), the same notation is adopted as in Salganik (1973). The coordinate system is connected with the crack, the first and second axes are directed in the crack plane, the third axis is perpendicular to the crack plane. Given the gas pressure in the crack, we have

$$V_i = \frac{8\pi(1-\nu_0^2)}{3E_0} a^3 C_{(i)}(\sigma^/_{i3} + P\delta_{i3}), \quad i = 1, 2, 3 \quad (2)$$

here ν_0, E_o are Poisson's coefficient and Young's module of the material, a is the crack radius, $C_{(i)}$ are coefficients calculated in a known way (Salganik 1973; Green and Sneddon 1950), δ_{i3} is Kronecker's symbol P is gas pressure in the crack.

The contribution $V_i^{\prime P}$ that gas pressure gives into the components of vector V_i is equal to

$$V_i^{\prime p} = \frac{8\pi}{3}\frac{(1-\nu_0)^2}{E_0} a^3 C_{(i)} P\delta_{i3} \quad (3)$$

In the lab coordinate system

$$V_k^P = V_3^\prime n_k \quad (4)$$

Taking into account that for a disco crack $C_{(3)} = \frac{2}{\pi}$ and entering a designation $\alpha = \frac{16}{3}\frac{(1-\nu_0^2)}{E_0}$, we get

$$V_k^P = \alpha\, a^3 P\, n_k$$

Since

$$V_k = V_k^e + V_k^P$$

$$S_{iklm} = S^0_{iklm} + \frac{1}{2}\int (n_i\frac{\delta V_k^e}{\delta\sigma_{lm}} + n_k\frac{\delta V_i^e}{\delta\sigma_{lm}})\ FdY + \frac{1}{2}\int (n_i\frac{\delta V_k^P}{\delta\sigma_{lm}} + n_k\frac{\delta V_i^p}{\delta\sigma_{lm}})\ FdY \quad (5)$$

The second term of the right part is defined in Kovalenko (1980), let us calculate the third term.

Let us assume that all processes in the system are slow enough, and the isothermal law is valid for gas. From (4) follows

$$\Delta V = \frac{16}{3}\frac{(1-\nu^2)}{E}(P+\sigma)a^3 \tag{6}$$

$$P_0 V_0 = P\left[V_0 + \frac{16}{3}\frac{(1-\nu^2)}{E}\right](P+\sigma)a^3 \tag{7}$$

P can be expressed from (7)

$$P = -\frac{1}{2}\left[\frac{V_0}{\alpha a^3} + \delta\right] + \frac{1}{2}\sqrt{(\frac{V_0}{\alpha a^3} + \delta)^2 + \frac{4V_0}{\alpha a^3}P_0} \tag{8}$$

Differentiating (8) by σ and considering that in the laboratory coordinate system $\sigma = \sigma_{lm} n_l n_m$ we find

$$\delta P = \frac{1}{2}\left[-1 + \frac{\frac{V_0}{\alpha a^3} + \delta}{\sqrt{(\frac{V_0}{\alpha a^3} + \delta)^2 + \frac{4V_0}{\alpha a^3}P_0}}\right]\delta\sigma_{lm} n_l n_m \tag{9}$$

From (9) and (4) we get

$$\frac{\delta V_k^p}{\delta \sigma_{lm}} = \frac{\alpha a^3}{2} B n_l n_m n_k \tag{10}$$

where

$$B = -1 + \frac{V_0 + \alpha a^3 \sigma}{\sqrt{(V_0 + \alpha a^3\sigma)^2 + 4\alpha a^3 P_0 V_0}}$$

By substituting (10) in (5), we can now find an equation to determine any component of the pliability tensor. In our case of isotropic crack distribution, all components of this tensor are expressed through some two. Let us find the equation for the following components: Young's module and Poisson's coefficient $E = \frac{1}{S_{3333}}$

$$S_{3333} = S^e_{3333} + \frac{\alpha B}{8\pi}\int n_3^4 a^3 \sin\theta F(a) da d\psi d\theta \tag{11}$$

In (11) the same designations as in Salganik (1973) are accepted: S_{3333} is pliability without taking into account gas pressure in cracks.

Given that

$$v = \int a^3 F(a) da = n\bar{a}^3,$$

we have

$$S_{3333} = S^e_{3333} + \frac{2(1-\nu_0^2)}{3\pi E_0} B 2\pi \frac{2}{5} v$$

Using the results (Salganik 1973), we obtain the first equation

$$E = E_0 \left\{ 1 - \left[\frac{16(1-\nu_0^2)(10-3\nu_0)}{45(2-\nu_0)} + \frac{8(1-\nu_0^2)}{15} B \right] v \right\} \tag{12}$$

Next,

$$G = \frac{1}{4S_{1313}} \tag{13}$$

$$G = \frac{E}{2(1+\nu)} \tag{14}$$

Similarly, calculating S_{1313} from (5) and (13) and (14) we find the equation for ν:

$$\nu = \nu_0 \left\{ 1 - \left[\frac{16(3-\nu_0)(1-\nu_0^2)}{15(2-\nu_0)} + \frac{8(1-\nu_0^2)(1+3\nu_0)}{45\nu_0} B \right] v \right\} \tag{15}$$

If we rewrite Eqs. (12) and (15) for small differences in the corresponding characteristics resulting from small increments in relative volume Δv, replace the ratio of the first to the second by derivatives, and the value ν_0 by the current value in the right parts, we get the differential equations:

$$\frac{d \ln E}{dv} = -\frac{16}{45}(10-3\nu)\,\frac{1-\nu^2}{2-\nu} - \frac{8}{15} B(1-\nu^2) \tag{16}$$

$$\frac{d \ln \nu}{dv} = -\frac{16}{45}(3-\nu)\,\frac{1-\nu^2}{2-\nu} - \frac{8}{15} B(1+3\nu)\frac{(1-\nu^2)}{\nu} \tag{17}$$

For initial conditions: at $v = 0$ must be $E = E_o, \quad \nu = \nu_o$.

Thus, having a system of differential equations that give a link between the change of effective characteristics of the material and the increase in the relative volume of cracks in the material, as well knowing the pattern of the dependence of the cracks growth on the external load, we can find the effective characteristics of the material for isotropic distribution cracks on directions, in the same way as it was done in (Kovalenko 1980) for the material with a system of parallel cracks. However, difficulties arose when trying to apply the computational procedure used in (Kovalenko 1980) to an isotropic case. Let us describe the procedure briefly.

At first, the external load at which the crack growth began was found using equation of the isothermal gas expansion, crack growth conditions, and equations to change the volume of the crack loaded. Figure 1 shows the crack growth beginning

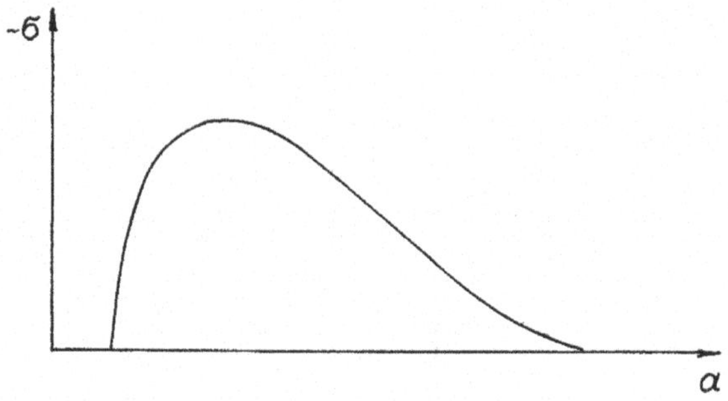

Fig. 1 The crack growth beginning curve

curve for cracks with an initial volume V_0, i.e. the dependence of the stress value at which the crack starts to grow on its size.

It was discovered that the growth of cracks begins with a jump when external stress achieves critical value σ_c. First, the effective characteristics were calculated for this value σ_c and the initial crack radius. Then, the transcendent equation for determining the crack radius after a jump, derived from the equation of the isothermal gas expansion, crack growth conditions, and the equations to change the volume of the crack was solved. After that, the new effective characteristics were found for the new radius of cracks by using the system of differential equations. The transcendental equation for determining the crack radius after the jump was solved again with these new values of characteristics. And so on, until the process has converged.

Thus, a new curve and actually a new value σ_c corresponded to each new values of effective characteristics, σ_c increased in absolute value at each step. The described iteration process is shown graphically at Fig. 2.

When trying to apply the specified computational procedure for the isotropic case, it turned out that σ_c decreases in absolute value at each step, and thus, σ_c determined at the previous step after recalculation of the effective characteristics of the medium with increased cracks will not be sufficient to begin the growth of cracks. This situation is shown graphically at Fig. 3.

Therefore, a new calculation scheme was proposed. The value σ_c and the effective characteristics of the material with cracks of the initial size are calculated. A small

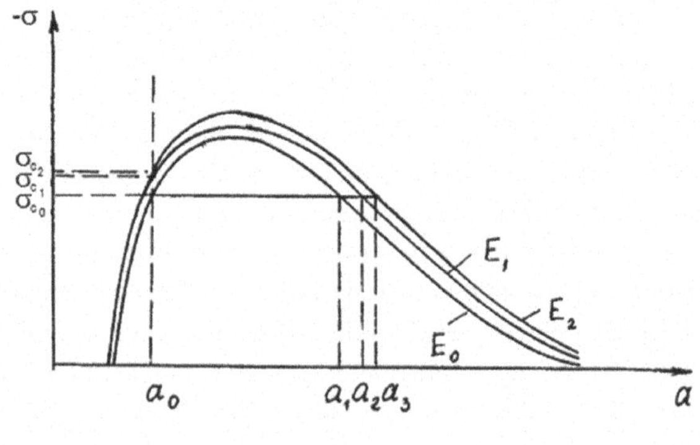

Fig. 2 An iterative process in calculating the dependence of the radius of cracks on the external applied load for a material with a system of parallel cracks. E_0, E_1, E_2—modules of material elasticity at zero, first and second iterations

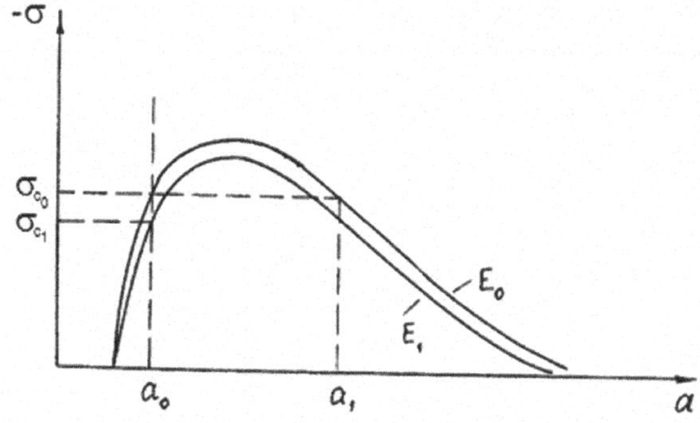

Fig. 3 The iterative process of calculating the dependence of the crack radius on the external applied load for a material with cracks chaotically distributed on direction is interrupted at the first step

increment to the radius of cracks is given. Gas pressure in the crack is determined from the isothermal equation and the equation to change the volume of the crack loaded. Now, if

$$(P + \sigma_c)\sqrt{2a} > K \tag{18}$$

where the K is clutch module at the end of the crack, then the same increment is given to the radius and the same procedure is done.

If

$$(P + \sigma_c)\sqrt{2a} < K$$

then the radius of the cracks is given half the increment with the opposite sign, again P is calculated and determined the sign of inequality (18). These calculations are carried out until $\left|(P + \sigma_c)\sqrt{2a} - K\right|$ will be less than some predefined value ε.

The calculation scheme is shown in Fig. 4.

This improved scheme was used to calculate once again the effective characteristics of the material with a system of parallel cracks, taking into account their growth.

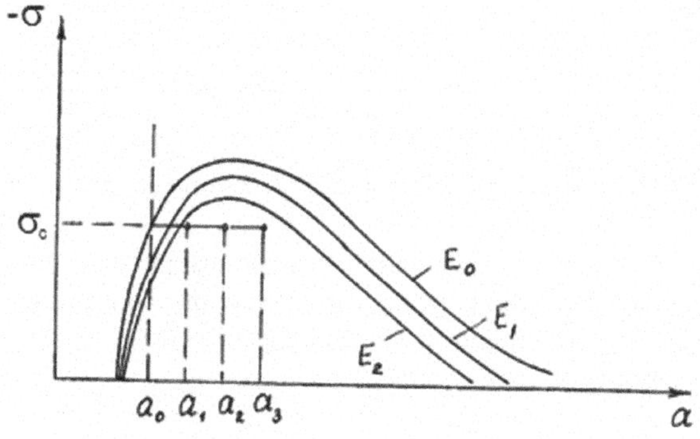

Fig. 4 A new iterative scheme to calculate the dependence of the crack radius on the external applied load, both for material with a system of parallel cracks, and for material with randomly distributed cracks

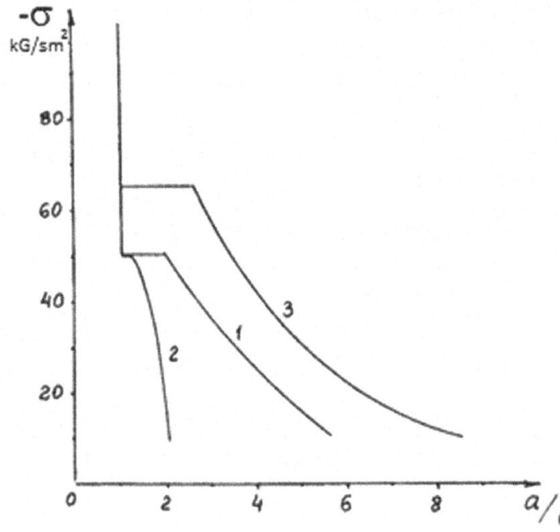

Fig. 5 Dependence of crack radius on external applied load at $\sigma_0 = -100$ kg/cm^2, $\frac{K}{\sqrt{2a_0}} = 44.7$ kg/cm^2, $a_0 = 0.1$ cm, $m_0 = 0.25$

At Fig. 5 the curves σ(a) are presented for a crack growing in an isotropic-elastic material without cracks (1), with isotropic distribution of cracks (2), with a system of parallel cracks (3). As it can be seen from the graphs, earlier than in other cases, the growth of cracks begins in a material with an isotropic distribution of cracks, in such a material the jump is the least and the cracks can, in principle, grow to a smaller value than in the other two cases.

How can it be explain this effect? Let us consider the case of isotropic distribution of cracks. As the cracks in this material grow, the pliability will increase. Therefore, as the compressive stress decreases, the volume of each crack in such material will increase more than the volume of the crack in an isotropic-elastic material without cracks (the pliability of which does not change) at the same increase in the radius of cracks in both cases, see (6). Since the volume of the crack and the gas pressure in it are related to the isothermal law, the gas pressure in the crack located in the material with isotropic crack distribution will be reduced accordingly to a greater extent than in the material without cracks. The condition for crack growth in isotropic material is

$$(P - |\sigma|) > \frac{K}{\sqrt{2a}}$$

Therefore, in order to achieve the same change in the crack radius in both cases, a higher degree of unloading is required in a material with isotropic crack distribution.

In the case of a parallel cracks system, this effect also occurs, but it is small compared to another effect that gives the opposite result. In this case, the material behaves effectively as transversally isotropic. In such material, the crack growth condition is written as follows

$$(P - |\sigma|) > \frac{\beta_N}{2I_{(3)}} \frac{K}{\sqrt{2a}}$$

where it is β_N and $I_{(3)}$ expressed through effective elastic material modules. The ratio $\frac{\beta_N}{2I_{(3)}}$ for isotropic material equaled to unity decreases with the parallel cracks growth. Thus, the growth of cracks in the material with the system of parallel cracks is provided with less pressure drop $(P - |\sigma|)$ than in isotropic-elastic material without cracks. Physically, this means that with increasing compliance in the direction perpendicular to the plane of cracks, the radius of curvature of cracks increases, which leads to an increase in the stress intensity factor on the edges of the crack that determines its growth. The gas pressure in cracks decreases to a greater extent than in an isotropic-elastic material without cracks, as well as in the case of isotropic distribution of cracks. But as the calculations show, the above effect is stronger, and to achieve the same increase in the crack radius in the material with the system parallel cracks it is required less reduction in compressive load than in an isotropic-elastic material without cracks.

Figure 6 shows the change in the gas pressure in the cracks P and the modulus of elasticity of the material E depending on σ. Curves *1* are obtained for a crack growing in an isotropic-elastic material without cracks; curves *2* are obtained for a crack in an isotropic-elastic material with a chaotic distribution of cracks; curves *3* are obtained for a material with a system of parallel cracks (in this case the elastic modulus in the direction perpendicular to the crack plane is considered).

Graphs presented at Figs. 5 and 6, obtained for the initial porosity of the material $m_0 = 0.25$. Figures 7 and 8 show the same dependences $a(\sigma)$ and $E(\sigma)$, $P(\sigma)$ for $m_0 = 0.1$.

From the comparison of graphs at Figs. 5, 6, 7 and 8 it is visible that at lower initial porosity of the material, the growth of cracks in the material with parallel system of cracks also occurs to a greater extent, and in the material with isotropic distribution of cracks to a lesser extent, than the growth of cracks in the material without cracks. However, these differences are less than for the material with higher initial porosity.

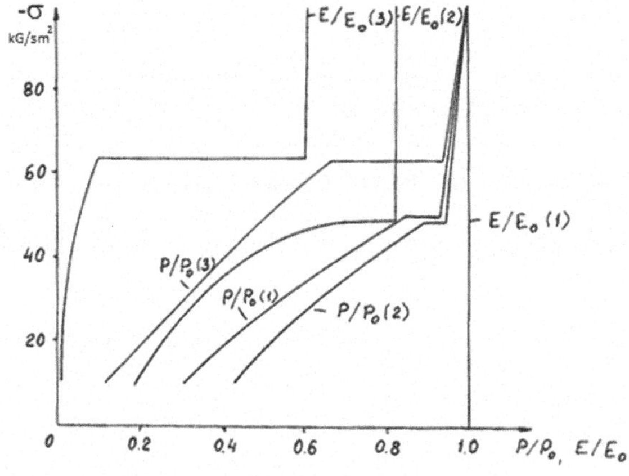

Fig. 6 The dependence of the gas pressure in the cracks and the effective elasticity modulus of the material (in the case of a material with a system of parallel cracks, the modulus of elasticity in the direction perpendicular to the plane of the cracks is considered) on the external applied load at $\sigma_0 = -100$ kg/cm^2, $\frac{K}{\sqrt{2a_0}} = 44.7$ kg/cm^2, $a_0 = 0.1$ cm, $m_0 = 0.25$

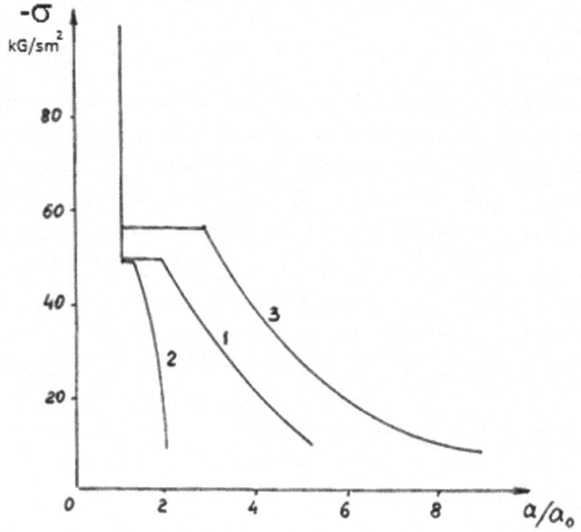

Fig. 7 Dependence of crack radius on external applied load at $\sigma_0 = -100$ kg/cm^2, $\frac{K}{\sqrt{2a_0}} = 44.7$ kg/cm^2, $a_0 = 0.1$ cm, $m_0 = 0.1$

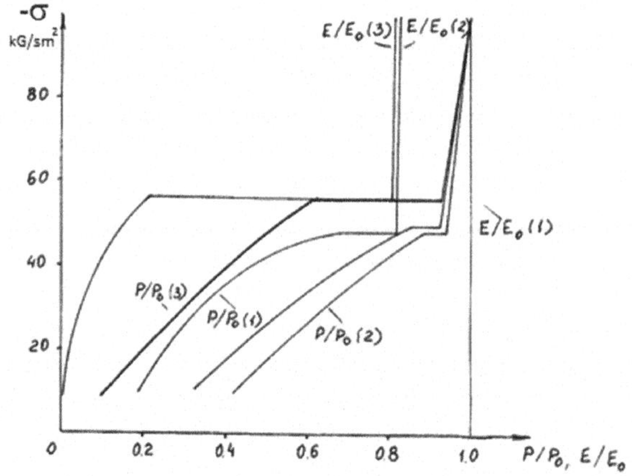

Fig. 8 Dependence of gas pressure in the cracks and the effective elasticity modulus of a material (in the case of a material with a system of parallel cracks the modulus of elasticity in the direction perpendicular to the plane of cracks is considered) on the external applied load at $\sigma_0 = -100$ kg/cm^2, $\frac{K}{\sqrt{2a_0}} = 44.7$ kg/cm^2, $a_0 = 0.1$ cm, $m_0 = 0.1$

Since the initial size of the cracks was always considered the same ($a_0 = 0.1$ cm), a lower initial porosity of the material means a lower concentration of cracks. The fact that the detected effect is smaller at a lower concentration of cracks confirms that this effect is due to the mutual influence of cracks.

Acknowledgements The study was supported by the Russian Science Foundation No. 16-11-10235 P.

References

Green AE, Sneddon IN (1950) The distribution of stress in the neighborhood of a flat elliptical crack in an elastic solid. Proc Cambridge Philos Soc 46:59–165

Karev VI, Kovalenko YF (1988) Theoretical model of gas filtration in gas-saturated coal seams. J Min Sci 24(6):47–55

Khristianovich SA, Salganik RL (1980) Sudden outbursts of coal (rock) and gas. Stresses and strains. Preprint No. 153. Institute for Problems in Mechanics of the Academy of Sciences of the USSR, Moscow, 88 p

Kovalenko YuF (1980) Effective characteristics of solids with isolated gas-filled cracks. A wave of destruction. Preprint No. 145. Institute for Problems in Mechanics of the Academy of Sciences of the USSR, Moscow, 52 p

Salganik RL (1973) Mechanics of solids with a lot of cracks. Mech Sol (4):149–158

The Calculation of the Earth's Insolation for the Period 3000 BC–AD 2999

V. M. Fedorov and A. A. Kostin

Abstract The method for calculating the downward solar energy falling to the Earth without an atmosphere is developed. The calculations take into account the ellipsoidal shape of the Earth and variations of the orbital motion and the rotation of the Earth produced by the Moon and planets. The insolation integrals were prepared for Earth's surface latitudinal strips and slices of tropical years from 3000 BC to AD 2999.

Keywords Earth's insolation · Astronomical ephemerides · Ellipsoid · Orbital motion

1 Introduction

The solar radiation coming to the Earth (downward one and missing the Earth's surface one) is the determinative of the radiation and thermal balance of the Earth (Kondratyev 1980; Lorentz 1970; Monin and Shishkov 2000). The radiant energy of the Sun and its changes make foundations for hydro-meteorological and many other processes taking place in the Atmosphere, Hydrosphere and on the Earth's surface. The energy of the Sun is the most important factor in the development of life on Earth, providing the necessary thermal conditions for life and photosynthesis. The main part of coming solar radiation is falling to the Earth's surface. Therefore, the accurate calculation and study of spatial and temporal changes in insolation is important for the study of the processes taking place in the Earth's geographical shell for understanding the causes of the formation and changes in the climatic conditions of life on the planet.

In contrast to the few known insolation calculations covering the range of high-frequency variations (Berger et al. 2010; Bertrand et al. 2002; Borisenkov et al. 1983), our calculation method takes into account the ellipsoidal shape of the Earth, changes in the duration of tropical year (Fedorov 2013, 2015, 2019) and other small variations of the orbital motion and the rotation of the Earth produced by the Moon and planets.

V. M. Fedorov (✉) · A. A. Kostin
Faculty of Geography, Lomonosov Moscow State University, Moscow, Russia
e-mail: fedorov.msu@mail.ru

C. T. Olegovna (ed.), *Processes in GeoMedia—Volume I*,
Springer Geology, https://doi.org/10.1007/978-3-030-38177-6_20

The initial variant of our method (http://www.solar-climate.com/en/ensc/kost.htm) is corrected here in some details noticed below.

The initial astronomical data for the insolation calculations are the declination and ecliptic longitude of the Sun, the distance from the Earth to the Sun (Fedorov 2012, 2016), the difference in the march of evenly running time and universal corrected one.

The basic principles of our approach: the distance from the Earth to the Sun and the orientation of the Earth's axis are determined from the high-precision NASA ephemerides DE-406 (Giorgini et al. 1996; http://ssd.jpl.nasa.gov), the Earth's ellipsoid is divided into latitudinal strips, and each tropical year—into slices, and each pair (strip, slice) connects with the integral (J) of insolation (W/m^2) per strip from the beginning to the end of slice. We can divide the integral by the strip area to get a specific energy (J/m^2) collected by the strip within the slice. We can multiply the integral by the ratio of the strip fragment length to the strip length to obtain an estimate of the energy (J) collected by this fragment within the slice.

A number of theoretical simplifications have been applied to calculate these values. The main simplifications are: (1) that solar activity is considered constant, (2) the radiation comes from the center of the Sun, and (3) the influence of the Earth's atmosphere is not taken into account.

All theoretical simplifications are outlined in Sect. 2. Strict formulas for calculations are gathered in Sect. 3. The applied technology of approximate calculations and its typical errors are described in Sect. 4.

2 Selected Approach to Describing the Falling Solar Energy

The period of time from 3000 BC to AD 2999 is considered. The surface of the Earth approximates an ellipsoid rocking relatively to the geoid, hereinafter referred to as MRS80 (Moving Reference System 1980), with the length of large axes $p_1 = p_2 = A = 6{,}378{,}137$ m and small axis $p_3 = B = 6{,}356{,}752$ m. The small axis at each moment is aligned with the axis of rotation of the Earth, and the center of the ellipsoid—with the center of the masses of the Earth.

The lengths of the axes are rounded to one meter and correspond to the parameters of the Earth's GRS80 ellipsoid, which is fixed relative to the geoid. The parameters GRS80 were recommended for usage by the International Union of Geodesy and Geophysics in 1980.

The MRS80 swinging ellipsoid is equipped with an imaginary grid of parallels and meridians, a system of normals and geodetic coordinates, according to which vertical lines, horizontal planes and latitude zones of the Earth are defined. These lines, planes and zones together with the ellipsoid oscillate relatively to the geoid. Oscillations are coupled with deviations of the Earth's rotation axis from its average position in the Earth's body. Deviations have been recorded since the end of the nineteenth century in terms of geographical pole movement. Each of the geographical poles moves relative to the geoid along a multi-turn, unclosed curve that fits in a square with a side of 30 m. One round (Chandler cycle) lasts about 14 months.

The oscillating ellipsoid is chosen instead of the fixed ellipsoid for two reasons: first, not to complicate the calculations, and second, because of the absence of a reliable swinging model covering the entire period of time under consideration.

The model of solar radiation and its imaginary measurement on the surface of the Earth is considered, according to which

(1) isotropic radiation comes to Earth from the center of the Sun, the power density of solar radiation decreases inversely to the square of the distance from the center of the Sun,
(2) eclipses are ignored,
(3) density of radiation power at a distance of 1 a.u. from the center of the Sun at any time is equal to $U_0 = 1361$ W/m^2, where 1 a.u. = r_0 = 149,597,870,691 m,
(4) the dissipating effects of the atmosphere are not taken into account,
(5) the earth's surface is replaced by a rocking MRS80 ellipsoid.

The oscillating ellipsoid is divided into Δ—degree latitudinal stripes (the geodetic latitude is meant) where $\Delta \in \{1, 5\}$. Projection of each strip onto the geoid "floats" relative to the geoid (type of motion—various vibrations of the ring), moving away from its middle position by a maximum of 15 m according to Chandler's cycles.

The insolation integrals are calculated—the energy (in joules) falling to the Earth through each of the strips in each of the L lobes of each tropical year under study, where $L \in \{12, 360\}$, as well as linear combinations of these integrals (tropical decades, months, quarters, half-years, years).

Tropical years are selected instead of calendar years to exclude four-year calendar rhythms (Fedorov 2013, 2015, 2019). The number of the tropical year coincides with the number of the calendar year in which it begins.

A tropical year is a projective tropical year that is tracked by the movement of the Sun's projection onto the ecliptic. If L is the number of slices into which the projective tropical year is divided, then the n-th slice starts at the moment when the ecliptic longitude of the Sun takes on the value of $360(n - 1)/L$ (in degrees).

To take into account the lengthening of the day due to the gradual deceleration of the Earth's rotation, a calendar time scale differs, where each day corresponds to an array of 86,400 calendar seconds (excluding days, adjusted for 1 s), and a scale of evenly running time, according to which the daily intervals are measured in true seconds, and between themselves are not equal. Solar radiation integrals are calculated on a scale of evenly running time. The remark "excluding days, adjusted for 1 s" was omitted in http://www.solar-climate.com/en/ensc/kost.htm.

The imaginary clock, which counts evenly for the current time, is located in the center of the Earth. An event "a portion of solar radiation has arrived" on a small area of the Earth's surface is fastened to the axis of evenly running time as follows. The moment of imaginary start of the corresponding portions bunch from the center of the Sun is marked. In this bunch, a portion moving to the center of the Earth is picked out. The moment of imaginary arrival of this portion in the center of the Earth (if there were no obstacles in its way) is calculated. This arrival time is selected as the moment to which the event is fastened.

This method of binding results with small (0–20 ms) delays (different in different places of the strips). However, in terms of the impact on large-scale terrestrial processes, such systematic delays are insignificant. This delay are equivalent to the small (about 10 ms) vibrations of the boundaries of tropical years slices. The delay option is chosen because their exclusion would cause unnecessary complications in the calculations. The estimates "0–20 ms" and "about 10 ms" are better than ones in http://www.solar-climate.com/en/ensc/kost.htm.

3 Strict Calculation Formulas

According to the chosen model of solar radiation and its measurement, the calculation of the integrals (in joules) is based on the calculation of insolation $\Lambda(t, \varphi, \alpha)$ (W/m^2), which would be observed in the absence of the Earth's atmosphere in a given time at a given point of MRS80. Here t is the moment on the scale of evenly running time (s), φ and α—the geodetic latitude (relative to MRS80) and sliding longitude (hour-angle of the Sun at a given point of MRS80) of the imaginary insolation measurement point, expressed in radians. The remark "hour-angle of the Sun at a given point" was omitted in http://www.solar-climate.com/en/ensc/kost.htm.

An elementary fragment of a tropical year is a product of its fragmentation into 360 parts (Fedorov 2013). The energy falling to Earth through the strip limited by the latitudes φ_1 and φ_2 (in radians), in the *n-th* elementary fragment of the *m-th* tropical year, denote $I_{\mathrm{nm}}(\varphi_1, \varphi_2)$. The energy falling through the same strip in the q-th slice of the *m-th* tropical year, denote $J_{\mathrm{nm}}(\varphi_1, \varphi_2)$. Then

$$
\begin{aligned}
L = 360 &\Rightarrow J_{qm}(\varphi_1, \varphi_2) = I_{qm}(\varphi_1, \varphi_2),\\
L = 12 &\Rightarrow J_{qm}(\varphi_1, \varphi_2) = \sum_{n=30(q-1)+1}^{30q} I_{nm}(\varphi_1, \varphi_2)
\end{aligned} \tag{1}
$$

Let t_{nm1} and t_{nm2} be the beginning and the end of the n-th elementary fragment of the m-th tropical year on the scale of evenly running time (s). Then

$$
I_{nm}(\varphi_1, \varphi_2) = \int_{t_{nm1}}^{t_{nm2}} \left(\int_{\varphi_1}^{\varphi_2} \sigma(\varphi) \left(\int_{-\pi}^{\pi} \Lambda(t, \varphi, \alpha)\, d\alpha \right) d\varphi \right) dt \tag{2}
$$

$$
\sigma(\varphi) = q_1(\varphi) q_2(\varphi),\ q_1(\varphi) = \frac{A^2}{B} h(\varphi) \cos\varphi,\ q_2(\varphi) = \frac{A^2}{B} h^3(\varphi) \tag{3}
$$

$$
h(\varphi) = \frac{1}{\sqrt{1 + \varepsilon^2 \cos^2 \varphi}},\ \varepsilon^2 = \left(\frac{A}{B}\right)^2 - 1 \tag{4}
$$

where $\sigma(\varphi)$ is the area factor in the place of imaginary measurement of solar radiation. It is used to calculate $\sigma(\varphi)d\alpha d\varphi$—the area ($m^2$) of infinitely small trapeze on the ellipsoid MRS80. Length of the middle line of the trapeze (along the local parallel): $q_1(\varphi)d\alpha$, height of the trapeze (along the local meridian): $q_2(\varphi)d\varphi$.

Let $b(t)$ be the moment of start of the imaginary light pulse from the center of the Sun reaching the center of the Earth at the moment t. For the moment $b(t)$ let $r(t)$ be the distance (m) between the centers of the Sun and the Earth, $\gamma(t)$—the declination of the center of the Sun (radians) and $\lambda(t)$—the ecliptic longitude of the center of the Sun (degrees). Then

$$\lambda(t_{nm1}) = n - 1,\ \lambda(t_{nm2}) = n, \tag{5}$$

$$\Lambda(t, \varphi, \alpha) = u_0 \left(\frac{r_0}{r(t)} \right)^2 \max\left(\frac{D_0(t, \varphi) + D_1(t, \varphi)\cos\alpha}{(C_0(t, \varphi) - C_1(t, \varphi)\cos\alpha)^{3/2}},\ 0 \right) \tag{6}$$

$$D_0(t, \varphi) = \sin(\varphi)\sin\gamma(t) - \frac{B}{r(t)h(\varphi)},\ D_1(t, \varphi) = \cos(\varphi)\cos\gamma(t) \tag{7}$$

$$C_0(t, \varphi) = 1 - \frac{2Bh(\varphi)\sin(\varphi)\sin\gamma(t)}{r(t)} + \frac{B^2 + \varepsilon^2 A^2 h^2(\varphi)\cos^2\varphi}{r^2(t)} \tag{8}$$

$$C_1(t, \varphi) = \frac{2A^2 h(\varphi)\cos(\varphi)\cos\gamma(t)}{Br(t)} \tag{9}$$

$$A = 6{,}378{,}137,\ B = 6{,}356{,}752,\ r_0 = 149{,}597{,}870{,}691,\ u_0 = 1361 \tag{10}$$

Before integrating according to Formula (2), it is useful to ask the question: if t, φ are given, then at which α does the positive insolation exist? The values of α from the range $(-\pi, \pi)$ at which the positive insolation is observed, are determined by the inequality $|\alpha| < \alpha_M(t, \varphi)$, where $\alpha_M(t, \varphi)$ is the observability boundary at a given latitude φ.

If the specified latitude φ is close to 0, then α_M fluctuates in a small area of $\pi/2$, when t changes. When the module of the given latitude increases, the range of α_M oscillation increases too. If the module is close to $\pi/2$ then α_M oscillates from 0 to π, takes on extreme values and lingers on them. At times when $\alpha_M = 0$, there is a polar night at the given latitude. At times when $\alpha_M = \pi$ the polar day is observed at a given latitude.

If $\alpha_M = 0$ at a given t, φ, then the positive insolation is not observed. In this case, the insolation at $\alpha = 0$ equals to 0. If $\alpha_M > 0$, then with growth of $|\alpha|$ the insolation decreases from a positive maximum at $\alpha = 0$ to some minimum at $|\alpha| = \alpha_M$. If $\alpha_M < \pi$, the minimum is 0. If $\alpha_M = \pi$, the minimum is either 0 or greater than 0 (between the beginning and the end of the polar day). In the second case, the positive insolation does not exist only at $|\alpha| < \alpha_M$ but also at $|\alpha| = \alpha_M$.

One of the properties of the insolation is α parity: $\Lambda(t, \varphi, \alpha) = \Lambda(t, \varphi, -\alpha)$. Thus, Formula (2) can be converted to a more computable form

$$I_{nm}(\varphi_1, \varphi_2) = \int_{t_{nm1}}^{t_{nm2}} u_0 \left(\frac{r_0}{r(t)}\right)^2 \left(\int_{\phi_1}^{\phi_2} \frac{2\sigma(\varphi)}{C_0^{3/2}(t,\varphi)} \left(\int_0^{\alpha_M(t,\phi)} \frac{D_0(t,\varphi) + D_1(t,\varphi)\cos\alpha}{(1 - E(t,\varphi)\cos\alpha)^{3/2}} d\alpha \right) d\varphi \right) dt \quad (11)$$

$$\alpha_M(t,\varphi) = \arccos\left(\max\left(-1, \min\left(-\frac{D_0(t,\varphi)}{D_1(t,\varphi)}, 1\right)\right)\right) \quad (12)$$

$$E(t,\varphi) = C_1(t,\varphi)/C_0(t,\varphi),\ 0 < E(t,\varphi) < 10^{-4} \quad (13)$$

4 Approximate Calculations and Their Errors in Case Δ = 5, L = 12

4.1 Calculation Plan

Formulas (1), (3)–(13) cannot be calculated with absolute accuracy. Inaccuracies are inherent in the source data, interpolation and root-finding procedures of equations in the processing of source data, as well as integration procedures.

For the variant $\Delta = 5$, $L = 12$ we applied the following system of approximate calculations corresponding to formulas (1), (3)–(13).

The first stage is the operational markup of the time scales used, the access to the Internet service HORIZONS NASA (Giorgini et al. 1996; http://ssd.jpl.nasa.gov/?horizons_doc#specific_quantities) and obtaining the primary source data, tied to the beginning of the day by Greenwich. Primary data are the Earth-Sun distance (km), the Sun's declination and ecliptic longitude (degrees), and the difference in stroke (s) between the coordinate time (we take it as evenly running time) and universal corrected time.

The second stage is to calculate the displacements of the beginnings and ends of the elementary fragments of tropical years relative to the beginnings of the day by Greenwich (for this purpose—the roots of the equations with the participation of the ecliptic longitude are found) and on this basis to prepare the secondary source data linked to the beginnings, ends and intermediate moments of the elementary fragments of tropical years (for this purpose—interpolation of the primary data is made). Secondary source data—Earth-Sun distance (m), Sun declination (radians), and duration of fragments (s) on the scale of evenly running time.

The third stage is the calculation of insolation integrals from the secondary source data (for this purpose—calculation of auxiliary variables and their substitution in the integration procedures).

4.2 Three Time Scales and Their Working Markup

Three time scales are used: CT (Coordinate Time), UT1 (Universal Time Without Correction) and UT2 = UTC (Universal Time With Correction). The UT2 scale is derived from the UT1 scale by episodic (once every few years) movements of ±1 calendar second to align the UT2 watch with the CT clock (since 1962).

The unit of the CT scale is the true second. Two markups are entered on the CT scale: tropical and calendar.

Tropical markup is made up of the main moments $\{t_{nm1}, t_{nm2}\}$—the beginnings and ends of elementary fragments of tropical years—and intermediate moments $\{t_{nm4/3}, t_{nm5/2}\}$:

$$q \in \{4/3, 5/3\} \Rightarrow t_{nmq} = t_{mm} + (q - 1)(t_{nm2} - t_{nm1}) \tag{14}$$

Tropical markup extends from the first fragment of the tropical year 3000 BC to the last fragment of the tropical year AD 2999 (Fedorov 2013, 2019). The true duration (in true seconds) of the n-th fragment of the m-th tropical year is

$$d_{nm} = t_{nm2} - t_{nm1} \tag{15}$$

The calendar marking $\{t_k\}$ consists of the beginning of the day according to Greenwich: the date of 3000BC.02.23 corresponds to the zero day, then the through numbering is used till the date of AD3000.05.05. The moments $\{t_k\}$ selected on the CT scale correspond to the moments $\{T_k\}$ on the UT2 scale (calendar seconds are counted on this scale):

$$T_k - T_{k-1} - 86400 \in \{-1, 0, 1\} \tag{16}$$

The functions $\tau_1(\cdot)$ and $\tau_2(\cdot)$ are used:

$$\tau_1(t_k) = t_k - t_{k-1} \tag{17}$$

$$\tau_2(t_k) = T_k - T_{k-1} \tag{18}$$

Their meaning: $\tau_1(t_k)$ and $\tau_2(t_k)$—duration of a day with number k in true and calendar seconds.

Sequence $\{t_k – T_k\}$—part of the primary source data. The formula is used to obtain $\{\tau_1(t_k)\}$:

$$\tau_1(t_k) = (t_k - T_k) - (t_{k-1} - T_{k-1}) + \tau_2(t_k) \tag{19}$$

Formulas (16)–(19) are free of mistakes in Formulas (16)–(19) in http://www.solar-climate.com/en/ensc/kost.htm.

4.3 Extraction of Primary Source Data

The primary source data is an array of values of the form $r(t_k)/1000$, $(180/\pi)\gamma(t_k)$, $(180/\pi)\lambda(t_k)$, $t_k - T_k$. Here $r/1000$ is the distance between the centers of the Sun and the Earth in kilometers, $180\gamma/\pi$ and $180\lambda/\pi$—declination and ecliptic longitude of the Sun's center in degrees, $t_k - T_k$ is the difference between the CT and UT2 in seconds. As already noted in Sect. 2, the values $r(t)$, $\gamma(t)$, $\lambda(t)$ recorded at the time of t refer to the earlier moment $b(t)$ (an important correction for the run of the light pulse from the center of the Sun to the center of the Earth).

The primary source data is extracted from the NASA DE406 ephemerides with the help of the HORIZONS internet service. The following parameters were specified in the queries (for *Time Span* there is a sample):

$$\begin{gathered} Ephemeris\,Type = OBSERVER, \\ Target\,Body = Sun\,[Sol]\,[10], \\ Observer\,Location = Geocentric\,[500], \\ Time\,Span = Start = 2001-01-01,\ Stop = 2200-12-31,\ Step = 1\,d \\ Table\,Settings = QUANTITIES = 2, 20, 30, 31;\ extra\,precision = YES. \end{gathered}$$

4.4 Calculating Secondary Source Data

Secondary source data is an array of values like $r(t_{nmq})$, $\gamma(t_{nmq})$, d_{nm}, where $q \in \{1, 4/3, 5/3, 2\}$. Secondary data are calculated from primary data using smooth spline-interpolation (the spline itself, its first and second derivatives are continuous).

$$\begin{aligned} & f_M = f(x_M),\ f_0 = f(x_0),\ f_1 = f(x_1), \\ & f_2 = f(x_2), x_0 < x_1, u = (x - x_0)/(x_1 - x_0) \end{aligned} \tag{20}$$

$$(0 < x_0 - x_M = x_1 - x_0 = x_2 - x_1, x_0 \le x \le x_1) \Rightarrow f(x) \approx \mathrm{spl}(f_M, f_0, f_1, f_2, u) \tag{21}$$

$$spl(f_M, f_0, f_1, f_2, u) = d_0 + d_1 u + d_2 u^2 + d_3 pol(u) \tag{22}$$

$$d_0 = f_0,\ d_1 = \frac{f_1 - f_M}{2},\ d_2 = \frac{f_1 - 2f_0 + f_M}{2},\ d_3 = \frac{f_2 - 3f_1 + 3f_0 - f_M}{2} \tag{23}$$

$$pol(u) = \frac{-31u^3 + 35u^4 + 21u^5 - 35u^6 + 10u^7}{12} \tag{24}$$

Spline interpolation formulas:

$$\lambda_M(n,t) = \begin{cases} \lambda(t) - 2\pi, & (\pi \le \lambda(t))\ \&\ (n \le 3) \\ \lambda(t), & (\lambda(t) < \pi) \vee (3 < n) \end{cases} \tag{25}$$

$$\lambda_P(n,t) = \begin{cases} \lambda(t) + 2\pi, & (\lambda(t) \le \pi)\ \&\ (358 \le n) \\ \lambda(t), & (\pi < \lambda(t)) \vee (n < 358) \end{cases} \tag{26}$$

$$j = j(n,m,q), \lambda_M(n, t_j) \le \pi(n+q-2)/180 \le \lambda_P(n, t_{j+1}) \tag{27}$$

$$r(t_{nmq}) = spl(r(t_{j-1}), r(t_j), r(t_{j+1}), r(t_{j+2}), u_{nmq}) \tag{28}$$

$$\gamma(t_{nmq}) = spl(\gamma(t_{j-1}), \gamma(t_j), \gamma(t_{j+1}), \gamma(t_{j+2}), u_{nmq}) \tag{29}$$

$$\tau_1(t_{nm2}) = spl(\tau_1(t_{j-1}), \tau_1(t_j), \tau_1(t_{j+1}), \tau_1(t_{j+2}), u_{nm2}) \tag{30}$$

$$d_{nm} = ((j(n,m,2) + u_{nm2}) - (j(n,m,1) + u_{nm1}))\tau_1(t_{nm2}) \tag{31}$$

where u_{nmq}—root of equation

$$\lambda(t_{nmq}) = \frac{\pi(n+q-2)}{180} = spl(\lambda_M(n, t_{j-1}), \lambda_M(n, t_j), \lambda_P(n, t_{j+1}), \lambda_P(n, t_{j+2}), u_{nmq}) \tag{32}$$

The root of each equation of the form (32) is searched for by the method of successive approximations with an error of 10^{-9} (in a day).

4.5 *Calculating SAR Integrals from Secondary Source Data*

After the transition from (2) to (11), further simplification consists in an approximate analytical calculation of the integral by α. Applying decomposition

$$\frac{1}{(1-x)^{3/2}} = 1 + \frac{3}{2}x + \frac{15}{8}x^2 + \frac{35}{16}x^3 + \frac{315}{128}x^4 + Q_5(x) \tag{33}$$

$$0 < x < 10^{-4} \Rightarrow 0 < Q_5(x) < 2.71x^5 \tag{34}$$

and omitting, for the sake of brevity, the arguments t, φ introduced above the functions α_M, D_0, D_1, E, and auxiliary functions μ, F_0, h_0, …, h_5, we find

$$\int\limits_{0}^{\alpha_M} \frac{D_0 + D_1 \cos\alpha}{(1 - E\cos\alpha)^{3/2}} d\alpha = (1+\mu)F_0,\ 0 < \mu < 3 \times 10^{-20}$$

$$F_0 = h_0\alpha_M + \sum_{j=1}^{5} \frac{h_j}{j}\sin(j\alpha_M) \tag{35}$$

$$h_0 = \frac{3E}{4}\left(1 + \frac{35}{32}E^2\right)D_1 + \left(1 + \frac{15}{16}E^2 + \frac{945}{1024}E^4\right)D_0 \tag{36}$$

$$h_1 = \left(1 + \frac{45}{32}E^2 + \frac{1575}{1024}E^4\right)D_1 + \frac{3E}{2}\left(1 + \frac{35}{32}E^2\right)D_0 \tag{37}$$

$$h_2 = \frac{E}{4}\left(3 + \frac{35}{8}E^2\right)D_1 + \frac{15E^2}{16}\left(1 + \frac{21}{16}E^2\right)D_0,$$

$$h_3 = \frac{15E^2}{32}\left(1 + \frac{105}{64}E^2\right)D_1 + \frac{35}{64}E^3 D_0 \tag{38}$$

$$h_4 = \frac{35E^3}{128}\left(D_1 + \frac{9}{8}ED_0\right),\ h_5 = \frac{315E^4}{2048}D_1 \tag{39}$$

When integrating by φ and t, *the* $\mu(t, \varphi)$ is discarded because of its smallness:

$$I_{nm}(\varphi_1, \varphi_2) \approx \int\limits_{t_{nm1}}^{t_{nm2}} F_2(t)dt,\ F_2(t) = u_0\left(\frac{r_0}{r(t)}\right)^2 \left(\int\limits_{\varphi_1}^{\varphi_2} F_1(t,\varphi)d\varphi\right) \tag{40}$$

$$F_1(t,\varphi) = \frac{\sigma(\varphi)}{C_0^{3/2}(t,\varphi)} F_0(t,\varphi) \tag{41}$$

For integrating by φ we correct the limits (only values of φ, at which $\alpha_M > 0$) using pair $(\varphi_{1M}, \varphi_{2M})$ instead of (φ_1, φ_2) and form the integration step near $\pi/180$ (one degree).

Omitting the argument t of the above functions r, γ and auxiliary functions F_1, γ_H, γ_B, η, η_H, η_{H1}, η_{H2}, η_B, η_{B1}, η_{B2}, φ_{1M}, φ_{2M}, *set*, $\{\varphi_{MK}\}$, for short, we find

$$\gamma_B = \arcsin\frac{B}{r},\ \gamma_H = -\gamma_B,\ \eta(x) = \arcsin\frac{B\sqrt{1+\varepsilon^2\sin^2(\gamma+x)}}{r} \tag{42}$$

$$\gamma \le \gamma_H \Rightarrow \varphi_{1M} = \varphi_1,\ \gamma > \gamma_H \Rightarrow \varphi_{1M} = \max\left(\varphi_1, -\frac{\pi}{2} + \gamma + \eta_H\right),\ \eta_H = \eta(\eta_H) \tag{43}$$

$$\gamma \ge \gamma_B \Rightarrow \varphi_{2M} = \varphi_2,\ \gamma < \gamma_B \Rightarrow \varphi_{2M} = \min\left(\varphi_2, \frac{\pi}{2} + \gamma - \eta_B\right),\ \eta_B = \eta(-\eta_B) \tag{44}$$

$$\eta_H \approx \eta_{H2} = \eta(\eta_{H1}),\ \eta_{H1} = \eta(0),\ |\eta_H - \eta_{H2}| < 3.7 \times 10^{-18} \tag{45}$$

$$\eta_B \approx \eta_{B2} = \eta(-\eta_{B1}),\ \eta_{B1} = \eta(0),\ |\eta_B - \eta_{B2}| < 3.7 \times 10^{-18} \tag{46}$$

$$\varphi_{1M} < \varphi_{2M} \Rightarrow set = \left[\frac{\varphi_{2M} - \varphi_{1M}}{\pi/180}\right] + 1,\quad \varphi_{MK} = \varphi_{1M} + \frac{\varphi_{2M} - \varphi_{1M}}{set} K \tag{47}$$

$$\varphi_{1M} \geq \varphi_{2M} \Rightarrow \int\limits_{\varphi_1}^{\varphi_2} F_1(\varphi) d\varphi = 0,$$

$$\varphi_{1M} < \varphi_{2M} \Rightarrow \int\limits_{\varphi_1}^{\varphi_2} F_1(\varphi) d\varphi = \sum_{K=0}^{set-1} \left(\int\limits_{\varphi_{MK}}^{\varphi_{M(K+1)}} F_1(\varphi) d\varphi \right) \tag{48}$$

Each summand in the integral sum by φ is calculated applying the replacement of the integrand function with a polynomial of the 3rd degree:

$$\int\limits_{\varphi_{MK}}^{\varphi_{M(K+1)}} F_1(\varphi) d\varphi \approx \left(\frac{\varphi_{2M} - \varphi_{1M}}{set}\right)$$
$$\frac{1}{8}\left(F_1(\varphi_{MK}) + 3F_1\left(\frac{\varphi_{MK}}{3/2} + \frac{\varphi_{M(K+1)}}{3}\right) + 3F_1\left(\frac{\varphi_{MK}}{3} + \frac{\varphi_{M(K+1)}}{3/2}\right) + F_1(\varphi_{M(K+1)})\right) \tag{49}$$

Integral $\int_{t_{nm1}}^{t_{nm2}} F_2(t)dt$ is taken on an interval in which the ecliptic longitude of the center of the Sun grows on 1 degree. This increase is close to the latitude change at the latitude integration step. Therefore, it is natural to calculate the integral $\int_{t_{nm1}}^{t_{nm2}} F_2(t)dt$ as well as the integral $\int_{\phi_{MK}}^{\phi_{M(K+1)}} F_1(\varphi)d\varphi$ by the polynomial method of the 3rd degree:

$$\int\limits_{t_{nm1}}^{t_{nm2}} F_2(t)dt \approx \frac{d_{nm}}{8}\left(F_2(t_{nm1}) + 3F_2\left(t_{nm4/3}\right) + 3F_2\left(t_{nm5/3}\right) + F_2(t_{nm2})\right) \tag{50}$$

As a result, the practical calculation of insolation integrals is made on the basis of secondary input data on formulas (1), (40), (50), (41), (48), (49) using (42)–(47), (35)–(39), (12), (13), (7)–(10), (3), (4).

4.6 Final Calculation Errors

The final error in the calculation of each insolation integral is a few percent of the mean module of its change from year to year. The relative error does not exceed 0.005% for insolation integrals near the poles and 0.00005% for insolation integrals near the equator.

5 Conclusion

The method for calculating the downward solar energy falling to the Earth without an atmosphere is developed. A publicly available database of downward solar energy falling to all 5-degrees latitudinal strips of the Earth's ellipsoid for each astronomical month of each year for the period from 3000 BC to AD 2999 is formed [http://www.solar-climate.com/en/en/ensc/bazard.(htm)]. The data can be used for calculations of the Earth's radiation balance (Hansen et al. 2011; http://earthobservatory.nasa.gov/Features/EnergyBalance/page6.php; Trenberth et al. 2009). The data can also be used in numerical experiments as an input energy signal in the radiation blocks of climate models (Fedorov 2019).

References

Berger A, Loutre MF, Yin Q (2010) Total irradiation during any time interval of the year using elliptic integrals. Quat Sci Rev 29:1968–1982. https://doi.org/10.1016/j.quascirev.2010.05.07

Bertrand C, Loutre MF, Berger A (2002) High frequency variations of the Earth's orbital parameters and climate change. Geophys Res Lett 29(18):40-1–40-3. https://doi.org/10.1029/2002gl015622

Borisenkov EP, Tsvetkov AV, Agaponov SV (1983) On some characteristics of insolation changes in the past and the future. Climatic Change 5(3):237–244

Fedorov VM (2012) Interannual Variability of the Solar Constant//. Sol Syst Res 4(2):170–176. https://doi.org/10.1134/S0038094612020049

Fedorov VM (2013) Interannual variations in the duration of the tropical year. Doklady Earth Sci 451(Part 1):750–753. https://doi.org/10.1134/s1028334x13070015

Fedorov VM (2015) Spatial and temporal variation in solar climate of the Earth in the present epoch. Izvestiya Atmos Oceanic Phys 51(8):779–791. https://doi.org/10.1134/S0001433815080034

Fedorov VM (2016) Theoretical calculation of the interannual variability of the Earth's insolation with daily resolution. Sol Syst Res 50(3):220–224. https://doi.org/10.1134/S0038094616030011

Fedorov VM (2019) Variations of the earth's insolation and especially their integration in physical and mathematical models of the climate. Phys Uspekhi 62(1):32–45. https://doi.org/10.3367/ufne.2017.12.038267

Giorgini JD, Yeomans DK, Chamberlin AB et al (1996) JPL's on-line solar system data service. Bull Am Astron Soc 28(3):11–58

Hansen J, Sato M, Kharecha P, Von Schuckmann K (2011) Earth's energy imbalance and implications. Atmos Chem Phys 11:13421–13449. https://doi.org/10.5194/acp-11-13421-2011

http://ssd.jpl.nasa.gov—NASA, Jet Propulsion Laboratory California Institute of Technology (JPL Solar System Dynamics). Electronic resource of the US National Aerospace Agency

http://ssd.jpl.nasa.gov/?horizons_doc#specific_quantities

http://www.solar-climate.com/en/en/ensc/bazard.(htm)

http://earthobservatory.nasa.gov/Features/EnergyBalance/page6.php

http://www.solar-climate.com/en/ensc/kost.htm

Kondratyev KYa (1980) Radiation factors of modern changes in global climate. Hydrometeoizdat, Leningrad, 279 p

Lorentz EN (1970) Nature and the theory of general atmospheric circulation. Hydrometeoizdat, Leningrad, 260 p

Monin AS, Shishkov YA (2000) Climate as a problem of physics. Successes Phys Sci 170(4):419–445

Trenberth KE, Fasullo JT, Kiehl J (2009) Earth's global energy budget. Bull Am Meteorol Soc 90:311–323. https://doi.org/10.1175/2008BAMS2634.1

Timan-Ural Stone Province—Mineral Resource Base of the Gemstones of the Russian Federation

Vitaly Gadiyatov, Peter Kalugin, and Alexander Demidenko

Abstract The paper reviews the validity of the allocation of the so-called 'gemstones province' within Timan region. The conducted minerogenic zoning of the reported province distinguished Timan region, North-Ural, Middle-Ural and Southern Ural sub-provinces, which would also include Kanin-Indigskaya, Polar Urals, Lapinsky, Shabrovsko-Nizhnetagilskye, Murzinskaya-Aduyskaya, Uraltausskaya, Orsk-Magnitogorsk, and Sysertsko-Mugodzhar mineragenic zones. Given the thorough description of the above minerogenic zones, the detailed description of ore regions with commonly known deposits of emerald, alexandrite, amethyst, beryl, tourmaline, sapphire, ruby, topaz, jasper and other gemstones that comprise the mineral and raw material base of colored gemstones of the Russian Federation is provided.

Keywords Mineragеny · Zoning · Province · Sub-province · Plate · Platform · Field · Gemstones

1 Introduction

Many jewelry stones are known in the Urals, and about a thousand of the almost 5000 thousand minerals and their varieties discovered on Earth are found (Talancev 2000). Granite pegmatites contain deposits of aquamarine, topaz, beryl, tourmaline; granitoids contain deposits of amethyst, mica—emerald and alexandrite. Volcanogenic-sedimentary formations contain deposits of rhodonite and colorful jasper. Malachite is associated with the oxidation zones of copper deposits. Irizing feldspars, amazonite, corundum and chrysolite are also important (Kievlenko 2000).

In different years, attempts were made to divide mineragenicaly the territory of the USSR and the Russian Federation, including the Urals, by colored stones. Between 1965 and 1991, intensive prospecting and exploration activities resulted in the division of the country into a number of self-colored-stone provinces. The proposed location schemes were based on geotectonic and formational features. Thus, Samsonov and Turinge (1984) has identified 15 complex and 9 monomineral

V. Gadiyatov (✉) · P. Kalugin · A. Demidenko
Voronezh State Technical University, Voronezh, Russia
e-mail: gadiatovvg@mail.ru

C. T. Olegovna (ed.), *Processes in GeoMedia—Volume I*,
Springer Geology, https://doi.org/10.1007/978-3-030-38177-6_21

self-colored-stone provinces. Different geodynamic structures have been adopted as provinces and additional taxonomic units have been introduced. As a result, the North-Ural-Paikhoi, Middle-Ural, South Ural, as well as the Polar and East Ural provinces of the jewelry crystal were identified in the Urals.

At the beginning of the new millennium the author's team of FSUE "Tsentrkvarts" (Atabaev 2003) generalized materials on deposits of colored stones in Russia and together with the Central Research Institute of Geolnerud and VNIISIMS carried out mineralogenetic zoning of the territory of the Russian Federation at a scale of 1:5,000,000,000. The map of the Russian Federation shows 14 self-colored-stone provinces, including Timan and Ural provinces.

Later, using the data of the above work, the next mineralogenetic zoning of the territory of the Russian Federation was carried out, and the Consolidated Resource Map of Colored Stones of Russia was compiled (Kovalenko et al. 2004). The map of the Russian Federation shows 6 megaprovinces, 17 provinces and regions, 141 deposits and 265 manifestations of 29 colored stones with estimated resources. The map of the Urals shows the Timan-Paikhoi-Ural megaprovince, which includes the Paikhoi region and the Paikhoi-Ural province.

As can be seen from the brief historical excursion into the problem of mineralogenetic extraction of gemstone tucsons in the Urals and Timan, all authors were guided by their own vision of this problem. Therefore, the purpose of this study was to perform zoning based on the principles of mineralogenetic analysis.

2 Materials and Research Methods

In the course of mineralogenetic zoning taxonomic units of different hierarchical levels are taken as a basis. As the largest subdivision for the zoning of colored stone objects, the province is taken as the area of distribution of the totality of structural-mineragenic zones of different tectonic-magmatic cycles coinciding with the boundaries of large geotectonic structures of the Earth's crust. That is, the province corresponds to the large regional structures of the Earth's crust and corresponds mainly to the platforms and complex folded-orogenic areas. Next come subprovince, mineralogenetic region, mineralogenetic zone, ore district, ore field (Gadiyatov 2013a, b).

The Ural folded system is structurally a mega-anticlinorium consisting of meridionally elongated anticlinories separated by synclinories. Therefore, the territory of the Urals is undoubtedly a full-fledged stone and color province. And there are questions about Timan. Namely: should it be singled out as a separate gemstone province or can it be combined with the Ural Province?

According to the history of the geological development of the Urals, as a result of the Baikal fold on the site of the once united Timan-Ural geosyncline there was a Timan anticlinorium with the in the West Pre-Timan deflection associated with it. At present, fragments of three structures of the first order, including Mezensky syneclise, North-Timan anteclise and Pechora syneclise, are distinguished on Timan

(Gosudarstvennaya geologicheskaya karta Rossijskoj Federacii1985). The Mezensky syneclise represents the marginal zone of the Russian plate with the archean-lower-proteroic foundation, and the North-Timan anteclise and the Pechora syneclise form the Timan-Pechora plate with the reefish foundation (Timonin 1998; Dedeev et al. 1985).

Thus, the Timan-Pechora plate is located in the extreme North-East of European Russia, between the East European platform and the mountain structures of the Polar and Subpolar Urals. From the West and South-West, the Timan-Pechora plate is bounded by the West Timan Krai seam, from the east by the Ural folded system and the Paikhoi Rise. In the south-west, the West Timan joint adjoins the West Urals thrust fault, forming the Ural-Timan joint. That is, in geotectonic terms it is an independent structure, located on the border of the platform and folding area.

For all the mentioned above, there are reasons to separate the Timan-Pechora plate into a separate mineralogenetic taxon. At the same time, given history of geological development, Timan can be attached to the Ural Stone and Color Province. The latter is also stimulated by the insignificant variety of colored stones (almost only agates). Therefore, in order not to separate a relatively small region into a separate monominal province, the territory of Timan is merged with the Urals into the Timano-Uralic stonemasonry province.

3 Research Findings and Discussion

Timan-Ural Province is one of the longest stonemasonry provinces in Russian Federation. The province stretches in the meridional direction for more than 2 thousand kilometers from Novaya Zemlya Island—in the north to the southern tip of Mugodzhar—in the south, which is located on the territory of the Republic of Kazakhstan. From the West, Northwest, where the Timan region is located, the province stretches along the Timan ridge from the coast of the Barents Sea to the North Urals.

Administratively, the Timan-Ural Province is located in the Perm Territory, Khanty-Mansiysk Autonomous District, the Republic of Bashkortostan, the Sverdlovsk, Chelyabinsk, Kurgan and Orenburg Regions, as well as in the Nenets Autonomous District and the Komi Republic (Timan Region).

The geological position of the province is determined by the linearly elongated area of the Ural folded belt, which was formed over a long period of time—from the Upper Proterozoic to the Mesozoic.

The Timan-Ural province is divided into Timan, North Ural, Middle Ural and South Ural sub-provinces in terms of mineralogy. The province has nine mineralogenous zones, including 18 ore districts, according to its geological structure and mineral associations (Table 1).

Timan self-colored-stone region is located on the Timansky ridge, stretching to the South-East from the coast of the Barents Sea to the North Urals. Administratively—on the territory of the Nenets Autonomous District and the Komi Republic. Its position is determined by the spread of the Late Devonian trap formation, timed to

Table 1 Mineragenic zoning of the Timan-Ural province

Subprovince region	Mineragenic zone	Ore region	Main colored stones	Deposits, occurrences
Timan region	Kaninsko-Indigskaya	North Timan	Agate, Pomegranate	Teaichye, Ievskoe Chernorechenskoe, etc. (agate)
North Ural	Kara	Novaya Zemlya Paykhoi	Agate Jasper	Gromashor, Darka-Ruzshor and others (jasper), Silovaya (turquoise)
	Polar Ural	Hadatinsky Voikaro-Synyinsky	Jadeite, Jade, Demantoid, Chrysolite, Corundum, Malachite	Puseorka chalcedony, Novo-Kechpelsky (jadeite), Left Kechpel, Nyrdvomenshor and others (nephrite, uvarovite) Hadatinsky Voikaro-Synyinsky
	Lyapinskaya	Khasavarkinsky North Sosva	Amethyst, Citrine, Corundum, Almandine; Jasper, Agate, Rhinestone	Khasavarka (amethyst), Zhelannoe (horn. Crystal, citrine), Ulyatemiya (jasper)
Middle Ural	Shabrovsko-Nizhny Tagil	Sysertsky Nizhny Tagil	Demantoid Rhodonite; Demantoid, Sapphire, Uvarite, Malachite, Agate, Chrysoprase	Mednororjanskoe (Malachite), Poldnevskoe, Korkodinskoe (demantoid), Malo-Sedelnikovskoe (rhodonite), Bobrovskoe (demantoid, Saranovskoe (uvarovit)
	Murzinsko-Aduiskaya	Murzinsky Aduysky	Alexandrite, Emerald, Sapphire, Ruby, Demantoid, Topaz, Amethyst, Rhodonite, Serpentine, Jasper, Agate, Jade; Tourmaline, Amazonite, Iridescent	Vatiha feldspar (amethyst), Verbaniy Log, Kornilov Log, Polozhikha (ruby, sapphire), Murzinskoe (beryl) Malyshevskoe (emerald, beryl)
South Ural	Uraltau	Satkinsky Sabyrovsko-Nuralinsky	Ophiocalcite; Serpentine, Jadeite, Chrysoprase, Jade, Agate	Monastic, Chereshkovskoe (ofiocalcite), Taganayskoe, Sakmarskoe (aventurine), Bikilar, Khalilovsk (nephrite)

(continued)

Table 1 (continued)

Subprovince region	Mineragenic zone	Ore region	Main colored stones	Deposits, occurrences
	Orsk Magnitogorsk	Orsky Uchalinsky	Rhodonite, Jasper, Jasper, Jade, Agate	Severo-Kalinovskoe, mountain Colonel (jasper), Kushkuldinskoe, Tash-Kazgan (Jasper)
	Sysertsko-Mugodzharskaya	Sysertsko-Ilmenogorsky Shevchenkovsky	Emerald, Uvarovit, Serpentine, Chrysoprase, Corundum, Agate	Kuchinsky placer (corundum), Yuzhno-Fayzulinskoe, hodonite)
	Not selected	PredUchalinsky	Gypsum, Selenite	Bikulovskoe Grishinskoe, Yakovlevskoe (gypsum-selenite)

the activated northern part of the Russian platform. The Timan region is practically monominal agathonous.

Mineral association of agates is typical for the volcanogenic hydrothermal complex in basalts containing zeolites, agate, quartz, sometimes amethyst, calcite. In addition, in the Timan region there is a pomegranate of secondary importance.

The atonicity of the considered stone-samo-colored area is connected with the upper parts of the Upper Devonian tholeitic basalts' covers. Timan agates are contained in them in the form of almonds, coloring of light gray agates with a bluish hue, there is a thin striped concentric pattern, sometimes moire effect. Jewelry and finishing agates are considered to be one of the best in Russia in terms of the quality of raw materials.

Timan self-colored-stone region includes the Kaninsko-Indigian Mineralogical Zone.

The Kaninsk-Indigian Mineralogical Zone is located in the North Timan, in the upper reaches of the Pilma and Sula basins. In the Northern part of this zone is located the North-Timan agat-bearing area, which represents an area of 104 × 24 km, including all deposits and manifestations of agate. The main deposits of the agate-bearing region are (from North to South): Chaichie, Chernorechenskoe, Ievskoe, Malochnorechenskoe, Belorechenskoe, Sulsko-Belorechenskoe. Fields and agate manifestations in the North-Timan agate-bearing region are confined to the basalt fields of the Kumushkin Formation. Spatial connection of deposits and agate manifestations with volcanic apparatuses of central type is established. In the Ivskoye field, up to 5 agath-bearing lava with a capacity of up to 15–20 m are allocated.

The North Urals sub-province occupies the territory of the Polar, Subpolar and most part of the North Urals. It includes the Kara, Polar-Ural and Lyapin mineralogenic zones. In the sub-provincial area there are different age rocks of different composition and genesis, which are associated with colored stones. For example, in the Polar Urals among the dunite-garzburgite hyperbasites of the Vaikaro-Synyinsky massif is known Novo-Kechepleskoye and other deposits of jadeite, and in the Rai-izsky district—jade and jadeite. High alumina shale and gneiss contain collectible rubies, kyanite, almandine and staurolite.

The Kara Mineralogical Zone territorially covers Novaya Zemlya Island and the Paikhoi Ridge. The zone includes the Novozemelsk agate-bearing and Paikhoi Jasper-bearing regions. The appearance of agates on Novaya Zemlya is confined to volcanogenic rocks. Jasper and jasper manifestations of Gromashor, Darka-Rushor, Malaya Serya and Silovayakhskoye turquoise are known in the Paikhoi Jasper-bearing region. The occurrence of jasper is associated with Paleozoic volcanogenic-sedimentary rocks, Silovayakhskoye manifestation of turquoise, localized in the strata of siliceous shales and limestone jasper of the middle Devonian Northern wing of the Srednesilovskaya syncline structure.

The Polar-Ural Mineralogical Zone is located on the territory of the Polar Urals. It consists of Khodatinsky and Voykaro-Synyinsky districts. Mineragenic specialization of the zone is defined by nephritis and jadeite. There are deposits and manifestations of the Left Kechpel, Novo-Kechpelskoye, Pusyorka, Karovoye, Nyrdvomenshor, Nyagarneoshor, Yai-Yu, Montanelle, Pogranichnoye, associated with the ultramafic massifs of Syum-Keu, Rai-Iz, Khulginskoye, Pai-Er Voykaro-Khodatinskoye ophiolite belt of Ordovician age. There are also known manifestations of uvarovite, demantoid, ruby, lazurite, chrysolite, beryl, jasper and malachite.

In the Novo-Kechpelskoye deposit of jadeite, Pusyerka, Rai-Iz jadeite is localized in apoplagioclasitic metasomatites of officeolites. At the Pusyorka deposit there is jadeite of the highest grade—imperial, which is known in Russia only at the Borusskoye deposit in the South Siberian subprovince. The manifestation of Nyrdvomenshor grenade with respect to the Kershorskoye chromite deposit contains a collector's demantoid.

Lyapinsky mineralogenic zone is located in the Subpolar Urals. Includes Khasavarkinsky Amethystonosny and Severo-Sosvinsky jasper bearing areas. In Khasavarkinsky district, the most important deposit belongs to the deposit of amethyst Khasavark, which is located in the upper Kozhim River (Pechora River basin). Amethyst is localized in the crushing zones of late Proterozoic quartz-chlorite-sericite shales. The host intensively metamorphosed rocks contain corundum, pomegranate almandine, spinel, coil, and ophiocalcite.

The deposits and manifestations of Zhelannoye, Severnaya, Tsentralnaya, Bolshaya Lapcha, Dodo, Puiva, Pelingichiy-Tri, etc. are also localized in the characterized ore area. Citrine (Zhelannoe deposit), quartz with chlorite inclusions and inclusions in the aggregates with axinite (Puiva deposit); rutile, tourmaline and actinolite-containing quartz (manifestations of Upper Parnuk, Cheln-Iz, Pedy Shor) are found there together with rock crystal.

The Severo-Sosvinsk jasper-bearing region covers the structures of the Central Ural uplift and the Tagil-Magnitogorsk deflection, which have different mineralogenic specialization. Along with the Ulyatemye jasper deposit, the Olkhovochnoye deposit of rock crystal and citrine, chalcedony, agate and jasper manifestations are known here.

The Middle Urals sub-province, located in the Middle Urals, includes the Shabrovo-Nizhnetagilskaya and Murzinsko-Aduiskaya mineralogenic zones, in which there are deposits and manifestations of many colored stones. Among them are the stones that brought glory to the Urals: emerald, alexandrite, demantoid, topaz, malachite.

Most of the gems of the most studied **Sredne-Ural subprovince** are confined to granitoid complexes adjacent to the ultrabasic massifs of the hyperbasite belt. Within the limits of the recently separated Murzinsk-Aduyskaya gemstone strip (Malikov et al. 2007) there is a meridionally elongated emerald strip of the Urals. Emerald mica contains emerald and alexandrite, and topaz beryl, amethyst, tourmaline, kunzite, and almandine were extracted from pegmatite fields. Demantoid, chrysoprase, decorative breeds—larch, coil, talcochlorite—are associated with ultrabasic massifs.

The Shabrovo-Nizhny Tagil zone stretches submerionally from Sysert in the south-east to Nizhny Tagil in the north-west. The structure includes Sysert and Nizhny Tagil ore districts. The zone includes the geological structures of the northern (Tagil) part of a large linear deflection made by volcanogenic-sedimentary formations, which were broken through by numerous basite-ultrabasite and granitoid intrusions.

On the territory of the zone is known rhodonite of Malosidelnikovskoye deposit, malachite and demantoid. Malachite of the Mednorudyanskoye deposit is associated with the oxidation zones of the skarn copper ore deposits. The demantoid is contained in the placers and bedrock of the Bobrovskoye, Polnevnevskoye and Korkodinskoye deposits, as well as in the manifestation of Vershinnoye.

In Sysersky ore district the main importance belongs to demantoid and rhodonite. The source of the Ural demantoid is the Poldnevskoye (Sysertskoye) field. The Korkodinskoye field is also located here. The demantoid is contained in serpentinized pyroxenites and serpentinites of the Korkodinsky ultramafic massif located in the basin of the Chrysolitka River and in alluvial placers. Painting from pale apple to bright herbal.

Malo-Sedelnikovskoye, Borodulinskoye, Oktyabrskoye and Kurganovskoye rhodonite deposits are represented by manganese-silicate deposits in quartzite and jasper-green shale strata. Rhodonite mineralization is observed in quartzite shale rocks with predominance of biotite-quartz shales. Rodonite forms lenso and stockwork deposits.

In Nizhny Tagil ore district the leading colored stones are ugrandite pomegranates—uvarovit, demantoid and topazolit. Uvarovit is contained in the Saranovskoye chromite field, located in the Perm region. Uvarovit forms brushes that perform cracks in the cracks and ruptures in hyperbasites.

In the Bobrovskoye (Nizhny Tagil) field, located near Nizhny Tagil, the demantoid is contained in serpentinite among dunites and pyroxenites of the Nizhny Tagil ultramafic massif. The color of the demantoid is green with different saturation. The Bobrovskoye field contains about 50% of Russia's total demantoid reserves (Lyashenko 2004).

The Murzinsk-Aduysk mineralogenic zone is located in a band of meridional strike, from the upper reaches of the Pyshma River (south of the city of Asbestos in the Sverdlovsk region)—in the south to the middle reaches of the Tagil River—in the north. In the northeastern part of the zone there is a gemstone strip of the same name, including crystal-self-precious pegmatites with beryls, topazes, tourmalines and other precious stones, as well as phlogopite (mica) apoultramafic graysens with emeralds, beryles, alexandrites, chrysoberyles and phenacites.

The Murzinsky and Aduisky ore districts with colored stones were identified as part of the zone under consideration.

Murzinsky ore district is located in the northern part of the mineralogenic zone, on the area of the granite massif of the same name. Here are known amethyst, beryl, tourmaline, sapphire, ruby, demantoid, topaz, jasper, agate, amazonite, and feldspar irrigating. The amethystonous area covers the western contact zone of the Late Paleozoic Murza granite massif with Cambrian biotite and biotite-amphibole gneisses. Within this area there are the Murzinskoye berilla deposit, the Vatikha amethyst deposit, the Pridorozhnoye and Artemieva manifestations. The best amethyst is considered to be the amethyst of the Vatikha deposit, which is contained in close quartz veins and mineralized cracks developed in the granites of the endocontact zone of the Murzinsky massif. Balance reserves of raw materials make up 13.4% of the reserves of amethyst in the Russian Federation (Lyashenko 2004).

The ruby and sapphire from the Verbal Log, Kornilov Log, Polozhikha and Cheremshanka deposits are associated with alluvial placers (Kornilov Log is a deluvial placer). The sources of the placers are disseminated corundum mineralization in Paleozoic marbles and plagioclasites in metamorphosed ultramafiles.

In the Adui ore district near Asbestos there are Emerald mines of the Urals, which are a group of deposits of emerald and other gemstones that are in paragenesis (chrysoberyl, alexandrite, phenacite, beryl). The Ural emerald belt is traced for 25 km parallel to the Adua granite massif. Among the group of emerald deposits (Malyshevskoye, Sverdlovskoye, Krupskoye, Cheremshanskoye, Pervomayskoye, Krasnoarmeyskoye, etc.) the main emerald-beryllium deposit is Malyshevskoye, which is currently being developed. Rudovmeschayuschie mica, developed by rocks of ultrabasic composition, which are combined with pegmatites and quartz-plagioclase veins. As a rule, productive bodies lie along the contacts of ultrabasic rocks with Late Paleozoic rare-metal berylliferous granites and pegmatites. More than 90% of Russia's emerald reserves are concentrated in the Malyshevskoye field (Lyashenko 2004). The Ural emerald belt is the southern end of the Murzinsk-Aduysky zone.

South Ural sub-provinces: It includes the Uraltaus, Orsko-Magnitogorsk, Sysertsko-Mugodzhar Mineralogical Zones and the Ural Ore District.

In the South Ural Sub-province, more than 100 jasper fields are known to be located in the band of submeridional strike with a length of more than 500 km. The main source of jasper is the Mesozoic crust of weathering of Paleozoic Jasperiferous massifs. The alkaline pegmatites of the Ilmensky Mountains are associated with amazonite, topaz, zircon, corundum and rare collection minerals. Different rocks of Paleozoic-oic age contain quartz veins with jewelry rock crystal, citrine, aventurine, amethyst.

In the weathering bark of copper deposits of the Tagilsko-Magnitogorsk sagging revealed accumulation of malachite, and in the metamorphosed manganese siliceous shale formations—rhodonite manifestations of the Middle-Ural rhodonite region.

The exogenous group includes diamond placers, topaz, almandine, amethyst and agate deposits. Collection samples of pegmatite, hydrothermal, skarn, micaceous deposits are of particular value.

The Uraltau Mineralogical Zone is located in a north-eastern strike zone from the Ural River in the south near the city of Orsk to the upper Miass River in the north. Includes Satka and Sabyrovsko-Nuralinskiy ore districts adjacent to the metamorphic complexes of the Uraltaus and Bashkir anticlinoriums of the Central Ural Rise. The Monastyrskoye and Chereshkovskoye deposits of office-calcite, Kusinskoye, Malo-Berdyaushskoye and other deposits are located in Satka district. In addition to them, the Taganay and Sakmarsk deposits of aventurine quartzites are known. Sakmarskoye includes Novotroitskoye and Karayanovskoye manifestations, Yumaguzinskaya and Ivanovskaya squares.

The Bikilyarskoye and Khalilovskoye nephritic deposits are known in the Sabyrovsko-Nuralinskoye ore district. Jadeite, serpentinitis and chrysoprase are associated with metasomatites of ophiolites with jadeite and nephrite.

The Orsk-Magnitogorsk Mineralogical Zone is located within the southern (Magnitogorsk) part of the Tagil-Magnitogorsk Trough and adjacent areas of localization of hyperbasite massifs from the west. It stretches in a north-eastern direction from the state border of the Russian Federation with the Republic of Kazakhstan in the south to the city of Miass in the north for a distance of 500 km. Includes Orsk and Uchalinsk jasper-bearing regions.

The main colored stone of the ore zone is jasper of the Orsk and Uchalinsk group of deposits. All decorative types of jasper are known here—monochrome, striped, finely drawn (chintz) and landscape. The most valuable are the last two varieties of jasper, developed in the east of the Republic of Bashkortostan and near Orsk, respectively.

Orsk ore district (Orenburg region) is home to a significant group of jasper deposits, including the Severo-Kalinovskoye ribbon jasper deposit, the Kalinovskoye striped jasper deposit, and the Colonel's Colonel mountain colorful jasper deposit. Deposits of jasper are localized in gabbro-diabases and ancient crust of weathering, deposits of jasper—in volcanic-sedimentary and sedimentary rocks.

In the Uchalinskiy Jashmonosny district, the Kushkuldinskoye, Tash-Kazgan, Suleymanovskoye, Mamuinakskoye, Tungatorovskoye and other fields are localized. They are confined to the Bugulygyr horizon by jasper with volcanic rocks, and separate horizons are composed of solid jasper. Predominant are brightly colored red coffee and green jasper with sharp transitions of these colors.

The Sysert Mugodzhar ore zone, which used to stand out as the Kulikovsko-Shevchenkovskaya ore zone, is located on the eastern slope of the Southern Urals. A narrow north-eastern band stretches from the upper Tobol River in the south to the Miass River in the north. The zone includes SysertskoIlmenogorsk and Shevchenko ore districts.

The area is mainly specialized in noble corundum. There are deposits of corundum (Kuchinskaya placer, near Plast) and sapphire deposits in Sysert-Ilmenogorsk, Dzhabyk-Karagai and other complexes. In addition, a number of high-quality rhodonite deposits are known at the northern closure of the Sysert gneiss-migmatite complex. Numerous manifestations of rhodonite (South Faizulinskoye, Bikulovskoye, Kazgan Tash, Gubaidulinskoye) are known to occur in the ore zone area, lying in the same formations.

The Ural ore district is located in the Urals region, stretching a narrow band of submeridional direction along the river Sylva. It is timed to coincide with the Ufa-Solikamsk depression of the Urals Regional Trough (Perm Territory). In the ore district the deposits and manifestations of ornamental gypsum and gypsum-selenite Grishinskoye, Yakovlevskoye, Burovoye, Denisovskoye, Odinovskoye are localized.

4 Conclusion

Timan-Ural Province is one of the longest stonemasonry provinces in the Russian Federation. Metallogenically, it includes the Ural folded system and the Timan-Pechora plate (Timan). The province has deposits of emerald, demantoid, amethyst and other gemstones, which form the basis of the country's raw materials base. More than 90% of the emerald reserves are concentrated in the Malyshevskoye field. In the Timan self-colored-stone region the most valuable are coastal-sea beach and eluvial-deluvial placers of agate. Thin striped Timan agates are considered to be one of the best in the country. In the North Urals sub-provincial area of the Polar Urals, the Novo-Kechpelskoye and other jadeite deposits are known, and in the Rayizsky district—jade and jadeite. The most famous and rich in gems is the Middle Urals subprovince. Emerald mica contains emerald and alexandrite, and topaz beryl, amethyst, tourmaline, kunzite, and almandine are extracted from pegmatite fields. The South Ural subprovince is known for its unique design and color scheme of jasper. More than 100 fields are located in the zone of submeridional strike with the length of more than 500 km. The alkaline pegmatites of the Ilmensky Mountains contain amazonite, topaz, zircon, corundum and rare collectable minerals.

References

Atabaev KK (2003) Geologicheskaya izuchenie i ocenka mineralnyh resursov territorii Rossijskoj Federacii i eyo kontinental'nogo shelfa. FGUP Centrkvarc, Moscow (in Russian)

Dedeev VA, Zaporozhceva IV, Timonin NI, Yudin VV (1985) Vostochnoe ogranichenie Pechorskoj plity. Tektonika, magmatizm, metamorfizm i metallogeniya zony sochleneniya Urala i Vostochno-Evropejskoj platformy. Sverdlovsk, pp 12–14 (in Russian)

Gadiyatov VG (2013a) Mineragenicheskoe rajonirovanie territorii Rossijskoj Federacii na cvetnye kamni. Novye idei v naukah o Zemle. Dokl. In: 11th international conference, vol 1, pp 234–235 (in Russian)

Gadiyatov VG (2013b) Prostranstvennoe razmeshchenie mestorozhdenij cvetnyh kamnej na territorii Rossijskoj Federacii. Gemmologiya. Mater. VI Sci. Conf. Tomsk, pp 28–36 (in Russian)

Gosudarstvennaya geologicheskaya karta Rossijskoj Federacii (1985) Scale 1:200,000. Seriya Timanskaya, list Q-39-VII-VIII (in Russian)

Kievlenko EY (2000) Geologiya samocvetov. Zemlya, Ass. EKOST, Moscow 580p (in Russian)
Kovalenko IV, Kostelova TG, Shulyaeva LN et al (2004) Svodnaya resursnaya karta cvetnyh kamnej Rossii. Prosp Protect Min Res (1):2–6 (in Russian)
Lyashenko EA (2004) Mineralno-syrevaya baza cvetnyh kamnej Rossii. Prosp Protect Min Res (1):20–22 (in Russian)
Malikov AI, Polenov YA, Popov MP et al (2007) Samocvetnaya polosa Urala: educ. and method. manual. Sokrat, Ekaterinburg, 283p [Russian]
Samsonov YP, Turinge AP (1984) Samocvety SSSR. Nedra, Moscow, 335p (in Russian)
Talancev AS (2000) Znamenitye ural'skie samocvety. Izd. Dom Pakrus, Ekaterinburg, 168p (in Russian)
Timonin NI (1998) Pechorskaya plita: istoriya geologicheskogo razvitiya v fanerozoe. Ekaterinburg, 240p (in Russian)

Interannual Variability of the Heat Exchange of the Ocean and the Atmosphere in the Arctic

A. E. Bukatov and A. A. Bukatov

Abstract Research of annual total heat flow variability in the Arctic region formed by the interaction between Ocean and Atmosphere is executed. It's made on the basis of average monthly values of sea ice closeness and thickness, salinity and sea surface temperature and also air temperature at the level of weather shelter and skin temperature of a snow-ice cover, zonal and meridional wind velocity components during 1969–2012. Heat flow through a snow-ice cover is calculated in the assumption of linearity of a temperature profile between its upper and bottom boundaries. Heat exchange on the water surface is defined by the equation expressing proportionality of incoming and outgoing heat to product of wind velocity and difference between water and air temperature. The estimation error of the heat flow value at changing of the skin temperature of a snow-ice cover by the air temperature at the level of a weather shelter is made. The analysis of heat flow fluctuations correspondence with the Arctic Oscillation index is carried out.

Keywords Heat flow · Ocean and atmosphere heat exchange · Arctic regions · Heat exchange in arctic regions

1 Introduction

One of the important characteristics defining a thermodynamic regime of the global climatic system is heat exchange between Ocean and Atmosphere. The basic areas of the climatic changes influencing on climate of other regions are located in high latitudes (Frolov et al. 2010). In this connection climate change of the Arctic is one of actual directions of the modern climatic researches. Substantial factor of the Arctic climate is the snow-ice cover which affect climate processes on the Earth (Frolov et al. 2007) and as a component of global climatic system actively influences on interaction of the Ocean and Atmosphere in a wide range of time scales, restricting possibility of the heat exchange, weight and the moment of motion between Ocean and Atmosphere (Doronin 1969; Johannessen et al. 2005). At the same time for the

A. E. Bukatov · A. A. Bukatov (✉)
Marine Hydrophysical Institute RAS, Sevastopol, Russia
e-mail: newisland@list.ru

C. T. Olegovna (ed.), *Processes in GeoMedia—Volume I*,
Springer Geology, https://doi.org/10.1007/978-3-030-38177-6_22

Arctic regions yet it was not possible to gain the good consent between observations and the results of occurring climate changes calculated on the basis of global models (Alekseev et al. 2010). It indicates urgency of the further researches of change of climatic system components of the Arctic, including heat exchange between Ocean and Atmosphere, depending on dynamics of a floating ice concentration and coverage of open water areas. Heat losses by Ocean through open areas of the freezing seas differ significantly from losses through an ice cover (Doronin 1969; Baidin and Meleshko 2014; Doronin 1974; Eremeev et al. 2013) that is subsequently manifested and in change of atmospheric circulation. Such researches are necessary for the construction of the climate theory, improvement and development new methods of natural processes and the phenomena forecast. Experimental and theoretical aspects of the problem of heat balance of system Ocean-Atmosphere in the Arctic basin are observed in (Doronin 1969; Tryoshnikov and Alekseev 1991; Makshtas 1984). For the North European basin quantitative estimations of heat flux components on interface of Ocean-Atmosphere are carried out in (Smirnov and Koroblev 2010).

In the present paper interannual variability of heat exchange between Ocean and Atmosphere in the Arctic regions is observed. The estimation of a correlation of monthly average climatic heat flux from Ocean in the Atmosphere with Arctic oscillation index is given.

2 Method and Data

Research of interannual variability of heat exchange of Ocean and Atmosphere is executed on the basis of monthly average of the values of sea ice concentration and sea surface temperature during 1969–2012 (http://badc.nerc.ac.uk/data/hadisst/, http://nomadl.ncep.noaa.gov) at 1° grids, and also monthly average values of sea salinity (http://www.nodc.noaa.gov), air temperature on the skin surface and at the level of a weather shelter, an ice thickness, zone and meridional components of ground wind velocity (http://www.esrl.noaa.gov), modified to 1° grids with help of spline-interpolation. Heat flux through a solid snow-ice cover is calculated in the assumption of linearity of a temperature profile between its upper and lower boundaries (Makshtas 1984; Bespalov 1959) by formula

$$Q_1 = -\lambda_2(T_0 - T_2)/(h_2 + \lambda_2 h_1/\lambda_1),$$

where λ_1 and λ_2 are the coefficients of thermal conductivity of snow and ice; h_1 and h_2—the thickness of the snow and ice; T_0 is the skin temperature, T_2 is the temperature at the bottom surface of the ice, assumed to be equal to the freezing point of water for a given salinity S (‰) and determined in accordance with the empirical relation (Perry and Walker 1979)

$$T_2 = -0.053S$$

Heat conductivity coefficients $\lambda_{1,2}$ (kcal/(h m grade)) were found from the empirical formulas

$$\lambda_1 = 0.36(0.05 + 1.9\rho_1^4), \; \lambda_2 = 1.926\,(1 - 0.0048\; T_1)$$

from the papers (Khvorova and Ivanova 2007; Pounder 1967) accordingly. So in the Janson's formula for λ_1 dimensions of a snow density ρ_1 specified in g/sm^3, and expression for λ_2 defines close to linear dependence of heat conductivity of an ice on its temperature T_1 (°C) and is assumed to be equal to T_0. Heat flux through an open water surface is computed by V. S. Samojlenko's formula

$$Q_2 = 13(T_W - T_a)V/6,$$

expressing proportionality of the quantity of heat Q_2 (kcal/h m^2) arriving from water in the air and in the opposite direction (Shuleikin 1968), differences of temperatures between water T_W and air T_a, and also wind velocity V m/s. Heat flux through the surface of the one-degree calculation grid cell partially occupied by the snow-ice cover is calculated by the formula

$$Q = \delta Q_1 + (1 - \delta)Q_2,$$

where δ—an amount of ice concentration in a mesh grid. Positive heat flux is directed from Ocean to an Atmosphere.

Although in the open Ocean below 60° N sea ice does not play special role, however in the coastal waters of such more southern regions as Hudson Bay, St. Lawrence Bay, the Baltic and Okhotsk Seas it plays an extremely important role (Pounder 1967). Considering this circumstance, the heat flux calculations are executed not only for Arctic regions northward of 70° N, but also for the area enclosed between 50° N and 70° N.

Value of the general heat flux for each year from the interval 1969–2012 is the sum of monthly values of a flux through each one-degree grid with regard of latitudinal changes of grid areas.

3 Analysis of Results

Interannual distribution of the general (total on region) heat flux for the Arctic region northward of 70° N is presented in Fig. 1a. Distributions obtained with the ice thickness in the corresponding one-degree grid from the array NOAA-CIRES twentieth Century Reanalysis (http://www.esrl.noaa.gov) (line 1) and a constant ice thickness equal 1.5 m (line 2) at concentration from (http://badc.nerc.ac.uk/data/hadisst/, http://nomadl.ncep.noaa.gov) are represented. Both figures are received provided that on the ice surface there is present 0.15 m a layer of snow with density 0.35 g/sm^3. The solid line in this figure characterizes the distribution obtained at air temperature at

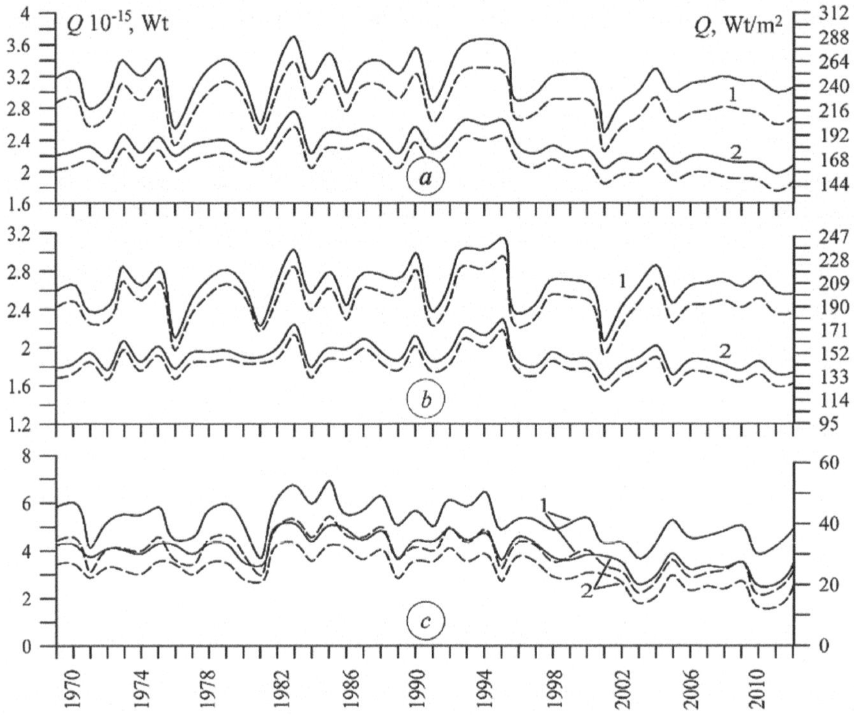

Fig. 1 Interannual variability of heat leaving from the Ocean to the Atmosphere in the Arctic area northward of 70° N

the skin surface level from the NCEP/NCAR Reanalysis (http://www.esrl.noaa.gov). The dashed line here shows the distribution in the case when the temperature at the level of the weather shelter is taken instead of the skin temperature.

The sea ice area changes in the annual cycle. In winter it has maximum, and by the end of the summer season it is considerably reduced. The end of the winter season is usually considered March, and the end of the summer season—September (Johannessen et al. 2005). Based on that for each year of the considered time interval the total flow for the six months of the cold season from October to March (winter) and the warm season from April to September is calculated. The results are shown on Fig. 1b, c in two sets of curves characterizing the distribution of Q over time. In these sets the solid and dashed lines denoted 1 and 2 correspond to the same conditions as in Fig. 1a.

The variability of the interannual distribution of the heat flux through the Ocean-Atmosphere interface section located in the latitudinal band between 50° N and 70° N is shown in Fig. 2 with the same notation as in Fig. 1.

The graphs in Figs. 1 and 2 show that the flow in time distributions of the total over the whole year (Figs. 1a and 2a) and over the cold period (Figs. 1b and 2b) are qualitatively similar to each other both in the region northward of 70° N and in the

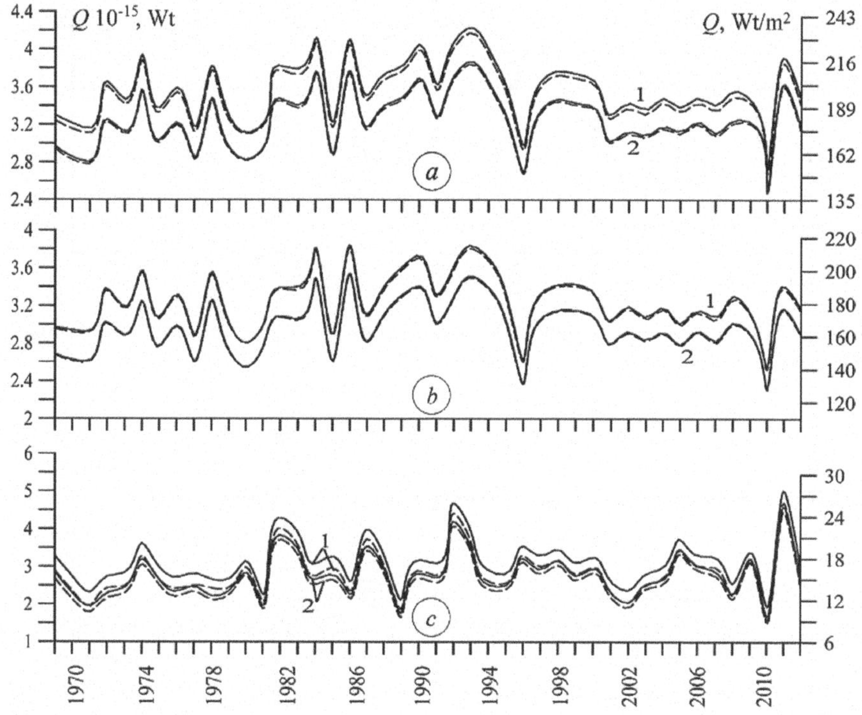

Fig. 2 Interannual variability of heat leaving from the Ocean to the Atmosphere in the area between 50° N and 70° N

latitudinal band between 50° N and 70° N. There are only quantitative differences in intensity of heat exchange. On the graphs in both considered regions the years of the burst and the attenuation of heat transfer with periods from 2 to 4 years, although not synchronously in time, are traced. Moreover, the intervals of bursts and attenuations are practically independent of thickness of the snow-ice. Apparently, as indicated in (Nikolaev 1977), the supply of heat to the Atmosphere is regulated by the Atmosphere itself. Extreme flow increases in the area northward of 70° N traced in 1983 and in the mid-90s. The increase of heat flow to the Atmosphere for this time was also noticed in (Smirnov and Koroblev 2010).

In the area between 50° N and 70° N the distribution of the flux over time both the total for the year and for the cold period differs notably from analogous distributions in the region northward of 70° N. Time intervals of extreme values of the flow are more clearly manifested here. The maximum of Q value is attained in 1993 and minimum—in 2010. Besides, here in 2011 (at the end of the considered period of time) there is a significant local maximum in time on the flow graphs, absent on the graphs for the area northward of 70° N. And the maximum is clearly traced on the graphs of interannual variability for a warm period.

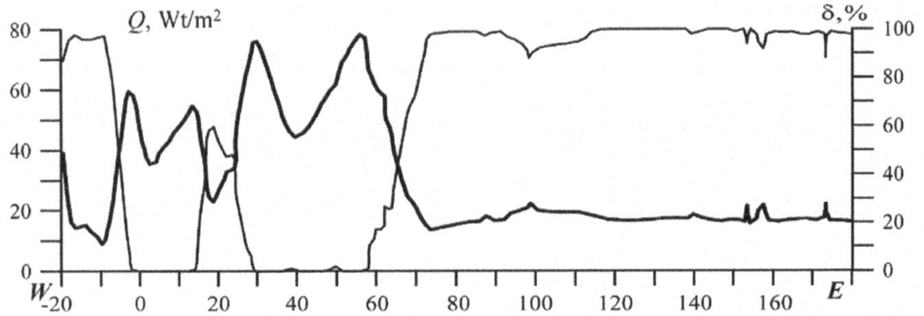

Fig. 3 The distribution of the heat flux on longitude in a one-degree band along 76° N for February 2009

Let's note that the total flow during considered time interval is directed from the Ocean to the Atmosphere, although in some months and for some areas southwards of 70° N the accumulation of heat by the Ocean can take place.

Interannual distribution of the heat flux is formed under the influence of changes in the components of the climate system, including the ice regime and hydrometeorological conditions. The role of a concentration of the snow-ice cover in the formation of the heat transfer of the Ocean is clearly illustrated by the graphs in Fig. 3, which shows the distribution of the heat flux Q (thick line) and the amount δ of concentration (thin line) along the 76° N longitudinal for February 2009 at ice and snow thicknesses equal to 2 m and 0.15 m respectively. It is visible that change of an amount of concentration leads to response in the heat transfer of the Ocean. The minimum quantity of heat goes into the Atmosphere at a longitude of 9° W where the ice concentration is 98%, and the flow is equal to 8.7 W/m^2. With small amounts of concentration the flow can reach 78.5 W/m^2, as for example, in region 56° E. After 73° E up to 180° E a high, more than 90%, concentration is maintained. The value of flux is in limits 15–27 W/m^2.

In the warm period (Figs. 1c and 2c), the heat transfer intensity is markedly weakened in comparison with the cold (March–October) period in the all region north of 50° N, which was noted in (Nikolaev 1977).

From an analysis of the interannual variability of the heat flux graphs, shown in Figs. 1 and 2 follows that the replacement of the skin temperature by the air temperature at the level of the weather shelter also leads to a decrease in heat flow from the Ocean to the Atmosphere. A quantitative estimate of the effect of such a substitution on the flux attenuation is given in Table 1, which shows the decrease (in %) of the quantity of heat leaving into the Atmosphere per unit area (m^2) for the latitudinal band between 50° N and 70° N (left column) and for the polar region northward of 70° N (right column). The upper line corresponds to the ice thickness from the reanalysis and the second and third from above—the ice thicknesses are 1 and 1.5 m, provided that a snow layer of 0.15 m is present on the ice surface. Data in Table 1 show that the attenuation of heat transfer is most essentially manifested in the warm period (April–September) in the polar region northward of 70° N. In

Table 1 Effect of substitution of the skin temperature by air temperature at the level of the weather shelter

Year	October–March		April–September	
1976	0.51	6.19	11.10	19.15
	0.38	6.48	8.65	20.45
	0.32	6.15	5.50	18.96
1993	0.77	6.07	6.65	26.16
	0.52	5.61	3.89	22.35
	0.43	5.17	3.19	20.25
2001	0.27	7.11	11.36	21.45
	0.21	7.33	6.53	21.78
	0.18	6.89	5.35	20.36
2010	1.46	8.39	18.49	44.08
	0.93	7.58	11.02	38.31
	0.77	7.02	9.09	35.81

the cold period (October–March), the flux attenuation here is less than 9% for the considered values of snow-ice cover parameters.

The estimation of the decrease in the heat flux leaving into the Atmosphere, obtained at the skin temperature in (http://www.esrl.noaa.gov), due to the increase in an ice thickness in the zones of its given real concentration and thickness of a snow layer covering the ice, is given in Table 2. Here gathered the characteristics of the decrease (in %) of heat outflow quantities with an increase in the ice thickness from 1 to 3 m with a snow thickness of 0.15 m, and with an increase in the thickness of snow from 0.15 to 0.5 m when an ice thickness of 1 m. The data presented in the

Table 2 The effect of changes in the ice thickness and the snow thickness covering the ice on decrease (in %) the quantity of heat per unit area (m^2)

Year	Season	In the band 50° N–70° N		In the region northward of 70° N	
		Decrease of heat flux	Snow thickness increase	Decrease of heat flux	Snow thickness increase
1993	October–March	4.57	5.20	28.74	33.04
	April–September	6.05	6.74	27.30	30.79
2000	October–March	4.58	4.54	29.43	33.65
	April–September	4.99	5.54	31.08	35.02
2002	October–March	4.81	5.13	30.63	35.03
	April–September	7.91	8.81	32.07	36.26
2010	October–March	4.81	5.43	29.82	34.09
	April–September	7.81	8.63	34.48	38.88

table indicate, as expected (Doronin 1969), that the snow effects on heat transfer by the Ocean more significantly than the ice cover of similar thickness.

To estimate the dependence of interannual variability of heat flow from the Ocean into the Atmosphere by the atmospheric circulation, a correlation analysis is made of heat flow anomalies at the region with a change of Arctic oscillation index. Significant correlation of 95% confidence interval is obtained. It is revealed that the quasisynchronous reaction in the Arctic region northward of 70° N between the anomaly of the monthly average values of the heat flux and the Arctic oscillation index takes place only for January and October with a correlation coefficient of 0.26 and 0.21 respectively.

In other months, a quasisynchronous reaction is not revealed. Significant correlations are found only on a shift from one year to a decade with a correlation coefficient not exceeding 0.42 in absolute value. The reaction of the total flow for the year in the region on the change in the Arctic oscillation index lags a year with a correlation coefficient of 0.13 over the period under review. More stable correlation is observed for the region northward of 50° N. Here for the period from 1969 to 2012, the correlation coefficient of the total flow for the year on region with Arctic oscillation index is 0.47. The correlation coefficient between the anomalies of the average monthly climatic values of the quantity of heat leaving the Ocean through the unit area in this region and Arctic oscillation index takes on the values 0.66, 0.52, 0.54, 0.10, 0.44, 0.39, 0.5 for January, February, March, April, October, November, December respectively. In the months of May—September (warm period) heat flow reaction to changes in the Arctic oscillation index appears not earlier than three years with a correlation coefficient of not more than 0.38.

4 Conclusion

Research of interannual variability of heat flux in the Atmosphere through Ocean surface in the latitudinal band between 50° N and 70° N and in the polar region northward of 70° N is performed. It is shown that interannual distributions of the total heat flux leaving the Ocean at region in the cold (October–March) period and total for whole year are qualitatively similar. There are only quantitative differences in the intensity of heat transfer.

The years of increase and attenuation of heat transfer are traced with periods from 2 to 4 years. Moreover, the time intervals of bursts and flux attenuations are practically independent of changes in the thickness of the snow-ice cover in the region of its real concentration.

The replacement of the skin temperature by the air temperature at the level of the weather shelter in calculating the flow leads to its weakening. The weakening is most essentially manifested in the warm (April–September) period in the polar region northward of 70° N.

The increase in the ice thickness in the areas of its preset real concentration and the thickness of the snow layer covering the ice leads to a decrease in heat transfer. In the warm period, the intensity of heat transfer is much lower than in the cold period.

Correlation between the heat flux to the Atmosphere and the Arctic oscillation index is more considerably manifested in the region bounded from the southward of 50° N, than in the Arctic region northward of 70° N.

Acknowledgements The present study is carried out within the framework of the State Order No. 0827-2019-0003.

References

Alekseev GV, Radionov VF, Aleksandrov EI, Ivanov NE, Kharlanenkova NE (2010) Climate change in the arctic and the northern polar region. Problemy Arktiki i Antarktiki. Arct Antarctic Res 84 (1):67–80 [In Russian]

Baidin AV, Meleshko VP: Response of the atmosphere at high and middle latitudes to the reduction of sea ice area and the rise of sea surface temperature. Meteorol Gidrol 6:5–18 (2014) [In Russian]

Bespalov DP (1959) On heat exchange between the atmosphere and the ocean in the central arctic. Trudi Arktiki i Antarktiki institut and Glav. Geophis. Observatoria. In: Proceeding of arctic and antarctic institute and main geophysical observatory, vol 226, pp 30–41 [In Russian]

Doronin YP (1969) Heat interaction of atmosphere and hydrosphere in the arctic. Gidrometeoizdat, Leningrad, p 300 [In Russian]

Doronin YP (1974) Effects of sea-ice cover on the atmosphere-ocean heat exchange. Problemy Arktiki i Antarktiki. Arct Antarctic Res 43–44, 52–60 [In Russian]

Eremeev VN, Bukatov AE, Bukatov AA, Babiy MV (2013) Interannual variability of the atmosphere-ocean heat exchange in the antarctic. Dokl. Nats. academii nauk Ukrainy. 1:96–104 [In Russian]

Frolov IE, Gudkovich ZM, Karklin VP, Kovalev EG, Smolyanitsky VM (2007) Scientific researches in Arctic. In: Climate changes in the ice cover of the Eurasian shelf seas, vol 2. St. Peterburg, Nauka, p 158 [In Russian]

Frolov IE, Gudkovich ZM, Karklin VP, Smolyanitsky VM (2010) Climate change in the arctic and antarctic—result of natural causes. Problemy Arktiki i Antarktiki. Arct Antarctic Res 85(2):52–61 (2010) [In Russian]

http://badc.nerc.ac.uk/data/hadisst/, http://nomadl.ncep.noaa.gov

http://www.nodc.noaa.gov

http://www.esrl.noaa.gov

Johannessen OM, Bobylev LP, Kuzmina S, Shalina E, Khvorostovsky KS (2005) Arctic climate variability in the context of global change. Computational technologies. Special issue. In: Proceedings of the international conference and the school of young scientists “Computational and informational technologies for environmental sciences” (CITES 2005), vol 10. Tomsk, pp 13–23 March 2005. Part 1. pp 56–62 [In Russian]

Khvorova LA, Ivanova OA (2007) Methodological bases of mathematical modeling of a hydrothermal mode of soil during the winter period. Izv. Altai. Gos. Univ. Altai State University. V. 1. 1–13 [In Russian]

Makshtas AP (1984) Heat balance of arctic ice in winter. Gidrometeoizdat, Leningrad, p 68 [In Russian]

Nikolaev YV (1977) Large-scale interaction between the atmosphere and the ocean and the problem of long-term meteorological forecasts. Trudi Arktiki i Antarktiki institut. In: Proceeding of arctic and antarctic institute, vol 347, pp 4–28 [In Russian]

Perry AH, Walker JM (1979) The ocean—atmosphere system. Gidrometeoizdat, Leningrad, p 196 [In Russian]
Pounder ER (1967) The physics of ice. Mir, Moscow, p 190 [In Russian]
Shuleikin VV (1968) Physics of the sea. Nauka, Moscow, p 1084 [In Russian]
Smirnov AV, Koroblev AA (2010) Interrelation between mixed layer properties and heat fluxes at the ocean-atmosphere interface in the north-European basin. Problemy Arktiki i Antarktiki. Arct Antarctic Res 86(3):79–88 [In Russian]
Tryoshnikov AF, Alekseev GV (eds) (1991) Interaction of the ocean and atmosphere in the northern polar region. Gidrometeoizdat, Leningrad, p 176 [In Russian]

Structural Features of the Dip Process of Thermics with Negative Buoyancy

A. A. Volkova and V. A. Gritsenko

Abstract The paper presents the results of a study of the process of immersion in fresh water of a finite volume of salt water (or thermics with negative buoyancy) using laboratory and numerical experiments. In a series of laboratory experiments were recorded already known structural features of the process: mushroom shape and transformation into a vortex ring. The calculated flows were obtained using a nonlinear 2d model of the dynamics of stratified in density fluid. The analysis of model fields: vorticity, current function, density and tracer of neutral buoyancy introduced to identify the water mass of thermic, allowed to identify the main physical factors responsible for the formation of the observed mushroom form. As it turned out, the key structure-forming factor of the evolution of thermic during its immersion is the baroclinic vortex generation. A parallel is drawn between the processes of immersion of thermic and the collapse of density inhomogeneity in a stable stratified fluid.

Keywords Convection · Buoyancy · Cooling · Numerical model · Thermic · Baroclinity · Instability of Rayleigh-Taylor · Computational mesh · Vorticity

1 Introduction

Termics of mushroom-shaped with non-zero buoyancy are well known from the works of various authors (Gershuni 1989; Gebhart 1995; Ingel 2016; Meleshko 1993; Scorer 1980; Turner 1977). There are two kinds, in the first, negative, when water is cooling down from the surface, and, in the second, positive, when the bottom of the tray is heated. Structural elements of motions of a non uniform in density fluid are also observed in the case of Rayleigh-Taylor instability (Belotserkovsky 2003; Inogamov 1999). The main difference between the process of immersion the termics

A. A. Volkova (✉)
Immanuel Kant Baltic Federal University, Kaliningrad, Russia
e-mail: a.al.volkova@gmail.com

V. A. Gritsenko
Shirshov Institute of Oceanology Russian Academy of Sciences, Moscow, Russia

C. T. Olegovna (ed.), *Processes in GeoMedia—Volume I*,
Springer Geology, https://doi.org/10.1007/978-3-030-38177-6_23

and a number of other phenomena is the volume of water with negative buoyancy is limited (Scorer 1980; Turner 1977).

The purpose of this work is to analyze the main factors of the formation of the structural features of the process of immersion of a finite volume of salt water (or a termics with negative buoyancy) in fresh water observed in laboratory experiments and field conditions.

2 Laboratory Experiment

Figure 1 shows photographs of two series of the immersion of termics generated by the supply heavier water (saline, $\Delta\rho_0 = 0.0001$ g/cm^3 and stained) to the surface of

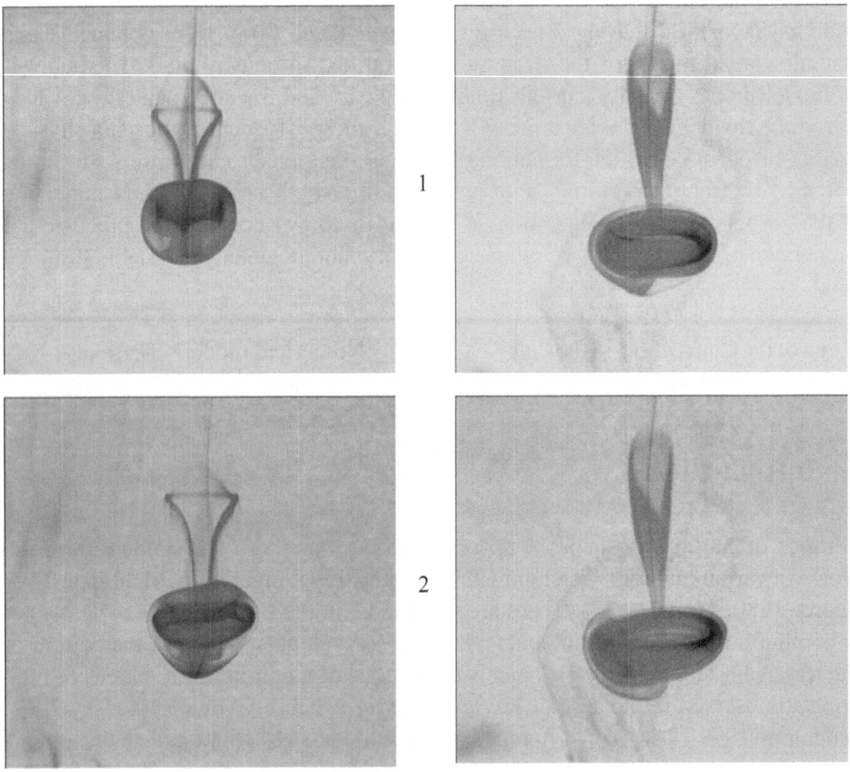

1

2

Series A

Series B

Fig. 1 Two series of three consecutive shots (A and B) illustrating the process of immersion the termics from 1 ml of salt water ($\Delta\rho_0 = 0.0001$ g/cm^3) in fresh water from the surface. The vortex character of inside movement the volumes of salted and colored water is clearly visible fresh water of a finite volume (~1 cm^3). All images clearly show the mushroom shape of the immersing volume of salt water and the existence "cometary tail" behind it.

The features of the form of sinking termics recorded in the experiment and the presence of a trace behind are known from various sources (Meleshko 1993; Turner 1977). R. Skorer also described processes that are similar by the physical phenomenon and structural features (Scorer 1980). Thus, the results of studies by various authors (Van Dyke 1986; Ingel 2016; Meleshko 1993; Scorer 1980; Turner 1977) allow us to state that the appearance of mushroom shape of the plunging final volume of salt water (termics) is a natural property of this unsteady flow stratified by density. Obviously, the problems of the collapse of density inhomogeneities in a stably stratified fluid are similar in the physics of the process and differ only in the absence of their mean vertical displacement (Barenblatt 1978; Popov 1986).

3 Numerical Model

In this work the structural features of the process of immersion of termics were studied using a nonlinear 2d model of the dynamics of a fluid that is inhomogeneous in density (Gritsenko and Yurova 1997; Gritsenko 1999). The traditional set of basic model variables—vorticity, current strength, density—was supplemented by two additional fields of tracers with neutral buoyancy introduced to identify volumes of water with specific properties and position in the model space. The system of equations of the model had a completely traditional view for the class of problems under consideration (Gebhart 1995; Roach 1980; Turner 1977):

$$\frac{d\omega}{dt} = \frac{g}{\rho_0}\frac{\partial\rho}{\partial x} + \nu_T \Delta\omega \tag{1}$$

$$\frac{d\rho}{dt} = D_T \Delta\rho \tag{2}$$

$$\Delta\psi = \omega, \tag{3}$$

$$\frac{dtras_K}{dt} = D_T \Delta tras_K, \quad K = 1, 2 \tag{4}$$

where $\omega = \partial u/\partial z - \partial w/\partial x$—vorticity, ψ—function of current, $g = 982$ cm/s^2, ρ_0 и $\rho = \rho_0 + \Delta\rho$—density of fresh and salt water, $\nu_T = \nu_0 + tras_1(x, z, t) \cdot \nu_{EF}$, $D_T = Sc^{-1} \cdot \nu_T$—turbulent exchange coefficients, $Sc = 2$—turbulent Schmiedt number, ν_{EF}—coefficient of effective viscosity, $\nu_0 = 0.01$ cm^2/s—coefficient of molecular viscosity, d/dt и Δ—full derivative operator and Laplace operator, $tras_K$, $K = 1, 2$—tracers with neutral buoyancy. Axis Ox is horizontal and coincides with the bottom of the model space, and the axis Oz is directed vertically upwards. To calculate currents of a density-stratified fluid, we used a grid by dimension 601×801 with dimensionless spatial discretization was used $\Delta x = \Delta z = 0.05$ and, thus the model space has a size of [0; 30] × [0; 40].

At the initial moment of time, fresh water in the model space assumed motionless. Termics in the form of a spot of salt water, "stained" with the help of the first tracer ($tras_1 = 1$) with an initial density $\rho_0 + \Delta\rho_0$ was set in the form of a horizontally extended rectangle with dimensions 31 × 16 grid nodes or 1.5 h_0 on 0.75 h_0. The characteristic scales of model currents was into the ranges: h_0 = 1–3 cm, u_0 = 0.5–2 cm/s, $\Delta\rho_0$ = 1–5 × 10^{-4} g/cm^3, ν_{EF} = 0.1–0.2 cm^2/s.

4 Calculated Currents

Calculated currents. Figure 2 shows the graphs of the initial distribution of density and the corresponding by him field of the horizontal pressure gradient. Calculate of the pressure field was calculated using an iterative fast upper relaxation algorithm (Gritsenko 1999; Roach 1980). Hereinafter, for better resolution, all the graphs in all the figures constructed only for a part of the model space that is directly adjacent to the termics.

The calculations showed that the density isolines very quickly (time = 2.5) acquire well-supposed signs of the future mushroom shape of the plunging termics (see Fig. 3a). The rotational nature of the movements with centers in the vicinity of the original right and left boundaries of the termics is clearly visible on the distribution of current lines (see Fig. 3b, the direction of rotation is shown by the arrows).

An obvious source of the appearance of the vorticity of different signs on the boundaries of the termics is the baroclinic member in the right side of Eq. (1). Immediately after the start of movement, due to the mismatch between isolines of density and pressure, in accordance with the Bjerknes theorem (Turner 1977), rotational motion of the fluid is generated at the left and right boundaries of the

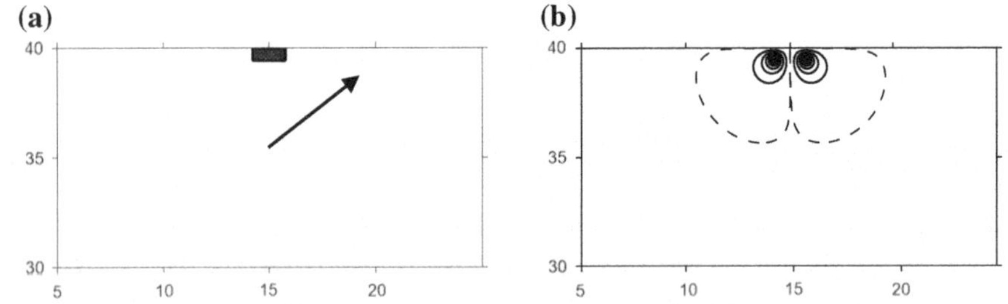

Fig. 2 The initial positions of the salt water stain (**a**, marked by an arrow) and distribution of isolines of horizontal pressure gradient values $-P'_x$ (**b**) for the initial state of one of the calculated currents with characteristic scales h_0 = 2 cm, u_0 = 0.5 cm/s, $\Delta\rho_0$ = 0.00025 g/cm^3, $T_0 = h_0/u_0 = 2.0$ s, ν_{EF} = 0.1 cm^2/s. Pressure gradient isolines "run through" values from −1.3 to 1.3 in increments of 0.2. Dashed lines indicate the isoline with the zero [or very small (±0)] value of the corresponding calculated parameter. Model time = 0

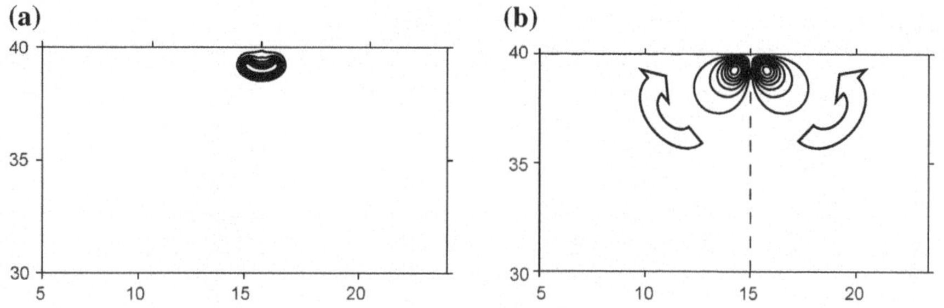

Fig. 3 The distribution of: **a** field of excess density ($\sigma = (\rho - \rho_0)/\Delta\rho_0$, isolines "run through" values from 0.05 to 0.65 with a step of 0.1), **b** field of function of current Ψ (from −0.4 to +0.4 through 0.05, the direction of rotation is indicated by arrows), for one of the initial phases of the model current. Estimated time = 2.5

density inhomogeneity (termics) clearly visible on the distribution of isolines of the function of current (see Fig. 3b). Note that it is possible to say that in the process of his immersion the termics also participates in the process of collapse of his heavier (salty) water, during which the vorticity of different signs is also generated (Barenblatt 1978; Popov 1986).

By the subsequent development of the model current (see Fig. 4), all the above features of structure of the water mass of the termics become more distinct.

The figure clearly shows the similarity of the geometric forms of the termics received in laboratory (see Fig. 1, series A, second shot) and of calculated termics (see Fig. 4a). It should be noted that the water with the highest density is located on the axis of the vortex tube (see Fig. 4b, a higher density of paint means a higher salt content) in the laboratory termics and in the center of the calculated maxima (see Fig. 4a).

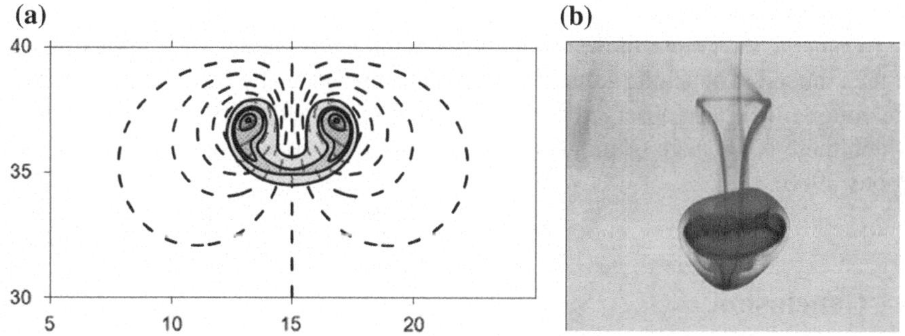

Fig. 4 The distribution of the field of excess density (**a**) ($\sigma = (\rho - \rho_0)/\Delta\rho_0$, isolines from 0.002 to 0.04 through 0.005 with gray fill in) on the background of the isolines of the current function Ψ (dashed lines, from −0.9 to +0.9 through 0.15) model time 12.5. Shot of one of the immersion phases of the termics (**b**) in the tray (Series A, see Fig. 1, left row of images, bottom image) clearly demonstrates the nucleation of a vortex ring

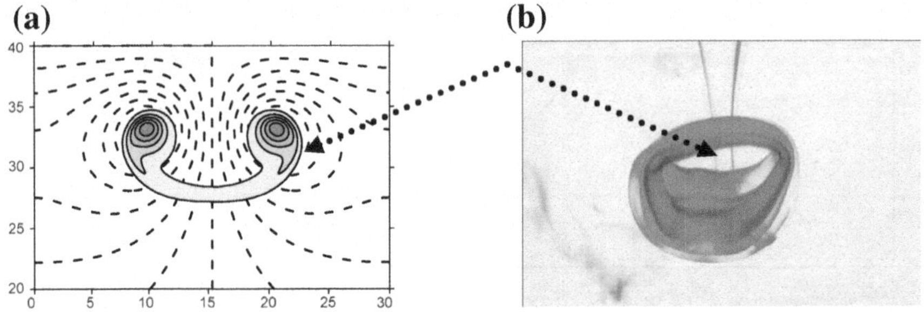

Fig. 5 The graph (**a**) of excess density [$\sigma = (\rho - \rho_0)/\Delta\rho_0$, isolines change from 0.002 to 0.04 with a step of 0.005 with gray fill on the background of isolines of the current function Ψ (dashed lines, from −1.6 to +1.6 through 0.2), model time = 50]. A snapshot (**b**) of one of the phases of immersion of the termics from the tray (series A), clearly demonstrating the nucleation of a vortex ring

The absence tail of a comet form at the computational termics is explained by the specifics of specifying the computational conditions for the excess density field. The salt water rectangle was formed only on the internal nodes of the computational grid, i.e. on the first subsurface layer. This approach removed the need to constructing a design situation for density at the upper boundary of the model space. Note that the vorticity is concentrated mainly inside the termics, while maintaining the natural symmetry only horizontally. It is the localization of the vorticity inside the termics that is responsible for the formation of a vortex ring at a later stage.

It is known (Gershuni 1989; Meleshko 1993; Chetverushkin 2003) that the next stage in the evolution of the finite volume of salt water (termics) plunging into fresh water is its transformation into a vortex ring (Meleshko 1993). Illustrative examples of vortex rings are given in the Album of Van Dyke Currents (Van Dyke 1986). Figure 5 shows the beginning of the process of transformation of the mushroom-like termics into a vortex ring.

In general, the graphs of the model fields in the figures indicate the vortex nature of the immersion of a finite volume of salt water surrounded by fresh water. Note that the vortex ring is also observed during the collapse of the density inhomogeneity, in which there is no stage of immersion of density inhomogeneity (Barenblatt 1978; Popov 1986).

5 Conclusion

A comparative analysis of the results of laboratory and numerical experiments suggests to a say that a crucial role for the mechanism of baroclinic eddy generation in the transformation of the initial state of a termics with negative buoyancy and him acquisition of a fungoid shape. The evolution of the vorticity fields during convective immersion of a finite volume of salt water in the surrounded by fresh water is well traced in the photographs and graphs. It is allows to state the similarity of their variability in laboratory and calculated termics.

Advection of heavier (saline) waters in the process of immersion and, at the same time, of the collapse of density heterogeneity, to the central part of well-localized vortex zones determines the transformation of the original saline water spot, first, into the usually mushroom-like shape of the termics, and then into the vortex ring.

The calculations showed the satisfactory ability of the chosen numerical model to qualitatively right reproduce the processes of immersion of a heavier and finite volume of water into a fresh water. The overall picture of the calculated flows, as well as the individual phases of the motion of termics in laboratory experiments, is quite consistent with the existing ideas about such processes (Van Dyke 1986; Gebhart 1995; Meleshko 1993; Scorer 1980; Turner 1977; Chetverushkin 2003).

Acknowledgements Model calculations were carried out in the framework of the state assignment number 0149-2019-0013, laboratory experiments—with the support of the RFFI grant No. 19-05-00717.

References

Barenblatt GI (1978) Dynamics of turbulent spots and intrusions in a stable stratified fluid. Izv USSR Acad Sci 14(2):195–206

Belotserkovsky OM (2003) Turbulence: new approaches. Nauka, Moscow, 286 p

Chetverushkin BN (2003) The use of high-performance multiprocessor computing in gas dynamics. In: Mathematical modeling: problems and results. Nauka, Moscow, pp 123–198 (2003)

Gebhart B (1995) Free convective flow, heat and mass transfer. Mir, Moscow, B.1. 678 p

Gershuni GZ (1989) Stability of convective flows. Nauka, Moscow, 492 p (in Russian)

Gritsenko VA (1999) Yurova AA (1999) About the basic phases of separation of near-bottom gravity currents from the slope of the bottom. Oceanology 39(2):187–191 (in Russian)

Gritsenko VA, Yurova AA (1997) About the propagation of the near-bottom gravity flow along the steep slope of the bottom. Oceanology 37(1):44–49 (in Russian)

Ingel LKh (2016) On the theory of convective jets and thermals in the atmosphere. Izv Atmos Oceanic Phys 53(6):676–680 (in Russian)

Inogamov N (1999) Development instabilities of Rayleigh-Taylor and Richtmyer-Meshkov in three-dimensional space: topology of vortex surfaces. Lett GETV 69(10):691–697

Meleshko VV (1993) Dynamics of vortex structures. Naukova Dumka, Kyiv, 279 p (in Russian)

Popov VA (1986) Development of a region of partially mixed fluid in a stratified medium. Izv USSR Acad Sci 22(4):389–394

Roach P (1980) Computational fluid dynamics. Mir, Moscow, 616 p

Scorer RS (1980) Environmental aerodynamics. Mir, Moscow, 549 p (in Russian)

Turner JS (1977) Buoyancy effects in fluids. Mir, Moscow, 431 p (in Russian)

Van Dyke M (1986) An album of fluid motion. Mir, Moscow, 184 p (in Russian)

Alternative Use of Satellite Maps of Yandex for Visualization of Geophysical Objects

N. Voronina and N. Inyushina

Abstract The paper presents the method of visualization of the variability of the geophysical parameters of the Black Sea region as an "overlay" on the Yandex satellite map as an alternative to the methods of visualization based on static images. The method is illustrated by the example of visualization of a numerical experiment on the propagation of tsunami waves. The advantages of using satellite maps in comparison with static images are shown. To achieve this goal, the existing modern graphic capabilities of JS, PHP, HTML software tools are used in conjunction with the Yandex.maps API map service tools.

Keywords Visualization · Geophysical parameters · Sea surface temperature · JS · PHP · HTML programming tools · Yandex.maps API · Google maps API mapping services

1 Introduction

With the development of space-based remote sensing tools, the flow of information about processes occurring in the atmosphere and the ocean has increased many times (Pustovojtenko et al. 2011; Pustovojtenko and Zapevalov 2012). The use of this information required the development of new tools for analyzing, storing, processing and visualizing sensing data.

Recently, considerable attention has been paid to the operational presentation for users of products, one way or another, tied to maps, comparing the results of satellite sounding and model calculations (Slepchuk et al. 2017). For these purposes, sites are created that display the results of operational data. The research results presented in this article are a development of the works started in 2009 in the framework of international projects MyOcean, then MyOcean2 of the European Union (http://www.myocean.eu/). The listed projects were devoted to the collection and display of the geophysical parameters of the state of the World Ocean (results of satellite monitoring, contact measurements, forecasts of the state of the marine environment,

N. Voronina (✉) · N. Inyushina
Marine Hydrophysical Institute RAS, Sevastopol, Russia
e-mail: voronina.nataly@mail.ru

C. T. Olegovna (ed.), *Processes in GeoMedia—Volume I*,
Springer Geology, https://doi.org/10.1007/978-3-030-38177-6_24

etc.). The authors of the article were the participants of these international projects as part of the team of the Marine Hydrophysical Institute of the Russian Academy of Sciences (MGI RAS).

Subsequently, the MyOcean projects, then MyOcean2, were transformed into Copernicus, an international service of operational marine monitoring of marine environment operating since April 2015 (https://www.copernicus.eu/en). This projects served a Copernicus starting foundation of collect, process, present information on the state of the marine environment as the entire oceans as a whole and individual marine components. In the service of Copernicus of the European Union, the latest achievements in the field of construction of mathematical models, as well as in the field of IT-technologies are applied. These Copernicus services are freely available. Displays of the geophysical parameters of the seas on satellite maps present in the service of Copernicus too, but such images are implemented using specialized technical means.

2 Tasks of Visualization and Methods of Their Solution

As part of the MyOcean project at MHI, the website of the Operational Oceanography Branch MHI was created and is currently operational (http://bsmfc.net) now, which provides a wide range of geophysical data on the Black Sea area on an operational basis. These data are represented by modern software and hardware visualization tools. It was during the work on the MyOcean project that the task was to obtain such images of geophysical fields that the user can instantly view with the maximum range of possibilities without the use of specialized technical resources. The work carried out within the framework of the RFBR grant, the results of which were presented in the article (Voronina et al. 2016), was devoted to solving the above problems by the example of the local region of the Black Sea basin. Data from the operational work of the Experimental Center for Marine Forecasts (ECMF) (http://innovation.org.ru/instan_param.php) of the Moscow State Institute of RAS was used. In the course of the work, emphasis was placed on the choice of freely distributed graphics-building software. The work was both scientific and practice-oriented, since its results were implemented and found their use up to the present time on-line when building images of daily marine forecasts.

When obtaining the results of work, namely, visualization of geophysical data in the Black Sea region, the Google Maps API map service was used with a powerful set of tools that allow using this service when developing your own software products. The JavaScript scripting language (JS), with which the Google Maps API toolkit functions, allows you to view images dynamically on the Web pages, which is necessary for a visual assessment of the variability of the geophysical parameters of the marine environment.

Later, in view of the well-known circumstances (the introduction of the sanctions regime in 2014 with the subsequent disconnection of some foreign software in Russia), it became necessary to switch to domestic products, in connection with which the

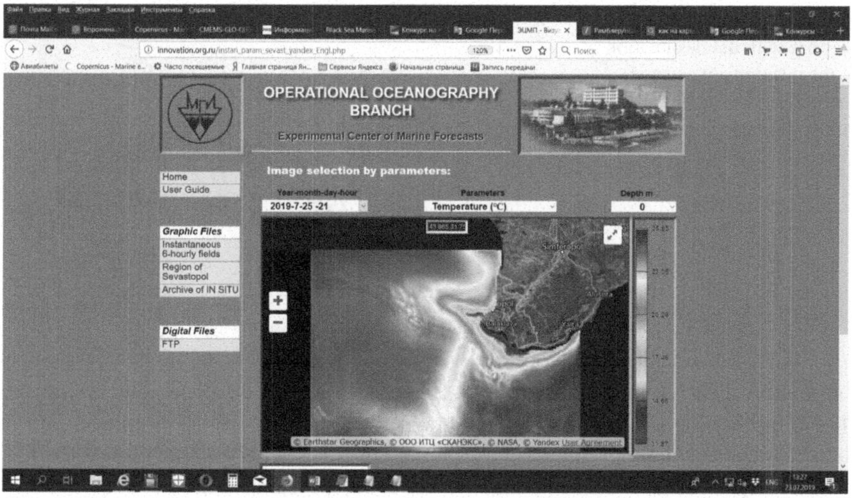

Fig. 1 Data visualization (temperature forecast at a depth of 0 m) using Yandex satellite maps

own software was developed using of standard programming tools JS, PHP, HTML, which resulted in the same images, but have already built no on Google maps, but on Yandex satellite maps (https://tech.yandex.ru/maps/), as shown in Fig. 1. The results of such a transition were presented in the materials of the Fourth International School of Young Scientists (Mamchur and Voronina 2018).

And finally, one of the objectives of this work became to study the broader use of the domestic Yandex satellite maps in research works of scientists, one way or another connected with geographic maps. This required the involvement of all currently known freely distributed software tools for Yandex satellite maps (https://tech.yandex.ru/maps/).

3 Material, Methods and Research Result

The results of numerical calculations of the evolution of tsunami waves depending on the characteristics of the shelf area (Docenko and Ingerov 2016) were taken as the source material for developing data visualization tools using Yandex satellite maps. The calculations were carried out for the Black Sea area with a different location of underwater seismic sources. In this article, the results of the analysis show the layout of the sections with the designation of the positions of seismic sources for conducting numerical experiments. The layout of the eleven sources and sections in which the wave field was calculated is shown in Fig. 2. The sections are designated A–K.

In the work (Docenko and Ingerov 2016), calculations were carried out for eleven cuts passing through the shelf. Here we presented an example of visualization using

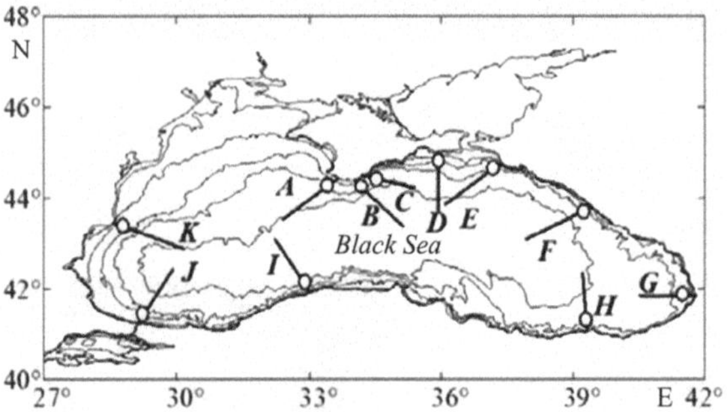

Fig. 2 The layout of the sections with the designation of the positions of seismic sources for conducting numerical experiments

the Yandex satellite maps service only for the first three cuts A–C. The results are shown in Fig. 3. Figures 2 and 3 show the same situation with a difference only in the visualization method—the scheme on the Fig. 2 is shown in static representation, whereas Fig. 3 presents the same scheme, but using the Yandex satellite map. Figure 4 is shown as a demonstration of one of the advantages of such a method of presenting data, namely the possibility of scaling (reducing or enlarging an image without "breakaway" from the map) with obtaining the coordinates of any point.

It should be noted all the possibilities of such a visualization method can be seen in the best way on the Web site. As we know Web sites pages allow you to view the object in Online mode. For example in our case, one can additionally see the coordinates of any point of interested to the researcher on the Yandex maps (see the panel on the top of Figs. 3 and 4) unlike any paper publication, which represents only the static state of the object of study.

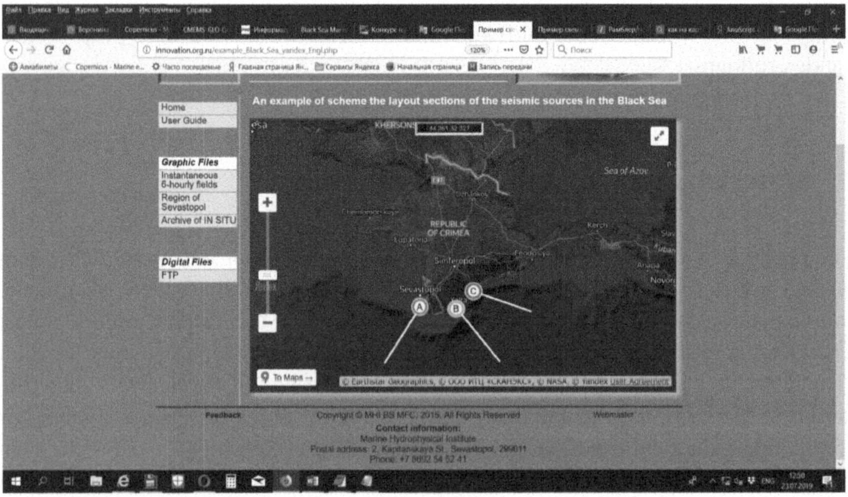

Fig. 3 Visualization of the location of the cuts with the designation of the positions of seismic sources for conducting numerical experiments

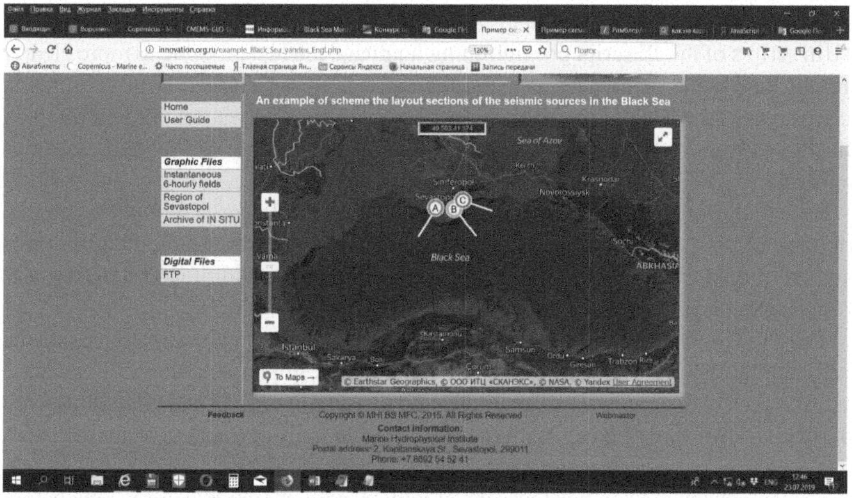

Fig. 4 The ability to scale when visualizing the location of the sections with the designation of the positions of seismic sources on the Yandex map for numerical experiments

This fundamentally distinguishes the images on satellite maps from ordinary graphic images, numerous examples of which we can see in almost every issue of some journal, in which materials are somehow related to geolocation.

And finally, Fig. 5 shows another possibility of using Yandex satellite maps—researchers can view the data in full screen mode (there is only a map on the screen and nothing more) by pressing just one button.

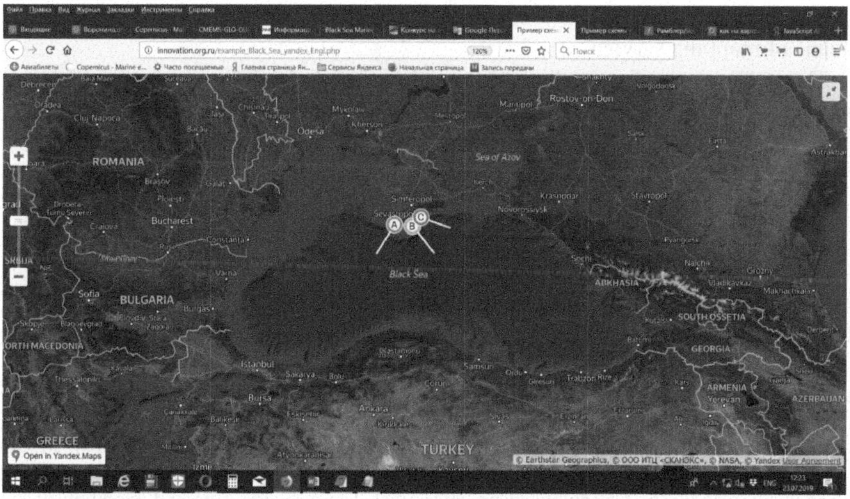

Fig. 5 Visualization in full-screen mode of the location of the cuts with the designation of the positions of seismic sources for conducting numerical experiments

The applying of the Yandex.maps API map service with a powerful set of tools for use in scripts of the JS script programming language made it possible to more clearly present the results of numerical experiments related to the location in seismically active areas of the Black Sea on the Web site page (http://innovation.org.ru/example_Black_Sea_yandex_Engl.php). And the construction of any Web-objects assumes implementation only with the help of languages PHP, HTML.

Note that non-commercial use of the Yandex.maps API service implies free distribution of both the toolkit and the developed application. In future, the use of this method of data visualization in the works will not require additional financial costs from researchers.

Thus, the proposed approach to solving the problem of data visualization using the cartographic service Yandex consists in the software implementation of displaying static images with geophysical parameters on the satellite map Yandex. It allows you to use your own static images with the full range of tools of Yandex. Maps API and, thanks to this, to leave from the disadvantages inherent in static images. Moreover, adding a mechanism to embed these developments into operational work, which does not require much time, will allow you to monitor regularly the current geophysical parameters of the Black Sea in the form of images on a satellite map.

4 Conclusion

Prospects for the use of satellite maps for the analysis and visualization of remote sensing data from spacecraft, as well as for data obtained in the framework of numerical models of processes occurring at sea, have being analyzed. It is shown the use of satellite maps has several advantages in comparison with static images. It was noted that in the current geopolitical situation, there is a need to move to the domestic means of mapping service.

The paper presents a method for visualizing the variability of the geophysical parameters of the Black Sea region by overlaying them on the Yandex satellite map, which in addition is a product of domestic developments as opposed to Google Maps. This method makes it relatively easy to perform such operations as zoning, taking the coordinates of any point on the map, scaling. The presented method of visualization perfectly shows the results of research in the operative mode, as it becomes possible to view the data in animation mode for a certain time. The applying of software tools for maps services allows you to use the above method of visualization of geophysical data in almost all areas of science, where geolocation finds its application.

Acknowledgements The research results presented in the paper are a continuation of a series of publications on the visualization of geophysical data, which began in the course of the development of the competition project No. 14-41-01635 with the support of the RFBR (Voronina et al. 2016).

This work was carried out as part of the state assignment on the subject No. 0827-2018-0003 "Fundamental studies of oceanological processes that determine the state and evolution of the marine environment under the influence of natural and anthropogenic factors, based on methods of observation and modeling".

References

Black Sea Marine Forecasting Centre. http://bsmfc.net
Copernicus. https://www.copernicus.eu/en
Docenko SF, Ingerov AV (2016) Spektry voln cunami v shel'fovoj zone CHernogo morya. Processes in GeoMedia 3(7):218–224
Eksperimental'nyj centr morskih prognozov. http://innovation.org.ru/instan_param.php
Mamchur NL, Voronina NN (2018) Sravnitel'naya harakteristika vizualizacii processov v geosredah na primere morskih prognozov regiona chernogo morya. V sbornike: Fizicheskoe i matematicheskoe modelirovanie processov v geosredah CHetvertaya mezhdunarodnaya shkola molodyh uchenyh: sbornik materialov shkoly. Institut problem mekhaniki im. A.YU. Ishlinskogo Rossijskoj akademii nauk; Moskovskij gosudarstvennyj universitet imeni M.V. Lomonosova. 179–180
Marine environment monitoring service. http://www.myocean.eu/
Pustovojtenko VV, Zapevalov AS (2012) Operativnaya okeanografiya: Sputnikovaya al'timetriya—Sovremennoe sostoyanie, perspektivy i problemy. Seriya. Sovremennye problemy okeanologii, Sevastopol': NPC "EKOSI-Gidrofizika" 11:218–224
Pustovojtenko VV, Zapevalov AS, Terekhin YV (2011) Sputnikovye sredstva operativnoj okeanografii: radiolokacionnye sistemy bokovogo obzora. Ekologicheskaya bezopasnost' pribrezhnoj i shel'fovoj zon i kompleksnoe ispol'zovanie resursov shel'fa 24:308–340
Slepchuk KA, Hmara TV, Man'kovskaya EV (2017) Uroven' evtrofikacii melkovodnyh akvatorij na osnove indeksa E-TRIX po model'nym dannym. Processes in GeoMedia 1(10):462–470
The example of the location scheme of the cuts. http://innovation.org.ru/example_Black_Sea_yandex_Engl.php
Voronina NN, Inyushina NV, Mamchur NL, Kryl' MV (2016) Analiz i sopostavlenie programmnyh sredstv vizualizacii morskih prognozov na osnove raschetnyh dannyh operativnoj okeanografii po chernomorskomu bassejnu na primerah sevastopol'skogo i krymskogo regionov. Nauchnaya vizualizaciya 1(8):146–155. http://sv-journal.org/2016-1/09.php?lang=ru
Yandex.maps API. https://tech.yandex.ru/maps/

Mechanism for Sediment Transport at Tidal Estuaries Under Bore Formation

E. N. Dolgopolova

Abstract Short review of recent research of propagation of tidal waves into an estuary and examples of estuaries where bore is formed are presented. Criterion of tidal bore formation and the cross section of its inception are discussed. In estuaries under consideration the bore is formed in cross sections at distances from the mouth site larger than the convergence length. Abrupt changes of water level and pressure gradient in a flow result in intensive sediment suspension and bottom erosion at the cross section of bore inception. Bottom deposits in the estuaries and low-landing marshes take a wave form with parameters corresponding to the bore waves. Bore waves contribute to the sediment accumulation in the upper parts of the estuary and its shallowing.

Keywords Estuary · Tide · Bore · Turbulent mixing · Sediment transport

1 Introduction

Bore is a reverse positive wave formed in tidal, funnel-shaped estuaries. The appearance of bore in the estuary is mainly conditioned by the ratio of the tide to the depth of the estuary and the rate of estuary narrowing (Zyryanov and Chebanova 2015). Studies of the formation and propagation of the bore wave at different estuaries have shown that the shape of the channel cross-section, the division of the channel into several ducts, and the high variability of the longitudinal profile of the macrotidal estuary bottom play an important role in the formation of bore in the shallow estuary (Dolgopolova 2017). Sediment movement in tidal estuaries is influenced by the estuarine circulation of water, which depends mainly on the flow of river water, tides and wind (Officer 1976; Omidvar and Nikeghbali 2015). The saline current moves to the bottom towards the land and the surface flow of fresh water towards the sea. A maximum turbidity zone (MTZ), migrating along the estuary in accordance with the tidal cycle, is formed near the saline water limit distribution site into the estuary, where the bottom flow velocity is zero (zero point). The length of the MTZ and the

E. N. Dolgopolova (✉)
Institute of Water Problems RAS, Moscow, Russia
e-mail: dolgopolova@gmail.com

C. T. Olegovna (ed.), *Processes in GeoMedia—Volume I*,
Springer Geology, https://doi.org/10.1007/978-3-030-38177-6_25

concentration of suspended sediments also vary depending on the tidal phase, there are areas of erosion and accumulation of sediments and the size of the sediments is sorted at the point of sediment deposition.

As a result of the propagation of the bore wave into the estuary, an intense turbulent mixing flow, moving upstream, increases the friction stress at the bottom, arising below the front of the bore at the bottom the turbulent vortices carry saline water and sediments upstream (Furuyama and Chanson 2008). The flow is accompanied by intensive washout of some parts of the channel, transfer of sediment and accumulation of sediment upstream.

The aim of this paper is to examine the characteristics of the sediment transport mechanism in the estuary region during the wave of wavy and collapsing bore, which are necessary to assess the effect of bore on the estuary morphology and deltaic armlets in the form of estuaries.

2 Influence of Bore on Current Structure and Sediment Transport in Estuaries

2.1 Salinity of Water

Saline water, which has a higher density than river water, also has a higher suspension capacity. The range of saline waters in the estuary serves to determine the position of its peak and length *L*. The possibility of bore formation in the cross section of the mouth is estimated by the Froude number Fr_b:

$$Fr_b = \frac{U + V}{\sqrt{g \cdot A/B}}, \tag{1}$$

where *U*, *V* is the velocity of river and tidal currents, *A*, *B* is the cross-sectional area and the width of the channel before the bore arrival.

The study of tidal wave propagation in macrotidal estuaries of different rivers showed that in armlets with estuarine expansions of the Amazon (Brazil), the Hooghly armlet of the Ganges River (India), in the estuaries of Gironde and Seine (France), bore is formed above the zone of saline water penetration into the estuary, and the salinity of the water does not affect the transport of sediment by bore (Dolgopolova 2013). In the estuaries of the Qiantang and Yangtze Rivers (North Hand) (China), Mercy and Severn (UK) and Mezen, the bore carries upstream saline waters. In the Mersey estuary, the bore is formed in the site with a vertically homogeneous distribution of salinity $S = 12‰$, which facilitates the transfer of water masses from this *S* upstream to the asteroid apex (16 km). In the Arlingem estuary of the Severn river, due to the transfer of saline waters by bore, which is formed in ~16 km downstream, there are large fluctuations in the salinity of water, where in summer during the tidal cycle they reach a maximum value of 15‰, with a maximum rate of

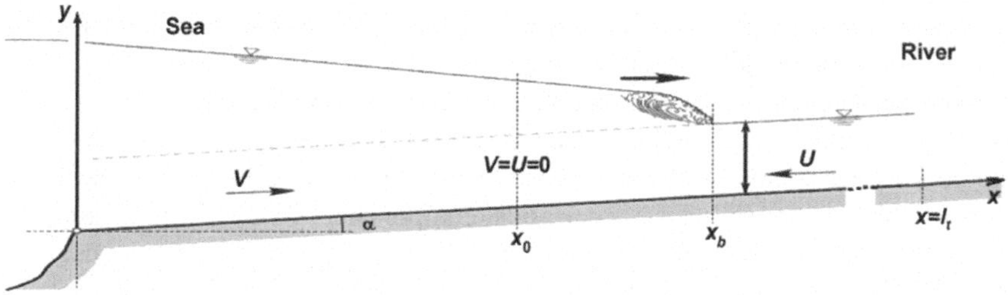

Fig. 1 Schematic diagram of the spread of bore into the estuary

change—15‰ per hour, and during the week the salinity of water S can change from 0 to 25‰. The top of the Mezen estuary, determined by the distance of saline water intrusion into the estuary L, is located near the town of Mezen at a distance of $L =$ 40 km from the estuary (Dolgopolova 2017). The maximum salinity values at the estuary site $S \sim 21$–22‰ (end of June) are observed during the period of changing the tidal current to the tidal one, and the minimum values $S \sim 17$‰ at the reverse change of currents. Formation of bore is possible at a syzygous tide in the area of 10–25 km below the city of Mezen, the bore waves are involved in the transfer of saline waters.

2.2 *Turning the Water Flow in the Estuary into a Unidirectional One*

The results of studies of bore in different estuaries indicate that the velocity of the current is directed to the apex of the estuary along the entire flow depth during the formation of bore (Dolgopolova 2017; Knezri and Chanson 2012; Tessier et al. 2016; Viero and Defina 2018). The latest numerical modeling results (Filippini 2018; Viero and Defina 2018; Uncles et al. 2006) convince us of necessity to evaluate the location of the bore formation site (Fig. 1) in order to investigate the movement of sediment in the macrotidal estuary.

2.3 *Changes in the Flow Structure of Bore Flow*

The formation of bore and its further spread into the estuary causes a sudden rise in water level. The unidirectional current formed in the tide phase moves up the estuary at a rapidly increasing rate. Immediately after a sharp rise in water level, the longitudinal component of the flow rate towards the estuary apex begins to grow and reaches $u = -1.5$ m/s in about 10 min (Tessier et al. 2016; Viero and Defina 2018).

Fluctuations in the vertical component of the flow rate v appear, reaching the value $v = 0.5$ m/s at the middle of the flow depth.

Large turbulent whirlwinds formed below the bore front are carried upstream of the bore near the bottom and can carry upstream sediment (Furuyama and Chanson 2008). The results of physical and numerical modeling show the presence of large oscillations of transverse and vertical components of velocity, indicating the intermittent currents behind the edge of bore. Evidences of turbulence focuses are obtained for wavy and collapsing bores, and the rate of turbulence production is proportional to $(\mathrm{Fr}_b - 1)^3$ (Hornung et al. 1995). Clouds of vorticity behind the bore are associated with secondary currents, and increase due to the non-prismatic form of the channel with complex bathymetry.

At formation of wavy ($1.4 < \mathrm{Fr} < 1.7$) and collapsing ($1.7 < \mathrm{Fr} < 2.1$) bores there are large fluctuations of longitudinal and transverse components of the flow rate, and the values of normal and tangential components of turbulent stresses below the edge of the overrun (Koch and Chanson 2009; Officer 1976). A fundamental difference in the structure of the flow has been found: with falling bore, large stresses are observed in the shear zone, i.e. in the area of high velocity gradients, while the reverse, eddy current was observed at the bottom; with wavy bore, large values of Reynolds' stresses are registered in the lower part of the flow up to the bottom, and the maximum normal and tangential components of turbulent stresses were observed at the wave ridges before the next ridge.

The comparison of dimensionless coefficients of turbulent longitudinal D_x and vertical D_y diffusion in the flow under the formed bore has shown that $D_x > D_y$ is an order of magnitude [for the collapsing bore $D_x = 0.12$, $D_y = 0.011$, for the undulating bore $D_x = 0.1$, $D_y = 0.018$ (Koch and Chanson 2009). Prolonged exposure of wavy bore to the free surface causes an increase in D_y. Values of vertical diffusion in the river flow vary in the range $D_y = 0.0026$–0.008 (Dolgopolova 2013), i.e. significantly less than D_y in the flow at the passage of the bore wave. The increase in turbulent pulsations of the flow rate leads to complete mixing in the bore site and facilitates the movement of sediments.

2.4 Movement of Sediment Particles During Bore Passage in the Estuary

The sudden increase in the depth and slope of the flow surface causes a change in the pressure gradient, one of the forces leading to particle displacement at the bottom. Experimental study of the regularities of the beginning of particles movement in the natural flow under the bore waves is extremely difficult, so the results of laboratory experiments with different restrictions and assumptions are used to develop the theory (Castro-Orgaz and Hager 2011; Chanson and Tan 2010; Knezri and Chanson 2012). In tidal estuaries, estuarine circulation forms a layer of fine sediment, which is periodically excreted and deposited. As a result of modeling the motion of small particles of practically neutral buoyancy, complex trajectories of suspended particles are found (Chanson and Tan 2010). Gravel particles are used to study the influence of different forces on the beginning of the bottom sediments movement in the horizontal direction during the passage of the collapsing bore (Knezri and Chanson 2012). Particles are assumed to be affected by three forces at the bottom of the flow: friction F_S, pressure gradient F_P and entrainment force F_V, the particle's velocity along the flow is described by the approximate equation:

$$m\frac{\partial U_s}{\partial t} \cong F_S + F_P + F_V \tag{2}$$

The friction force in (2) can be represented as:

$$F_S = 0.5C_d\rho(U - U_S)|U - U_S|A_S \tag{3}$$

where C_d is the hydraulic resistance coefficient, A_S is the cross-sectional area of the sediment particle perpendicular to the flow, U and U_S is the flow rate of the fluid and particles downstream.

The force of the longitudinal pressure gradient $\partial P/\partial x$ on a particle with a characteristic h_S size is:

$$F_P = -\frac{\partial P}{\partial x}A_S h_S \tag{4}$$

Let's denote the connected mass factor C_m, then the entrainment force is equal:

$$F_V = m_f C_m \frac{\partial(U - U_S)}{\partial t} \tag{5}$$

where the m_f is displacement coefficient, S—relative particle density.

Thus, at occurrence of bore at a tidal phase in a river mouth the small site is observed where velocities of river and return streams are equal to zero. Upstream, a reverse current ($u < 0$) is formed with an increasing longitudinal component of the velocity and large ripples of all three velocity components. At modeling of a wave of a bore in a tray growth of turbulent stresses in ~30 times at passage of a collapsing bore with c $Fr_b \cong 1.8$ (Knezri and Chanson 2012; Officer 1976) is registered. In this case, Reynolds' stresses are 1–2 orders of magnitude higher than the stress threshold that causes the movement of bottom and suspended sediments in the river flow. Thus, tidal bore is accompanied by a strong turbulence of the flow, which causes intensive mixing of water in the channel, which is not described by classical mixing theories.

3 Discussion of Results

3.1 Where Will the Bore Form?

The section in which a bore is formed during tidal wave propagation into the estuary, called the bore boundary, is determined by the condition $Fr_b > 1$. The use of (1) allows to determine more precisely the conditions of bore occurrence. The bore boundary shifts depending on the tidal phase: downwards at maximum tidal range and upwards along the river at comparatively lower tidal range. To describe the tidal wave motion in an estuary of length L, the narrowing of the river mouth in plan (width of the river B) and vertically (depth of the river H) is set by exponential functions of the species:

$$B = B_0 e^{-x\beta}, \quad H = H_0 e^{-x\gamma}, \tag{6}$$

where x is the distance along the estuary from the river mouth ($x = 0$ RM), B_0 and H_0 is the width and depth of the flow at $x = 0$; $\beta = 1/l_b$ and $\gamma = 1/l_h$—are the coefficients of narrowing of the channel as a result of reduction of its width and depth with the dimension of 1/length unit.

Table 1 shows the hydrological characteristics for some rivers. The coefficients β and γ in (6) show how quickly the cross-sectional area of the channel changes. At a distance of l_b in (6) from RM the channel width decreases by 35% and this parameter can be estimated by Google Earth maps. Most researchers believe that tidal slopes at the mouths of the tidal estuaries are small, and changes in depth have little effect on changes in A in (1). In mathematical and numerical flow models, the bottom is considered horizontal $\alpha = 0$ (Filippini 2018; Knezri and Chanson 2012; Omidvar and Nikeghbali 2015; Reineke 1860). However, it is shown in Dolgopolova (2017) that the parameter γ characterizing the change in depth for shallow estuaries is of

Table 1 Characteristics of flows at river mouths: average depth of estuary H, tidal value h, parameters β and γ calculated for full and small water (if available), X_b—bore formation site according to the observations

Estuaries, L (km)	Q (m^3/s)	h (m)	H (m)	β	l_b (km)	γ	Bore, x_b (km)	x_b/l_b
Mezen[a], 40	780	7.8	FW—7.4	0.0538	18.6	0.022	15–18	~1.0
			LW—1.7	0.057		0.015	35	1.9
Qiantang[a], 122	1000	8.9	FW—7.0	0.0263	38.0	0.013	50	1.3
			LW—2.3	–		0.018		
Seine[a], 50	430	7.5	8	0.0318	3.2	–	100	~31
Garonne[a], 64	600	6.0	5	0.0381	26.3	–	126	4.8

[a]Estuaries are described in detail in Dolgopolova (2017, 2013)

the same order of magnitude as β and an order of magnitude greater than γ for deep estuaries.

When considering tidal wave transformation in shallow estuaries, the ratio $\beta \gg \gamma$ is not applicable, and the change in depth should be taken into account. The distance from the RM to the site, which is reached by tidal water level fluctuations, l_t determines the position of the wellhead top. In general, $l_t > L$. At numerical modeling of tidal wave transformation in estuary for obtaining physically reasonable results a limit on distance *xb* from RS is introduced, on which a bore, $x_b/l_b \leq 3$ is formed (Filippini 2018), which is not always true (Table 1). At the mouth of the Mezen River in the nineteenth century, the bore was observed in the section of the Cape Tolstik—v. Okulovsk, 15–18 km from the RM (Reineke 1860; Tessier et al. 2016), and now the bore is fixed at the site of the v. Kamenka, 35 km from the RM (Dolgopolova 2017). A significant change in the section distance, in which bore is formed, from RM is also noted in the Seine estuary (Dolgopolova 2017). In the nineteenth century, one of the highest hogs in the world (maximum ridge amplitude up to 7.5 m) was observed in the estuary of the Seine River, extending over a distance of 80 km from the RM. The collapse of the banks and regular dredging of the Seine Estuary resulted in the disappearance of this huge bore and the appearance of a subtle low-frequency wavy bore with low-slope waves 45 km upstream of its original location.

3.2 *Influence of the Pressure Gradient on Sediment Transport*

The sudden change in the free surface of the flow during the bore wave causes changes in the longitudinal pressure gradient, which plays a significant role in the beginning of the movement of particles upstream. An assessment of the forces of F_S (3), F_P (4) and F_V (5) in the experiments showed that the force of F_V is low and its effect prevents the beginning of particle movement (Knezri and Chanson 2012). The resistance force did not affect the particle's separation from the bottom. When the force of the F_P pressure gradient weakened, the role of the resistance force increased, and as a result of the combined action of these two forces, the particles moved up the estuary for a long time. The highest pressure gradient is reached at the moment of bore passage over the particle. Most of the particles begin to move at this moment. The resistance force at the front of the bore decreases due to the slowing down of the flow, and there is a negative direction of F_S during the recirculation period caused by hydraulic jump, which plays a major role in the intensification of particle movement. The motion of the particles stops with the action on the particles of both forces.

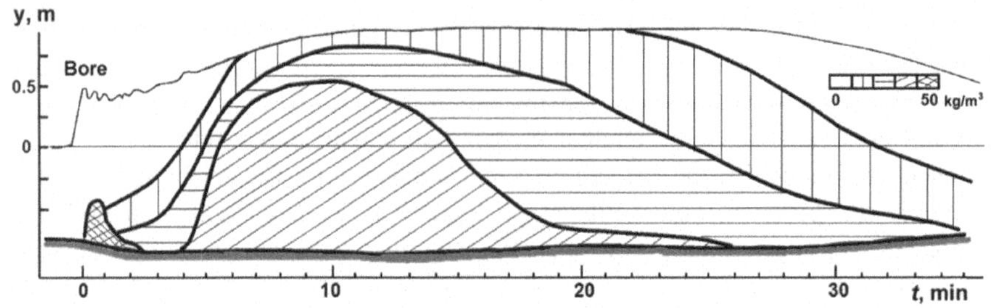

Fig. 2 Changes in suspended sediment concentration per tidal cycle from measurements in the estuary of the See River, hall. Mont-Saint-Michel (Viero and Defina 2018)

3.3 Suspended Sediment Concentration in the Estuary

The time of direction change for the entire flow is a variable, depending on the development of tidal bore. Sharp changes in velocity cause very large accelerations that reach a maximum before the bore front for longitudinal velocity and below the bore front for vertical velocity. There is intensive processing of the streambed accompanied by a very high concentration of suspended sediment near the estuary bed beneath the bore front. Measurement results in the estuary of the River C, hall. Mont-Saint-Michel showed a sharp increase of C at the bottom of the bore front site (Fig. 2) (Tessier et al. 2016; Viero and Defina 2018). The analysis of the sediment transport process has shown that an increase in pressure associated with a sharp rise in the water level, an increase in friction stress and an acceleration of the vertical component of the velocity under the front of the tidal bore lead to the filling of the estuary with sediment that is periodically excited and deposited in most of the estuary, with the exception of the upper estuary.

Sandy bottom processing during bore passage causes the formation of a sequence of sand rolls formed on the erosion surface. Modern studies of water dynamics in estuaries have shown that these shafts develop into chains reflecting the passage of tidal bore formed from sediment of different diameters. The occurrence of higher concentrations of suspended sediment from medium to maximum diameters than in estuaries without bore is a sign of tidal bore dynamics. Wave-shaped deformations of tidal deposits along the estuary in which the bore passes have a curved structure with characteristic dimensions: the amplitude of several centimeters—few decimeters, the wavelength of one to several centimeters. Some of them are located on dehydration, showing that the process of changing the surface of the bottom shifts from plastic deformations of the layered underlying surface of the bottom to liquefaction of mud and sandy fractions when the bore passes. The geometry of the curved bottom in the plane (2D) with normal axes to the flow agrees with the direction of bore propagation in the three-dimensional image.

Modeling of stable hydraulic jump in a straight-line tray with the ratio of depth H to width $B - H/B = 0.15$ (Castro-Orgaz and Hager 2011) has shown a close connection between the form of vibration of the flow surface and sand forms formed

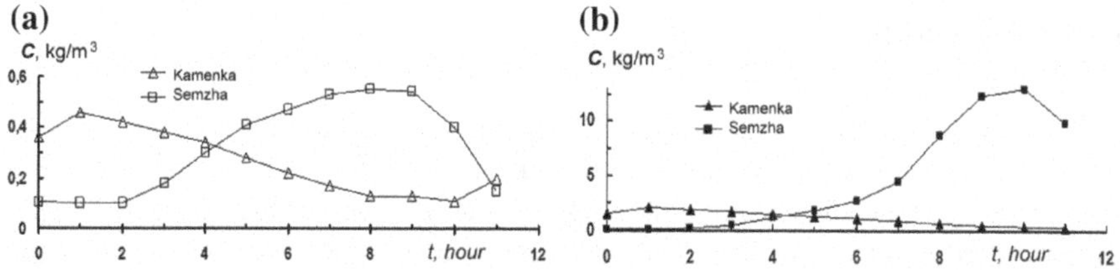

Fig. 3 Changes in surface (**a**) and bottom (**b**) flow at Semzha (6 km from RS) and Kamenka (35 km from RM) along the Mezen river estuary during the syzygous tide according to Polonsky et al. (1992). The countdown is full water, 6–8 is small water

on the bottom. Fluctuations in friction stress in the moving-bottom flow under the hydraulic jump cause the formation of bottom shapes: intensive transport of sediment under the basement of the wave and the deposition of sediment under the crest of the wave on the surface of the flow. The results of modeling (2D model based on the Boussinesq-Karman equations) explain the friction stress fluctuations under the crest and basement of the waves on the flow surface by periodic growth and reduction of the boundary layer thickness at the bottom. Thus, the relationship between the oscillations of the boundary layer thickness and the friction coefficient at the bottom is theoretically obtained. These results are consistent with the conclusions of Dolgopolova (2003), which examines the behavior of the boundary layer on the eroded bottom and the emergence of sand ridges in the natural flow at small Froude numbers. The oscillation period of the viscous sublayer thickness is described by a model based on the Navier-Stokes equations.

Studies of C change at the mouth of the Mezen River showed that in full water (FW: 0–2, 10–0 in Fig. 3a), high C values at the water surface are observed near v. Kamenka, at the site of bore formation (Polonsky et al. 1992). During the tide at Semzha (6 km from the RM), the C value at the bottom reaches 13 kg/m^3 (Fig. 3b). As a result of active mixing and high return flow rate in the estuary the water with increased C reach v. Kamenka site by the time the tidal currents change direction.

Measurements of the turbidity of water in the estuary during the passage of bore are very time consuming and therefore not numerous. A detailed study of the formation and migration of the maximum turbidity zone in the Hammer Estuary (GB) during the tidal cycle (Uncles et al. 2006) measured the vertical distribution of C in different sections of the Hammer Estuary and the estuaries of the Ouse and Trent rivers. The authors note that vertical transport at low tide longitudinal transport reduces of ~10% the transport of sediment downstream and contributes to its accumulation in upstream areas. It has also been noted that the 0.1–0.2 m tidal range generated by the syzygous tide at the upper sites of the Ouse estuary does not cause an increase in suspended sediment concentrations in the turbidity zone. However, the application of Lagrangian description of the flow of water with sediments as a two-phase medium in the modeling of bore has allowed the authors Omidvar and Nikeghbali (2015) to show the formation of a sediment accumulation zone in the form of a transverse sand bar with a steeper slope facing the bore front.

4 Conclusions

In the bore formation site, the suspended sediment concentration increases several times as a result of a sharp change in the pressure gradient and turbulent friction stress. Increased turbulent mixing in the flow under the crest of bore leads to a rapid spread of sediment to the surface of the flow and its movement upstream of the estuary. Bore waves contribute to the accumulation of sediment in the upper estuary and its shallowing. Variability in the shape of the estuary cross-sections due to the passage of bore or the channel correction leads to changes in the localization of bore formation, for example, in the estuaries of Mezen and Seine. The wave-shaped of deposits in estuaries and on coastal dehydration corresponds to friction stress fluctuations under hydraulic jump. In numerical modeling of bore propagation in estuaries, it is necessary to estimate the distance of the site where the bore is formed from the estuary and the salinity of water in this site.

Acknowledgements This study is supported by state program No. 0147-2019-0001 (registration No. AAAA-A18-118022090056-0).

References

Castro-Orgaz O, Hager WH (2011) Observations on undular hydraulic jump in movable bed. J Hydraul Res 49(5):689–692

Chanson H, Tan K-K (2010) Turbulent mixing of particles under tidal bores: an experimental analysis. J hydraul Res 48(5):641–649

Dolgopolova EN (2003) About the interaction of the flow with the eroded bottom. Water Resour 30(3):297–303

Dolgopolova EN (2013) Bore formation conditions and its impact on the transfer of saline water at river mouths. Water Resour 40(1):1–17

Dolgopolova EN (2017) Tidal waves at the mouth of the Mezen River and conditions of bore formation. Water Resour 44(6):628–640

Filippini AG et al (2018) Modeling analysis of tidal bore formation in convergent estuaries. Eur J Mech 1–14

Furuyama S, Chanson HA (2008) Numerical study of open channel flow hydrodynamics and turbulence of the tidal bore and dam-break flows. Hydraulic Model Report № CH66/08. University of Queensland, Brisbane

Hornung HG, Willert C, Turner S (1995) The flow field downstream of a hydraulic jump. J Fluid Mech 287:299–316

Knezri N, Chanson H (2012) Sediment inception under breaking tidal bores. Mech Res Commun 41:49–53

Koch C, Chanson H (2009) Turbulence measurements in positive surges and bores. J Hydraul Res 47(1):29–40

Officer CB (1976) Physical oceanography of estuaries (and associated coastal waters). A Wiley Interscience Publication, New York

Omidvar P, Nikeghbali P (2015) The study of breaking bore on movable bed using smoothed particles hydrodynamics. In: Proceedings of the 36 IAHR World Congress, The Hague, The Netherlands, pp 2559–2572

Polonsky VF, Lupachev YV, Skriptunov NA (1992) Hydrologic and morphological processes in river mouths and methods of their calculation. Hydrometeoizdat, St. Petersburg, 383p
Reineke M (1860) Hydrographic description of the northern coasts of Russia. First Part. WHITE SEA. Imprimerie Administrative de Paul Dupont, Paris
Tessier B, Furgerot L, Mouaze D (2016) Sedimentary signatures of tidal bores: a brief synthesis. Geo-Marine Lett 1–8
Uncles RJ, Stephens JA, Law DJ (2006) Turbidity maximum in the macrotidal, highly turbid Hamber Estuary, UK: flocks, fluid mud, stationary suspensions and tidal bores. Estuar Coast Shelf Sci 67:30–52
Viero DP, Defina A (2018) Consideration of the mechanisms for tidal bore formation in an idealized planform geometry. Water Res Res 1–23 (in press)
Zyryanov VN, Chebanova MK (2015) Tidal waves in estuary. Process Geosph 3(3):21–33

Intra-annual Variability of the Diurnal Water Temperature Variations on Sambian Plateau (South-Eastern Baltic Sea) in 2016

V. F. Dubravin, M. V. Kapustina, and S. A. Myslenkov

Abstract Paper presents the analysis of the thermistor chain data, installed in the South-eastern Baltic Sea on D-6 platform 22 km from the coast of the Curonian Spit at depth of 29 m. The thermistor chain consist of 9 temperature sensors 'Starmon mini' which installed at different horizons. Recording interval is 1 min and accuracy of sensors is ±0.025 °C. A series of data with averaging step of 1 h for the period from 06.08.2015 to 31.12.2016 is considered, and only data for 2016 have been used for the correct assessment of time components. Dispersion, harmonic and correlation analysis of the data was carried out. Diurnal water temperature variability along the vertical is analyzed. On the basis of the author's time series model, the intra-daily, diurnal and seasonal component contributions to the total time variability of the water temperature for 2016 are estimated. The intra-annual variability of the contribution and amplitude of the diurnal variability component is also estimated.

Keywords Baltic Sea · Coastal zone · Water temperature · Thermistor chain · Thermal structure of water · Diurnal water temperature variations · Synoptic variability · Variance analysis · Harmonic analysis · Correlation analysis

1 Introduction

The physical processes taking place in the World Ocean and the Atmosphere lead to the formation of heterogeneities in the distributions of different parameters on different spatial and temporal scales, from micro- to macro-scale. It should be noted that there are many models of both spatial (Ozmidov 1965; Kamenkovich et al. 1982) and temporal variability (Monin 1969; Cargo and Rankova 1980) of the Earth's climatic system. Sufficiently detailed descriptions of such classifications can be found

V. F. Dubravin · M. V. Kapustina (✉) · S. A. Myslenkov
P.P. Shirshov Oceanological Institute RAS, Moscow, Russia
e-mail: kapustina.mariya@yandex.ru

S. A. Myslenkov
Hydrometeorological Research Centre of Russian Federation, Moscow, Russia

Lomonosov Moscow State University, Moscow, Russia

C. T. Olegovna (ed.), *Processes in GeoMedia—Volume I*,
Springer Geology, https://doi.org/10.1007/978-3-030-38177-6_26

in Lappo et al. (1990), Dubravin (2014). These heterogeneities arise under the influence of processes with their own time scales, which allows to correlate spatial and time scales (Lappo et al. 1990; Gulev et al. 1994). This task has its difficulties due to the different spatial and temporal scales of processes in the Atmosphere and the Ocean.

In the Baltic Sea (Mediterranean, inland or intercontinental) (Terziev et al. 1992), as a result of atmospheric western transport, river runoff and water exchange with the North Sea (Antonov 1987), a stable density stratification is created, which determines the hydrological and hydrochemical regimes of the sea. This leads to the formation of a two-layer structure of hydrological and hydrochemical parameters: the surface structural zone (SZ) or the active layer (AS) and depth. In the seas, the thermal structure of DC is characterized by a great variety (Dietrich 1962). Previously, according to Rozhkova et al. (1992), for the Baltic Sea within the DC the upper quasi homogeneous layer, seasonal thermocline and cold intermediate layer (CRM) are allocated, later in Dubravin (2017) the upper part of the main thermocline was added. At the same time, according to our calculations, the depth of thermal convection gradually increases from 25 to 55 m in the west of the sea to 65–85 m in its central part, to 160–175 m in the Åland Sea and to 85–105 m in the Gulf of Bothnia, and the average depth of thermal convection for the Baltic States is about 100 m (Dubravin 2017). For the eastern slope of the Gdansk DC basin, DC penetrates to the bottom (70–75 m) in most of the Russian sector of the South-Eastern Baltic Sea (Dubravin 2017). According to estimates in Krechik and Gritsenko (2016) for the coastal zone of this sector, the depth of convection is 40–45 m.

This study analyzes the unique data of the thermistor chain installed on the platform D6, in order to obtain quantitative indicators of the daily, synoptic and seasonal component of the temporal variability of the thin thermal structure with the help of harmonic, dispersion and correlation methods of analysis (Brooks and Carousers 1963).

2 Material and Method

This work uses data from the garland of temperature sensors installed on a fixed offshore platform D6 in 22 km from the coast of the Curonian Spit (Myslenkov et al. 2017a), the depth at the installation site is 29 m. Nine Starmon mini temperature sensors are located on the horizons: −1 (uppermost horizon is in the air at a height of about 90 cm from the water and was used as air temperature data); 1; 3 [In the event of severe disturbance, the first 3 sensors were periodically in the air and in the water. Emissions in the data have been filtered out based on a sharp increase in temperature variance (Myslenkov et al. 2017b)]; 5; 8; 10; 13; 24 and 28 m. Measurements are made in increments of 1 min and are accurate to ±0.025 °C. A number of data were analyzed with an averaging step of 1 h for the period from 06.08.2015 to 31.12.2016. Observation time of the 4th time zone (UTC+04).

Earlier in Myslenkov et al. (2017b) the initial analysis of the thermocosmic data for a shorter period (06.08.2015–25.05.2016) was performed, where different time scales of the phenomena—from inertial to seasonal—were identified.

In the present paper, the time series model (Dubravin 2014) was used: the initial series (IR) is composed of short-period (SP) (high-frequency) and long-period (LF) (low-frequency) variability. The control consists of irregular intra-day variability (IIDV), regular intra-day variability (RT, the IDV of hydrometeorological elements according to Cargo and Rankova (1980), Lappo et al. (1990), Gulev et al. (1994), Woods (1980) refers to synoptic variability, and according to Kamenkovich et al. (1982), Monin (1969), as in our case, to mesoscale, i.e. T = 1 day—the boundary between mesoscale and synoptic components) and synoptic variability (SI). The DP is composed of irregular intra annual variability (IRAV), regular seasonal variations (RSV) and inter-annual variability (IAV):

$$IR = SP + DP \tag{1}$$

$$IR = IIDV + IDV + SI + IRAV + RSV + IAV \tag{2}$$

After the IR was smoothed by the moving monthly averaging

$$DP = IRAV + RSV + IAV, \tag{3}$$

and the series members were assessed within the time series model (Lappo et al. 1990). Similar time series models were proposed by R. V. Abramov. For short-period variability, the general dispersion is considered, which consists of the dispersions of daily, synoptic and "random" changes (Abramov 1982). For long-period variability, the general dispersion is considered, which consists of annual variations, long-term changes and "random" changes (Abramov 1988). The RSV is the monthly averaged values of the multiyear SP series, IAV is calculated from the multiyear series after averaging for each year and IRAV as a residual value.

Subtracting the DP from the IR will give a short-period variability of the control

$$SP = IR - DP, \tag{4}$$

or

$$SP = IIDV + IDV + SI \tag{4a}$$

Smoothing over IR with a daily period makes it possible to obtain the implementation of the SI (Dubravin 2014), the averaging of the IR for each hour—IDV, IIDV is a residual component.

However, taking into account the small length of the studied series (17 months) and large inter annual differences between the second half of 2015 and 2016, the data from January 1 to December 31, 2016 had to be used to correctly assess the time components. In this case, the DP will only consist of Sezh, i.e.

$$IR = IIDV + IDV + SI + RSV \tag{5}$$

3 Results

Structure of time series of water temperature

Using our proposed time series model, we analyzed the data on water temperature variability on the D6 platform (discreteness of the series is 1 h) (Table 1).

The specific contribution (relative share) of the SP dispersion to the dispersion of the initial series for water temperature (T_w) at different levels varies from 5% at the horizon of 8 m to 14% at the bottom. At the same time, the contribution of DP (represented only by RSV) to the total dispersion varied from 95% at the horizon

Table 1 Top row—dispersion (°C^2), bottom—relative proportion (%) of water temperature variability at D6: short-period (regular daily run—IDV, irregular intra-day—IIDV and synoptic SI components) and long-period (regular seasonal run—RSV) 2016. Dispersion is shown in bold (°C^2)

Element depth	Dispersion				
	General	Short-period			Long-period
		IDV	IRAV	SI	RSV
TW (1 m)	**34.35**	**0.009**	**0.11**	**2.16**	**32.07**
	100	0.025	0.33	6.29	93.35
TW (3 m)	**33.99**	**0.005**	**0.09**	**2.09**	**31.81**
	100	0.014	0.27	6.13	93.58
TW (5 m)	**33.09**	**0.002**	**0.09**	**1.86**	**31.13**
	100	0.006	0.28	5.62	94.09
TW (8 m)	**31.10**	**0.001**	**0.14**	**1.40**	**29.56**
	100	0.003	0.44	4.50	95.06
TW (10 m)	**30.08**	**0.001**	**0.09**	**1.49**	**28.50**
	100	0.002	0.30	4.96	94.73
TW (13 m)	**29.09**	**0.0003**	**0.09**	**1.68**	**27.31**
	100	0.001	0.31	5.79	93.90
TW (24 m)	**26.08**	**0.0004**	**0.13**	**2.23**	**23.72**
	100	0.001	0.50	8.55	90.95
TW (28 m)	**22.15**	**0.0008**	**0.24**	**2.82**	**19.10**
	100	0.004	1.07	12.72	86.21

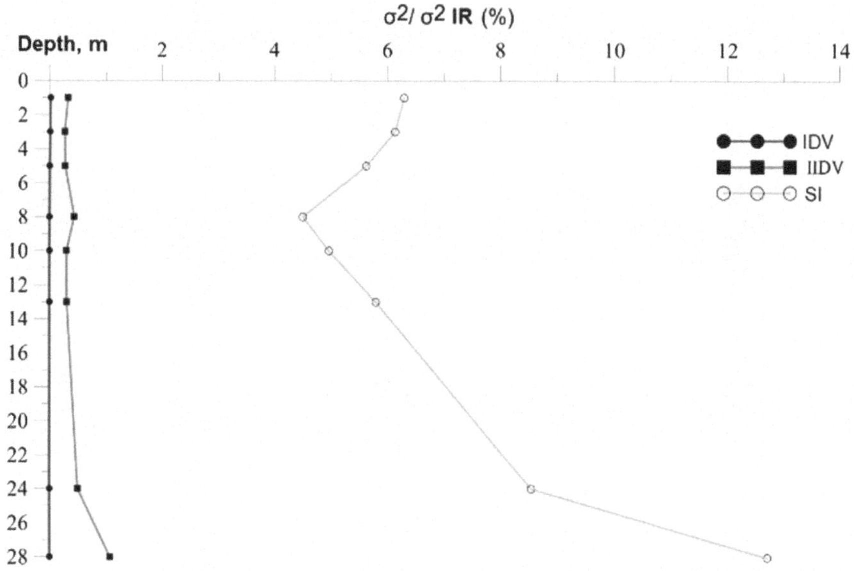

Fig. 1 Vertical distribution of the specific contribution (%) of short-period components, on D6 station, averaged for 2016

of 8 m to 86% at the bottom, respectively. The specific contribution of IDV to the total dispersion at all horizons is minimal (≤0.02%). The maximum contribution to the dispersion of the initial series of short-period components is made by SI, first decreasing with a depth of 6–4% by 8 m, and then increasing to 13% at the bottom (Fig. 1).

4 Regular Daily Variation

Figure 2 and Table 2 give an idea of the nature of the daily vertical water temperature variability on average station D6 for 2016. As can be seen, with the depth of IDV does not remain constant. The most correct average annual daily variability, characterized by the prevalence of a daily wave with high stability, horizons 1, 3, 5, 10 and 13 m have the quota of daily harmonic T_w is $q_I = 0.92$–0.99 and somewhat smaller—horizons 8 and 24 m (its contribution decreases to $q_I = 0.79$ and $q_I = 0.74$). However, at the 28 m horizon, the dominance of the daily harmonic is replaced by the dominance of the half-day harmonic ($q_I = 0.43$; $q_{II} = 0.48$) (Table 2).

If we consider the daily variability averaged on a monthly basis, it is as follows. At the horizon of 1–5 m, the dominance of daily harmonics is observed throughout the year, with the maximum quotas observed in July (decreasing from $q_I = 0.99$ to $q_I = 0.94$), and the minimum in December (increasing from $q_I = 0.58$ to $q_I = 0.61$). At the horizon of 8–28 m the daily wave prevails for most of the year with the

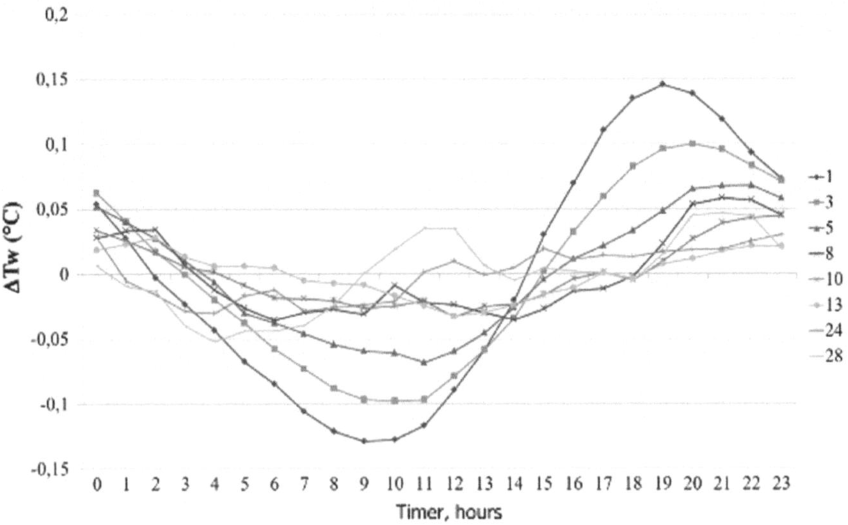

Fig. 2 Average anomalies of regular daily water temperature variations at D6 station for 2016

maximum contribution in February ($q_I = 0.94$–0.97), but in the warm season (May–September) two or three months at these depths the dominance of the daily harmonic is replaced by half a day ($q_{II} = 0.53$–0.85) (Table 2).

For the 2016 average IDV T_w on the horizon of 1 m maximum occurs at 19 h (GMT+4), and at least at 0900 h; growth time—10 h, fall time—14 h. This corresponds to the overall change of water and air temperature near the surface of the Baltic Sea (Dietrich 1962). Average for 2016 IDV T_a at a height of 1 m is characterized by a daily harmony ($q_I = 0.88$) with a maximum of 16 h, a minimum of 8 h, a growth time of 8 h and a daily variation of 1.2 °C.

With the depth of the onset of extrema in the average annual IDV shifts to a later time of day so that the maximum at a depth of 10 m is noted at 23 h, the minimum—falls on 12 h, the growth time has not much changes 10–11 h.

At the 13 m horizon, the maximum is shifted by 2 h, at least at noon, and the growth time has increased to 14 h. At a depth of 24 m, the maximum returned to 23 h, the minimum moved to 04 h, the growth time reached 19 h. Near the bottom (28 m), since the average annual IDV is dominated by a half-day wave (Fig. 2), there are two maxima (11–12 and 21 h) and two minima (4 and 19 h). The range of average daily variations decreases from 0.27 °C (near the sea surface) to 0.08 °C (by 10 m) and 0.06 °C (by 13 and 24 m). At the bottom, with the predominance of semi-daily harmonic, the range of daily variation slightly increases to 0.10 °C (Fig. 2).

Table 2 Average and limit values of harmonic constants of regular daily temperature of water, averaged monthly for 2016, at D6 station

Element depth	Harmonics						A_I/A_{II}	A_0
	I (daily wave)			II (semi-daily wave)				
	Amplitude	Phase	Quota	Amplitude	Phase	Quota		
T_W (1 m)	0.13	146.9	0.968	0.02	81.5	0.031	5.6	10.27
	0.49	163.2	0.985	0.06	81.5	0.132	9.2	
	0.01	18.4	0.581	0.00	123.5	0.012	2.4	
T_W (3 m)	0.10	135.2	0.984	0.01	88.5	0.015	8.0	10.23
	0.33	158.0	0.984	0.05	161.9	0.107	35.4	
	0.01	18.4	0.594	0.01	142.9	0.001	2.4	
T_W (5 m)	0.07	125.9	0.992	0.00	133.0	0.001	35.4	10.15
	0.24	156.6	0.937	0.09	156.3	0.150	8.5	
	0.01	18.8	0.610	0.00	155.9	0.012	2.1	
T_W (8 m)	0.04	115.5	0.791	0.02	142.0	0.143	2.4	9.91
	0.17	159.9	0.945	0.27	169.7	0.744	8.5	
	0.01	172.5	0.162	0.00	170.0	0.013	0.5	
T_W (10 m)	0.03	110.5	0.916	0.01	135.4	0.049	4.3	9.72
	0.18	162.4	0.955	0.12	163.2	0.605	8.4	
	0.01	160.5	0.316	0.00	151.9	0.013	0.7	
T_W (13 m)	0.02	88.1	0.937	0.00	126.4	0.014	8.3	9.45
	0.18	178.3	0.948	0.06	137.8	0.682	8.1	
	0.01	15.8	0.091	0.00	128.8	0.014	0.4	
T_W (24 m)	0.02	166.8	0.744	0.01	117.5	0.075	3.2	8.54
	0.21	161.5	0.959	0.08	134.8	0.531	5.8	
	0.01	163.9	0.154	0.00	167.2	0.029	0.5	
T_W (28 m)	0.03	174.2	0.431	0.03	137.7	0.480	0.9	7.95
	0.15	167.1	0.973	0.31	135.2	0.847	7.2	
	0.01	176.8	0.081	0.00	168.2	0.019	0.3	

5 Seasonal Variability of the Specific Contribution of the Daily Rate

On Fig. 3 the intra-annual variability of IDV and the vertical scale of water temperature is presented at D6 station. The greatest similarity of the intra-annual variability of the specific contribution of the IDV is found for the horizons 1, 3 and 5 m (closeness of the connection r = 0.86–0.97) (Table 3), characterized by the first two harmonics (decreasing with the depth) $q_I = 0.65$ to $q_I = 0.51$ and increasing from $q_{II} = 0.17$ to $q_{II} = 0.29$). With yearly maxima decreasing with depth, primary ($\sigma^2_{IDV}/\sigma^2_{IR}$ from 3.75 to 1.71%), coming in March–April and secondary ($\sigma^2_{IDV}/\sigma^2_{IR}$ from 1.83

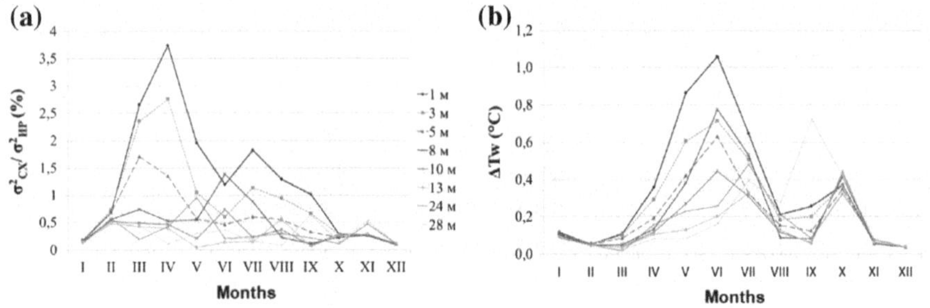

Fig. 3 Average for 2016 intra-annual variability of the specific contribution (**a**) and range (**b**) of the daily component of water temperature at D6 station

Table 3 Correlation matrixes between the intra-annual variability of the specific contribution of the daily component of water temperature on the horizons of D6 station in 2016

Depth (m)	1	3	5	8	10	13	24	28
1	1.00	0.97	0.86	0.42	0.36	0.26	0.29	−0.09
3		1.00	0.95	0.33	0.39	0.42	0.20	−0.04
5			1.00	0.39	0.51	0.52	0.15	0.10
8				1.00	0.75	−0.03	0.10	0.10
10					1.00	0.51	0.07	0.12
13						1.00	0.02	−0.05
24							1.00	0.24
28								1.00

to 0.60%)—in July and lows: main $\sigma^2_{IDV}/\sigma^2_{IR} = 0.10\%$, coming in December and secondary (decreasing $\sigma^2_{IDV}/\sigma^2_{IR}$ from 1.20 to 0.46%), falling in June (Fig. 3a). The horizons of 8 and 10 m ($r = 0.75$) are slightly less similar (Table 3). With yearly highs decreasing in depth, the main ($\sigma^2_{IDV}/\sigma^2_{IR}$ from 1.39 to 0.73%) coming in June and the secondary ($\sigma^2_{IDV}/\sigma^2_{IR}$ from 0.74 to 0.53%) in February–March and lows: main $\sigma^2_{IDV}/\sigma^2_{IR} = 0.11\%$ (December) or $\sigma^2_{IDV}/\sigma^2_{IR} = 0.08\%$ (September) and secondary (descending $\sigma^2_{IDV}/\sigma^2_{IR}$ from 0.53 to 0.20%), falling on April–May. At a depth of 13 m, the intra-annual variability of the specific contribution of IDV is characterized by the prevalence of a semi-annual wave ($q_{II} = 0.35$), with maximums in August ($\sigma^2_{CX}/\sigma^2_{ИР} = 0.56\%$) and February ($\sigma^2_{CX}/\sigma^2_{ИР} = 0.48\%$) and the lows of May ($\sigma^2_{IDV}/\sigma^2_{IR} = 0.04\%$) and September ($\sigma^2_{IDV}/\sigma^2_{IR}$ from 0.07%), on the horizons of 24 or 28 m the prevalence of a quarter-year wave ($q_{IV} = 0.55$ and 0.36), with major highs in May ($\sigma^2_{CX}/\sigma^2_{ИР} = 0.95\%$) or September ($\sigma^2_{CX}/\sigma^2_{ИР} = 0.64\%$), and secondary—in February ($\sigma^2_{IDV}/\sigma^2_{IR} = 0.58\%$) or November ($\sigma^2_{IDV}/\sigma^2_{IR} = 0.53\%$) and the main lows in December ($\sigma^2_{IDV}/\sigma^2_{IR} = 0.11\%$) or August ($\sigma^2_{IDV}/\sigma^2_{IR} = 0.09\%$) and the secondary lows in March ($\sigma^2_{IDV}/\sigma^2_{IR} = 0.20\%$) or December ($\sigma^2_{IDV}/\sigma^2_{IR} = 0.12\%$) (Fig. 3a). The

Table 4 Correlation matrices between the intra-annual variability of the daily component of water temperature on the horizons of station D6 for 2016

Depth, m	1	3	5	8	10	13	24	28
1	1.00	0.99	0.96	0.92	0.81	0.49	0.67	0.14
3		1.00	0.97	0.93	0.87	0.60	0.75	0.20
5			1.00	0.96	0.92	0.63	0.70	0.16
8				1.00	0.94	0.71	0.78	0.19
10					1.00	0.86	0.81	0.26
13						1.00	0.88	0.42
24							1.00	0.40
28								1.00

ratio between maximum and minimum values in the seasonal variability of the specific contribution of IDV is as follows: first, it decreases from 38 (1 m) to 9 times (10 m), then increases to 13 times (13 m) and decreases again to 7 times (28 m).

Let's consider the intra-annual variability of water temperature IDV range vertically on D6 station. The greatest resemblance of the intra-annual variability of the IDV range is found in the horizons of 1–8 m (closeness of the connection r = 0.92–0.99) (Table 4) with maximum annual progress: The main [from ΔTw = 0.63 °C (5 m) to ΔTw = 1.06 °C (1 m)] occurring in June and the secondary [from ΔTw = 0.35 °C (5 m) to ΔTw = 0.42 °C (8 m)] occurring in October and the minimums are the main, coming in December [ΔTw = 0.04 °C (1–8 m)] and secondary in August [ΔTw = 0.21 °C (1 m) to ΔTw = 0.18 °C (3 m)] or September [ΔTw = 0.12 °C (5 m) to ΔTw = 0.08 °C (8 m)] (Fig. 3b).

The horizons of 10–24 m differ somewhat less (closeness of connection $r = 0.81$–0.88) (Table 4), with maximums: the main one in June–July ($\Delta T_w = 0.45$–0.48 °C 10 and 24 m) or in October $\Delta T_w = 0.43$ °C 13 m), secondary one in October ($\Delta T_w = 0.44$ °C 10 m and $\Delta T_w = 0.33$ °C 24 m) or in July ($\Delta T_w = 0.33$ °C—13 m) and minimums: the main one coming in March ($\Delta T_w = 0.02$–0.04 °C) and the secondary one coming in September ($\Delta T_w = 0.06$–0.08 °C). The lowest similarity with other horizons is found in the horizon of 28 m (closely related r = 0.14–0.42) (Table 4), where the annual variability of the IDV range is characterized by the dominance of the annual harmonic ($q_I = 0.64$), with the maximum ($\Delta T_w = 0.72$ °C) occurring in September, and the minimum ($\Delta T_w = 0.02$ °C) in March (Fig. 3b).

The relation between maximum and minimum values of seasonally variable IDV range with depth varies as follows: decreasing from 28 (1 m) to 18 times (5 m), then increasing to 20 times (8 m) and decreasing again to 12 times (10–13 m), and increasing again to 28 times (24 m) and 30 times (28 m).

6 Conclusion

The analysis of the structure of time series of water temperature T_W on the Curonian-Sambian Plateau (South-East Baltic Sea) in 2016 showed the following.

The specific contribution (relative fraction) of SP to the initial series dispersion for T_W vertically varies from 5% at the horizon of 8 m to 14% at the bottom. The maximum contribution to the dispersion of the initial series of short-period components is made by SI, first decreasing with a depth of 6–4% by 8 m, and then increasing to 13% at the bottom. The specific contribution of IDV to the total dispersion at all horizons is minimal ($\leq$0.02%).

The stability of the regular daily flow of T_W vertically into the IR dispersion is confirmed by the results of the harmonic and correlation analysis.

The maximum contribution of the intra-annual variability of the daily component for T_W in the vertical direction does not remain constant: it decreases in depth and shifts from March–April (in the upper 5 m) to May–June (on the intermediate horizons) and September (at the bottom). The minimum specific contribution of IDV from December (1–8 m) is shifted to September (10–13 m) and August (near the bottom). The ratio between the maximum and minimum values in the seasonal variability of IDV decreases from 38 (1 m) to 7 times (28 m), with an intermediate maximum of 13 times (13 m).

The maximum variation of the intra-annual variability of the daily component for T_W in the upper layer of 1–10 m is observed in June, shifting to July (24 m) and September (near the bottom). The minimum range of IDV from December (1–10 m) is shifted to February (13–28 m). The ratio between the maximum and minimum values in the seasonal variability of IDV range first decreases from 28 (1 m) to 12 times (10–13 m), with an intermediate maximum of 20 times (8 m), and then increases to 28 times (24 m) and 30 times (at the bottom).

Acknowledgements The authors would like to thank OOO LUKOIL-KMN for the opportunity to install the thermistor chain and for assistance in organizing the monitoring processes on the D6 platform. The work is carried out within the framework of the state assignment of the Institute of Oriental Studies of the Russian Academy of Sciences (topic No. 0149-2019-0013).

References

Abramov RV (1982) Variability of meteorological fields in the equatorial Atlantic. In: Variability of the Ocean and Atmosphere in the equatorial Atlantic (studies under the PGEP program). Science, Moscow, pp 211–241

Abramov RV (1988) About the evolution of the Icelandic minimum. 52 c. Dep. in VINITI No. 7294—B88

Antonov AE (1987) Large-scale variability of the hydrometeorological regime of the Baltic Sea and its influence on the fishery. Hydrometeoizdat, L., 248 p

Brooks C, Carousers N (1963) Application of statistical methods in meteorology. Hydrometeoizdat, L., 416 p

Cargo GV, Rankova EY (1980) Structure and variability of the observed climate. In: Northern Hemisphere air temperature. Hydrometeoizdat, L., 72 p

Dietrich G (1962) General oceanography. Institute of Engineering Lit., Moscow, 465 p

Dubravin VF (2014) Evolution of hydrometeorological fields in the Baltic Sea. Capros, Kaliningrad, 438 p

Dubravin VF (2017) Evolution of thermohaline structure of Baltic Sea waters. Pen Publishing House, Moscow, 438 p

Gulev SK, Kolinko AV, Lappo SS (1994) Synoptic interaction of the Ocean and Atmosphere at middle latitudes. Hydrometeoizdat, St. Petersburg, 320 p

Kamenkovich VM, Koshlyakov MN, Monin AS (1982) Synoptic vortices in the ocean. Hydrometeoizdat, L., 264 p

Krechik VA, Gritsenko VA (2016) Thermal structure of the coastal waters of the Baltic Sea near the northern coast of the Kaliningrad region. Process Geomedia 5:77–84

Lappo SS, Gulev SK, Rozhdestvensky AE (1990) Large-scale thermal interaction in the ocean-Atmosphere system and energy-active regions of the World Ocean. Hydrometeoizdat, L., 336 p

Mamaev OI (1995) About the spatial-temporal scales of the oceanic and atmospheric processes. Oceanology 35(6):805–808

Monin AS (1969) Weather forecast as a task of physics. Science, Moscow, 184 p

Myslenkov SA, Krechik VA, Soloviev DM (2017a) Water temperature analysis in the coastal zone of the Baltic Sea on the basis of the satellite data and thermistor chain measurements. In: Proceedings of the Hydrometeorological Research Center of the Russian Federation, vol 364, pp 159–169 (in Russian)

Myslenkov SA, Krechik VA, Bondar AV (2017b) Daily and seasonal variability of water temperature in the coastal zone of the Baltic Sea according to the data of the thermistor chain on the platform D-6. Ecol Syst Instrum 5:25–33

Ozmidov RV (1965) About some peculiarities of the energy spectrum of the Ocean turbulence (in Russian). DAN USSR 2161(4):828–832

Rozhkova VA, Smirnova AI, Terzieva FS (eds) (1992) Hydrometeorological conditions. Project “Seas of the USSR” T. III. Baltic Sea. Come on, let’s go. 1, Under edn. Hydrometeoizdat, St. Petersburg, 450 p

Terziev FS, Rozhkov VA, Smirnov AI (eds) (1992) Hydrometeorology and hydrochemistry of the USSR seas. Volume III Baltic Sea. Issue I. Hydrometeorological conditions, Hydrometeoizdat, St. Petersburg, 449 p

Woods JD (1980) Do waves limit turbulent diffusion in the ocean? Nature 288(5788):219–224

On the Analogy of Pore Pressure and Temperature Effects in the Elastic and Elastoplastic Problems Solutions

K. B. Ustinov and E. V. Stepanova

Abstract The problems of thermoelasticity and poroelasticity with boundary conditions are considered. The disjoint formulation of thermoelasticity problem is compared with the poroelasticity problem in two different formulations with different choice of basic kinematic variables. An analogy between the equations for both cases when considering full strains and full stresses is shown. The ratio allowing to overwrite the thermal expansion coefficient is obtained so that for the transition of the equations for poroelasticity problem into the equations for thermoelasticity problem formal replacement of pressure by temperature is sufficient.

Keywords Theory of elasticity · System of equations · Isotropy · Boundary conditions · Thermoelasticity · Poroelasticity · Total deformations

1 Introduction

During the formation of the basic principles of the thermoelasticity and poroelasticity theories, these two areas developed separately from each other. The theory of thermoelasticity originates with the works of Duhamel (1838, 1837) and Neumann (1885), which appeared in the mid-nineteenth century. The development of the poroelasticity theory Terzaghi (1925) has engaged among the one of the forerunners at the beginning of the twentieth century. Lately the ideas were developed in works of Biot (1935, 1941). The systems of equations written taking into account the effects of thermo- or poroelasticity, in contrast to the classical theory of elasticity equations, contain at least one additional variable—temperature or pressure, respectively. The existence of an analogy between the two theories has been repeatedly noted and emphasized, but for the transition of equations into each other simple overwriting of symbols is not enough. The completeness clarification and identification of the limitations of this analogy requires more detailed consideration of the basic equations of both theories.

K. B. Ustinov · E. V. Stepanova (✉)
A. Ishlinsky Institute for Problems in Mechanics of the RAS, Moscow, Russia
e-mail: step@ipmnet.ru

C. T. Olegovna (ed.), *Processes in GeoMedia—Volume I*,
Springer Geology, https://doi.org/10.1007/978-3-030-38177-6_27

2 Thermoelasticity Equations

One of the methods of recording the system of decoupled thermoelasticity equations for small deformations corresponds to the possibility of representing complete deformations in the form of decomposition into elastic ε_{ij}^{E}, obeying Hooke's law, and inelastic ε_{ij}^{P} components (De Wit 1973) of deformations:

$$\varepsilon_{ij}^{T} = \varepsilon_{ij}^{E} + \varepsilon_{ij}^{P} \tag{1}$$

It should be emphasized that in general case it is assumed that the conditions of compatibility are complied for only complete deformations, and not for the decomposition components (1) separately, which corresponds to the assumption of the existence of the initial state in which both elastic and inelastic deformations are zero. And it is regarding this initial state inelastic deformations were caused first, and then the appearance of elastic deformations is allowed to remove the incompatibility. It is also important that the inelastic distortion in the decomposition (1) can be generally represented as a sum of symmetric, which corresponds to inelastic deformation, and asymmetric, corresponding inelastic rotation, parts (De Wit 1973). However, within the constraints of the considered problems it is possible to consider only symmetrical part of inelastic distortion.

Thus, it is the total deformation associated with the displacement vector u_i by means of Cauchy relations (tensile stresses and strains are considered positive)

$$\varepsilon_{ij}^{T} = \frac{1}{2}\left(u_{i,j} + u_{j,i}\right) \tag{2}$$

In the presence of isotropy of both elastic properties and thermal expansion coefficients, inelastic deformations are associated with temperature change ΔT by relation

$$\varepsilon_{ij}^{P} = \frac{\alpha_T}{3}\Delta T \delta_{ij}, \tag{3}$$

where $\alpha_T/3$—coefficient of linear thermal expansion, α_T—coefficient of volumetric thermal expansion. Whereas elastic deformations ε_{ij}^{E} are related to stresses σ_{ij} by the generalized Hooke law

$$\sigma_{ij} = 2\mu\varepsilon_{ij}^{E} + \lambda\varepsilon_{kk}^{E}\delta_{ij} \tag{4}$$

Here, $\lambda,\ \mu$ are the constants of Lamé. The components of the stress tensor σ_{ij} are bridged by the equilibrium equations

$$\sigma_{ij,i} + f_j = 0, \tag{5}$$

where f_j are components of the volume forces density vector, these components are usually absent in standard formulations of thermoelasticity problems.

Equations (1)–(5) in case with given temperature field form a closed system, which should be supplemented only by boundary conditions, which are usually stresses or displacements conditions. The temperature distribution is usually determined from the solution of stationary

$$T_{,ii} = 0, \tag{6}$$

or non-stationary heat conduction problem with certain boundary conditions.

For the convenience of further calculations, it is convenient to convert the system (1)–(5) to a form similar to the Lame equations system. To this end, elastic deformations ε_{ij}^E should be expressed in terms of the displacement vector u_i and the temperature change ΔT should be expressed by Eqs. (1)–(3)

$$\varepsilon_{ij}^E = \frac{1}{2}\left(u_{i,j} + u_{j,i}\right) - \frac{\alpha_T}{3}\Delta T\delta_{ij} \tag{7}$$

The obtained relations, when substituting in the equations of Hooke's law (4), lead to

$$\sigma_{ij} = \mu\left(u_{i,j} + u_{j,i}\right) + \lambda u_{k,k}\delta_{ij} - K\alpha_T\Delta T\delta_{ij}, \tag{8}$$

where $K = \lambda + 2\mu/3$ is the triaxial compression module. The obtained ratios are Duhamel-Neumann proportions.

Substitution of (8) in the equilibrium equation (5) gives

$$\mu\left(u_{i,ji} + u_{j,ii}\right) + \lambda u_{k,ki}\delta_{ij} - K\alpha_T\Delta T_{,i}\delta_{ij} + f_j = 0, \tag{9}$$

which transits into

$$(\mu + \lambda)u_{i,ji} + \mu u_{j,ii} - K\alpha_T\Delta T_{,j} + f_j = 0 \tag{10}$$

The relations (8)–(10) point out that thermoelasticity can be described by classical Lamé equations if the volume forces are formally supplemented by the value

$$\Delta f_j = -K\alpha_T\Delta T_{,j} \tag{11}$$

The system of Eqs. (10) and (6) in combination with boundary conditions for stresses (displacements) and temperatures is frequently mentioned as a system of decoupled thermoelasticity equations. The decoupling is due to the fact that temperature T is included in Eq. (10) as an external variable, its distribution is not affected by the determined displacement. Therefore, the solution of the problem is split in two stages, and the problem of thermal conductivity (6) is solved at the first stage. Then the obtained temperature distribution is used to solve the elasticity problem (10).

The obtained relations (10) and (6) are assumed to be compared with the poroelasticity problem formulation. When summarizing and taking into account the time effects, this system can be written in the form [see Novackiy (1975)]:

$$(\mu+\lambda)u_{i,ji}+\mu u_{j,ii}-K\alpha_T\Delta T_{,j}+f_j-\rho\ddot{u}=0 \tag{12}$$

$$T_{,ii}-\frac{1}{\kappa}\dot{T}-\frac{K\alpha_T T_0}{\lambda_T}\dot{u}_{i,i}=0 \tag{13}$$

Here ρ is the density, κ is the thermal diffusivity coefficient, λ_T is thermal conductivity coefficient, T_0 is the reference temperature, the dots over variables denote partial derivatives in time.

The system (12)–(13) becomes coupled: temperature T and displacements u_j are included in both equations. Additional components in these equations may arise when considering dissipative processes. If the last component in (13) is neglected, which is legitimate for a number of problems, the system ceases to be coupled. It should be pointed out that coupling is due to the presence of time derivatives.

3 Poroelasticity Equations

For the further comparison with the problem of thermoelasticity it is possible to choose the equations of poroelasticity in the form proposed by Khristianovich and Zheltov (1955). The postulate that the stresses are transmitted partially through the ground skeleton and through the fluid pressure, but the magnitude of the induced deformation is influenced only by the stresses transmitted through the ground skeleton is the feature.

The stress components acting on the ground skeleton are called effective stresses s_{ij} as a result of the considered problem of grain interactions, the connection of total stresses σ_{ij} with effective stresses s_{ij} and pore pressure p is obtained

$$\sigma_{ij}=s_{ij}-\alpha_P p\delta_{ij} \tag{14}$$

Here $0\le\alpha_P\le 1$ is a parameter of the pore space structure: the higher the value α_P the more permeable the rock. Sometimes $\alpha_P=1$ is assumed to simplify calculations. The general form of (14) in this approach allows to decompose the full stress tensor into two components: the effective stress tensor and the pressure with the reverse sign. In case $\alpha_P\ne 1$, the decomposition of the full stress tensor into the effective stress tensor and the component $-\alpha_P p\delta_{ij}$ can be made.

The pressure p distribution, if it is not pre-defined, is deduced from the filtration problem solution in a stationary

$$p_{,ii} = 0 \tag{15}$$

or non-stationary form with corresponding boundary conditions.

Only the full stresses obey the equilibrium equations (5), which should be taken into account to obtain a complete system of equations, whereas substitution of (14) in (5) for effective stresses gives

$$s_{ij,i} - \alpha_P p_{,j} + f_j = 0 \tag{16}$$

For effective stresses the Hooke's law is

$$s_{ij} = 2\mu\varepsilon_{ij}^E + \lambda\varepsilon_{kk}^E\delta_{ij} \tag{17}$$

In this case, there are no inelastic deformations, and therefore the elastic deformations are equal to the total, defined as the symmetric part of the displacement gradient

$$\varepsilon_{ij}^E = \varepsilon_{ij}^T = \frac{1}{2}\left(u_{i,j} + u_{j,i}\right) \tag{18}$$

When substituting of (18) into the equations of Hooke's law (17), it is obtained

$$s_{ij} = \mu\left(u_{i,j} + u_{j,i}\right) + \lambda u_{k,k}\delta_{ij} \tag{19}$$

Relations (14) with the committed changes and the relation between full and effective stresses can be written in the form

$$\sigma_{ij} = \mu\left(u_{i,j} + u_{j,i}\right) + \lambda u_{k,k}\delta_{ij} - \alpha_P p\delta_{ij} \tag{20}$$

In two cases, after substituting the expressions for total stresses (20) into the equilibrium equations (5) or the expressions for effective stresses (19) into the equilibrium equations for effective stresses (17), an analogue of Lamé equations is obtained

$$(\mu + \lambda)u_{i,ji} + \mu u_{j,ii} - \alpha_P p_{,j} + f_j = 0 \tag{21}$$

4 Biot's Poroelasticity Equations

The other way to solve the poroelasticity problem is to choose another set of basic kinematic variables. For example, it is possible to take the total strain, determined by the displacement vector using the relations (2), and the change in the volume of the pore space V as the main kinematic variables. The power variables (generalized forces) corresponding to these generalized displacements are stress σ_{ij} and pore pressure p.

The further construction of the solution is associated with the determination of the connection between kinematic and force variables. One of the methods is a direct indication of the linear relationship between the kinematic and corresponding force variables in the general form. The other is to assign the expression for the energy as an arbitrary quadratic form including introduced kinematic variables with subsequent variation. Both methods lead to the following constitutive relations:

$$\sigma_{ij} = \mu'(u_{i,j} + u_{j,i}) + \lambda' u_{k,k}\delta_{ij} - k_2 V \delta_{ij}, \tag{22}$$

$$p = -k_2 u_{k,k} + k_1 V, \tag{23}$$

where the equality of the k_2 coefficients follows from the assumption of the energy potential existence. The constant values indicated as μ', λ' are different from the constant values μ, λ in the previous equations. The representation form of (23) presumes the positive pressure to correspond negative stresses, hence the right side has negative sign.

The total stresses σ_{ij} included in the governing relation (22) have to obey the equilibrium equations (5). The pressure p remains an independent variable and can be found from the filtration equation (15) with corresponding boundary conditions. Substitution of (23) in (22) allows to exclude the change of the pore space volume V

$$\sigma_{ij} = \mu'(u_{i,j} + u_{j,i}) + \lambda' u_{k,k}\delta_{ij} - \frac{k_2^2}{k_1} u_{k,k}\delta_{ij} - \frac{k_2}{k_1} p\delta_{ij} \tag{24}$$

In this form, the governing ratio corresponds to the variant obtained above of the poroelasticity equations (20) with the following replacement

$$\mu = \mu', \lambda = \lambda' - k_2^2/k_1, \alpha_P = k_2/k_1 \tag{25}$$

The substituting of (24) and (25) into the equilibrium equations (5) leads to the relations in the form similar to (21), which allows to reduce both formulations of the poroelasticity problem to the similar notation.

5 Analogy Between Poroelasticity and Thermoelasticity

The equation of poroelasticity (21) coincides with the equation of thermoelasticity (10), up to the notations if we put

$$K\alpha_T = \alpha_P$$

$$\Delta T = p \tag{26}$$

There is also a complete analogy for the equations reflecting the full stresses connection to full deformations (20) and (8).

The equations in obtained form allow to use the solutions of thermoelasticity problems for poroelasticity and vice versa. That respectively facilitates the use of software packages that contain the thermoelasticity solvers, for calculations in poroelasticity problems based on formal replacement of the temperature by pressure and overwriting the thermal expansion coefficient according to (26).

An important point to pay attention to is that in the problems of poroelasticity, boundary conditions are usually set for effective stresses s_{ij}, while the boundary conditions should be set for full stresses while using developed solutions for thermoelasticity or software packages. The stresses transition can be easily made with relation (14).

Several difficulties are to be faced in the critical states evaluation (strength calculations) and, especially, in the inelastic deformation calculations using the theory of plastic flows. Difficulties stem from the fact that the criteria for the transition to non-elastic state and plastic potentials, according to the concept used, should be written for the effective stress. The expression of those criteria through the full stress via (14) leads to the additional parameter p manifestation. When using the considered analogy, the flow criteria parameters and plastic potentials become formally dependent on the pressure (or its analog—temperature).

The complexity of the transition from the poroelasticity to the thermoelasticity theories arise due to differences in the construction methods of these theories. When considering the phenomenon of thermoelasticity subjects of decomposition are elastic and inelastic parts to the deformation field, and when considering the poroelasticity the stress field subjected to decomposition into parts corresponding to the effective stresses and fluid pressure. Therefore, it is necessary to require the correspondence of total stresses and strains for both cases to preserve the complete analogy in the form of the final equations. It should be emphasized once again that all the above considerations and calculations relate exclusively to the tasks in decoupled formulation.

Returning to an overall formulation of the Biot's poroelasticity problem, it is necessary to pay attention to Eq. (23), linking the pore pressure and the change in the pore space volume. Indeed, if the dependence of the change in the pore space volume with permeability is considered

$$\kappa = \kappa(V) \tag{27}$$

then the equation of filtration should be denoted as

$$\left[\kappa(V)p_{,i}\right]_{,i} = 0 \tag{28}$$

The complete system of equations thus includes three scalar equations (21), (23) and (28) with five variables: three components of the displacement vector u_i, change in the pore space volume V and pressure p.

If the dependency (27) can be neglected and $\kappa(V) = \kappa_0$ is put, the system does not become coupled. In this case, the value V can be determined after finding the solution to the problem. If the dependency (27) is significant, the problem becomes coupled and sequential solution for filtration and elasticity does not bring the necessary accuracy of calculations. In this case, it is possible to consider a larger number of degrees of freedom, complementing the kinematic variables with V value, and also to solve filtration and elasticity problems by successive approximations.

It is also necessary to point out the nonlinearity of the coupled problem by virtue of (28) and its coupling in the stationary formulation, which leads to significant differences from the coupled thermoelasticity problem, as a result of which the analogy is also violated.

Acknowledgements Authors express their deep gratitude to V. G. Markov for a number of ideas on the presentation of the material, as well as useful tips, comments, which allowed to better understand the similarities and differences in the considered theories.

The work is carried out under financial support of the Russian Science Foundation, project No. 16–11–10325–п.

References

Biot MA (1935) Le problème de la consolidation des matières argileuses sous une charge. Ann Soc Sci Brux Ser B 55:110–113

Biot MA (1941) General theory of three-dimensional consolidation. J Appl Phys 12:155–160

De Wit R (1973) Theory of disclinations: IV. Straight disclinations. J Res Nat Bur Stand A Phys Chem 77A(5):607–658

Duhamel JMC (1837) Second mémoire sur les phénomènes thermo-mécaniques. J l'École Polytech 15, Cahier 25:1–57

Duhamel JMC (1838) Mémoire sur le calcul des actions moléculaire développées par les changements de tempé rature dans les corps solides. Mém Présentées Diver Savant l'Acad Sci 5:440–498

Khristianovich SA, Zheltov YP (1955) O gidravlicheskom razryve neftenosnogo plasta. Izv AN SSSR OTN 5:3–41 (in Russian)

Neumann F (1885) Vorlesung über die Theorie des Elasticität der festen Körper und des Lichtäthers. Teubner, Leipzig

Novackiy V (1975) Teoriya uprogosti. Mir, Moscow, 256 p (in Russian)

Terzaghi K (1925) Erdbaumechanik auf Bodenphysikalischer Grundlage. F. Deuticke, Leipzig

Total Duration of Arctic Air Outbreaks Over the Azov-Black Sea Region in 2000–2018

A. V. Kholoptsev, S. A. Podporin, and L. E. Kurochkin

Abstract In this paper we consider the impact of Arctic air outbreaks on the weather and the performance of certain sectors of the economy of the Azov-Black Sea region. Special emphasis is made on risks these phenomena can pose to the operation of transport complexes and marine navigation. We propose a new technique to detect Arctic outbreaks that enables more reliable and precise identification of these phenomena. Its performance is tested by means of estimation of the total duration change trends for the outbreaks that occurred in the region in question during 2000–2018 period. Unlike the conventional detection technique, the proposed one is shown to significantly lower the probability of overlooking the phenomena in question, as well as minimize cases of other weather phenomena being mistaken for the Arctic air outbreaks. The trial use of the proposed technique has revealed that during the period from 2000 to 2018 the total duration of Arctic air outbreaks in the Azov-Black Sea region had a tendency to increase in January. The opposite trends have been observed in April. The most affected part of the region has been found to be the North-Western region of the Sea of Azov while the least affected is the West and the South-West of the Crimean Peninsula. Based on the results obtained, we conclude that the operational conditions for marine transport complexes functioning in the area tend to somewhat worsen in winter, whereas during other seasons they become more favorable.

Keywords Arctic air outbreaks · Total duration · Azov-Black Sea region · Detection · Safety of navigation

1 Introduction

Arctic air outbreaks (hereafter referred to as AAO) are fast streams of cold and dry air from the high latitudes of the Northern Hemisphere penetrating temperate and subtropical latitudes. Since such air is very dense, the atmospheric pressure (P_{at}) and

A. V. Kholoptsev · S. A. Podporin (✉) · L. E. Kurochkin
Sevastopol State University, Sevastopol, Russia
e-mail: sapodporin@sevsu.ru; s.a.podporin@gmail.com

A. V. Kholoptsev
N.N. Zubov State Oceanographic Institute, Moscow, Russia

C. T. Olegovna (ed.), *Processes in GeoMedia—Volume I*,
Springer Geology, https://doi.org/10.1007/978-3-030-38177-6_28

air temperature (T_a) exhibit rise on all parts of the Earth's surface above it. Due to the abovementioned features of AAOs, the hydrometeorological conditions and the state of ecosystems in most extratropical regions of the world, as well as the safety of navigation carried out therein, are significantly affected.

In continental regions, AAOs tend to intensify winter frosts and summer droughts. Over oceanic regions, these phenomena may manifest themselves as strong cold northerly winds.

The impact of AAOs on the weather in a particular region is the stronger the longer the total duration (TD) of their existence during the year or month in question. The values of this indicator should be taken into account when planning the functioning and development of numerous sectors of the economy, including the transport sector. Therefore, the assessment of trends and average values of this indicator for the current period corresponding to a particular region is a pressing problem of physical geography, oceanography, meteorology, as well as the operation of transport complexes.

The solution of this problem is of theoretical and practical interest for marine and oceanic regions with intense transport communications. The Azov-Black Sea region, which includes the Sea of Azov and the Black Sea with the adjacent territories, is one of them. The main waterways of the region connect the ports of Russia, Ukraine, Bulgaria, Romania, Turkey, Georgia and exits from the mouths of the navigable rivers—the Danube and Dnieper—with the Kerch Strait and the Bosporus. Taking into account the features of the regional geography, we have adopted the meridians of 25° W and 47.5° W as its Western and Eastern boundaries.

Despite the presence of numerous meteorological stations in the region, the current trends in TD of AAOs over different parts of the region, as well as the corresponding average values of these indicators have not been revealed before. The main reason for this is the imperfection of the methods used to detect these processes.

Arctic air outbreaks are large-scale fast-progressing processes that lead to formation of high atmospheric pressure bands. These bands link the Arctic and Subtropical anticyclones with each other. In order to detect these phenomena, direct information from all weather stations located in the corresponding sector of the Northern Hemisphere or corresponding weather maps where this information is plotted must be available. Therefore, the existing method of AAO detection is based on detection of such baric heterogeneities (Dzerdzeevskiy et al. 1975) by analyzing short-term and daily weather maps or the results of the corresponding reanalyses of P_{at} (Central Institute of Forecasting 1946).

Such sources contain information that enables identification of the aforementioned high P_{at} bands (and, therefore, AAO-events) in any region of the world. The information is available for any date for the last few decades.

At the same time, the mentioned bands of high P_{at} may have different origins (Central Institute of Forecasting 1946; Kholoptsev et al. 2018; Salby 1996). In such cases the use of the mentioned technique leads to errors—so-called "false alarms". Another situation when an AAO remains undetected is also possible (it is referred to as "an AAO miss"). As a result, the current views on the trends in the TD of AAOs in the Azov-Black Sea region need further development.

In this research we address the following problems:

1. development of an alternative and more precise technique for detection of Arctic air outbreaks, and
2. revealing trends in variations of their total duration for different sectors of the Azov-Black Sea region in 2000–2018.

2 Material and Research Methods

The method of detecting Arctic air outbreaks is based on three stages. At the first stage, the origins of errors arising from the detection of AAOs by use of the method described in (Dzerdzeevskiy et al. 1975) are considered. The second stage involves development of an alternative AAO detection technique that takes into account current understanding of the responses to these processes that may be observed in the P_{at}, H_P and T_a fields. The third stage compares frequencies of correct detection of AAOs propagating over the Crimean Peninsula obtained by use of the traditional and the proposed techniques.

The methodology for the trend assessment includes calculation of TD of AAOs crossing different sectors of the Azov-Black Sea region in a particular month, as well as assessment of trends for interannual variations of these indicators for the period from 2000 to 2018.

As source data, the results of NCEP/NCAR reanalysis of daily means of P_{at}, T_a at the absolute height corresponding to 1000 hPa geopotential (H_{1000} level), as well as the H_{700} and H_{300} levels, for 2000–2018 period were used (Mankin 2011). The data from the same reanalysis for 1972–1988 period was used to carry out the verification of the proposed technique.

3 Results of the Study and Their Discussion

In accordance with the detection technique described in (Dzerdzeevskiy et al. 1975), the fact of an AAO-event is established if a band of high P_{at} linking the Arctic anticyclone with a subtropical one is present on a daily (or short-term) weather map. At the same time, P_{at} values are considered "high" if they exceed 1015 hPa level.

This level is chosen empirically, partly because the corresponding isobar is often displayed on weather maps compiled for the Northern Hemisphere. Hence, computers are not required to use this technique, which was implemented long before their invention.

In addition to frequent false alarms and AAO "misses", this technique features another significant drawback. Its implementation is possible only with the participation of a person with sufficient experience in dealing with weather maps. As a result, "anthropogenic" errors may arise when AAOs are being looked for in such a way. It is also worth noting the high labor intensity of the above methodology.

In order to prevent an AAO detection technique from anthropogenic errors, full automation is required. The probability of "misses" and false alarms can be reduced if known properties of these processes are properly accounted for when choosing the decision making criteria.

It is an established fact that at the onset of an AAO, the high P_{at} band is formed almost 24 h in advance, and, as a result, in any of its internal regions

$$P_{aT}(\lambda, \varphi, t) - P_{aT}(\lambda, \varphi, t-1) > 0,$$

where λ is longitude, φ is latitude, t is date.

The Arctic air involved in an AAO is not only dense, but also cold, hence the "response" to this process occurs not only in the P_{at} field, but also in the T_a field, where it manifests itself as a band of low T_a extending from the Arctic to the subtropics (ftp://ftp.cdc.noaa.gov/Datasets/ncep.reanalysis.Dailyavgs/surface). The geographical position of this band on weather maps nearly coincides with the high P_{at} band (Central Institute of Forecasting 1946). In any of its internal areas, the following relationship is true:

$$T_B(\lambda, \varphi, t) - T_B(\lambda, \varphi, t-1) > 0.$$

It is also known that the air involved in an AAO moves southward at a relatively high speed. This behavior arises because the baric gradient vector in the area where this current is formed has a significant westerly zonal component (Kholoptsev et al. 2018).

In addition, if the speed of this air is high, its flow is most likely turbulent, and, consequently, a turbulent boundary layer is inevitably formed above it (Salby 1996). The air involved in an AAO is partially mixed with the surrounding air, resulting in an increase in the density of the mixture. Therefore, within this layer the corresponding values of H_p exhibit rise as well. The turbulent boundary layer above the AAO is capable of reaching the tropopause at least.

This suggests that the following criterion applicable for detection of AAO-events including those crossing the Icelandic Low will be feasible: the fact of an AAO occurrence can be established if there is a band-shaped baric heterogeneity linking the Arctic anticyclone with a subtropical one, in all points of which the following conditions are met:

$$\begin{aligned}
&P_{aT}(\lambda, \varphi, t) - P_{aT}(\lambda, \varphi, t-1) > 0, \\
&T_B(\lambda, \varphi, t) - T_B(\lambda, \varphi, t-1) < 0, \\
&P_{aT}(\lambda, \varphi, t) - P_{aT}(\lambda + 2.5^\circ, \varphi, t) > 0, \\
&H700(\lambda, \varphi, t) - Hcp700(\varphi, t) > 0, \\
&H300(\lambda, \varphi, t) - Hcp300(\varphi, t) > 0,
\end{aligned}$$

Table 1 Frequencies of P_{at} rises and T_a drops observed at weather stations during the periods corresponding to AAO-events detected by use of the proposed technique (A) and the traditional technique (B)

Weather station	Technique		Weather station	Technique	
	A	B		A	B
Black Sea	0.95	0.73	Yalta	0.92	0.75
Evpatoria	0.96	0.77	Alushta	0.95	0.74
Sevastopol	0.96	0.68	Theodosius	0.90	0.76
Chersonesus lighthouse	0.95	0.69	Kerch	0.93	0.72
Mysovoye	0.94	0.72	Opasnoe	0.95	0.70

where Hcp(φ, t) is the arithmetic mean of Hp(λ, φ, t) values for longitude values λ corresponding to some latitude φ and date t.

To automatically identify all possible trajectories of arctic air incursions into lower latitudes, we applied the full search technique that guaranteed us the result for any AAO configurations.

To estimate frequencies of correct AAO detection by use of the traditional and the proposed techniques, the frequencies with which P_{at} rise and T_a drop were recorded on the corresponding dates at Crimean weather stations were analyzed. Information on the actual results of P_{at} and T_a variations was obtained from observations carried out during 1972–1988 period. The archived observation data was provided by the Sevastopol branch of the FSBI "N.N. Zubov State Oceanographic Institute" (Kholoptsev and Kurochkin 2018). The verification results are shown in Table 1.

As can be seen from Table 1, the frequency of detection of local signs of AAOs at the specified weather stations during the considered time periods by use of the proposed detection technique is significantly higher than that shown by the traditional method. Therefore, a conclusion can be made that the frequency of "misses" and "false alarms" when using the proposed technique is significantly lower.

As a result of the solution of the second problem by use of the proposed methodology, all AAO-events that occurred over the Azov-Black Sea region in the period 01.01.2000–31.12.2018 were identified. For each event, the dates and longitude ranges within which they occurred were established.

When calculating AAO TDs for the sectors of 2.5° width, all the events that occurred above them during the specified time interval were considered. From the estimates of annual and monthly TDs of AAOs obtained in this way, the corresponding time series were formed for all the sectors of the Azov-Black Sea region.

Table 2 shows the values of angular coefficients of linear trends of interannual variations in the monthly TD of AAOs obtained for different sectors of the region in 2000–2018 period.

As follows from Table 2, variations in mean monthly TD of AAOs in January of 2000–2018 period exhibited rising trends in any sector of the Azov-Black Sea

Table 2 Angular coefficients of linear trends of interannual variations in the TD of AAOs for each month and each sector of the Azov-Black Sea region in 2000–2018 period

No. λ	25	27.5	30	32.5	35	37.5	40	42.5
1	0.02	0.08	0.27	0.13	0.09	0.15	0.21	0.19
2	−0.03	0.07	0.14	0.13	0.19	0.16	0.18	0.26
3	−0.03	0.03	−0.00	−0.03	0.04	0.15	0.17	0.18
4	−0.05	−0.09	−0.12	−0.17	−0.22	−0.27	−0.23	−0.25
5	−0.12	−0.18	−0.25	−0.29	−0.24	−0.14	−0.18	−0.05
6	−0.23	−0.06	−0.02	0.06	0.04	0.00	0.09	0.05
7	0.07	−0.06	−0.02	0.00	−0.09	−0.05	−0.09	−0.04
8	−0.10	−0.15	−0.07	−0.06	0.01	−0.03	0.04	−0.06
9	0.12	0.12	−0.02	−0.04	−0.03	0.05	0.08	0.09
10	0.02	−0.06	−0.14	−0.16	−0.21	−0.25	−0.19	−0.18
11	0.14	−0.10	−0.15	−0.29	−0.29	−0.30	−0.28	−0.23
12	−0.07	−0.10	−0.11	−0.05	−0.03	0.00	0.10	0.00

region. The processes under consideration exhibited similar trends in February in all sectors of the region, except for the sector corresponding to 25°.

Decreasing trends for TD of AAOs were revealed in April in all sectors of the region. In October, the same trends were established for all sectors, except for 25°.

Over the year, the trends in the decline in TDs in 2000–2018 period have been identified for all sectors except for the most Eastern one. For the winter season, the tendencies for the increase of AAO TD prevail almost in the whole region. The spring and autumn seasons are characterized by decreasing trends everywhere. The same is true for summer months for all the sectors except those corresponding to the Southern coast of Crimea.

Since the cyclonic activity in the Azov-Black Sea region in winter has been on its rise (as shown in Dzerdzeevsky 1968), it is obvious that wind patterns during this season have become more variable. It is worth noting that due to the higher frequency of AAOs (and, therefore, of strong cold northerly winds), the risks of ice accretion on marine vessels tend to rise. As a result, winter operation of water transport in the region becomes more of a challenge. In particular, this is the case for the ferries servicing the Kerch Strait crossing, for which a strong lateral wind can significantly complicate mooring operations. Operational risks are also increasing for other vessels with limited seaworthiness.

In contrast, in spring-summer season, operating conditions are becoming more favorable for the local fleet, since the frequencies of cyclonic events as well as Arctic air outbreaks exhibit decreasing trends at this time of the year.

4 Conclusions

The proposed technique for detection of Arctic air outbreaks has been shown to operate more effectively by reducing the probability of arising errors and showing more robust performance compared to the traditional method. Its accuracy has been verified by cross-checking of the results obtained by its use with the actual monitoring data collected from weather stations.

In 2000–2018 period, total durations of Arctic air outbreaks over the entire Azov-Black Sea region have exhibited rising trends in January, whereas the opposite trends have been revealed in April. On average, the most intensive decline in total durations of these phenomena has been established in the sectors corresponding to the Western and South-Western Crimea. These indicators, on the contrary, have shown rising trends in the North-Eastern part of the Sea of Azov and the South-Eastern part of the Black Sea. The research results indicate that the influence of Arctic outbreaks on weather conditions in the region under study in the period from March to November has weakened.

Conditions for water transport operation in the Azov-Black Sea region tend to worsen in winter period, whereas the tendencies for their improvement are observed in other seasons.

References

Central Institute of Forecasting (1946) Synoptic meteorology. M.; L.: Hydromethyzdate, Come on, let's go. 21, 80 p

Database. Results of the reanalysis of mean daily values of atmospheric pressure and air temperature at the absolute height of geopotential 100 hPa. [Electronic resource]. ftp://ftp.cdc.noaa.gov/Datasets/ncep.reanalysis.Dailyavgs/surface

Dzerdzeevskiy BL, Kurganskaya VM, Vitvitskaya ZM (1975) Typification of circulation mechanisms in the Northern hemisphere and characterization of synoptic seasons (in Russian). In: Proceedings of the main directorate of hydrometeorology of the USSR Council of Ministers. Sir

Dzerdzeevsky BL (1968) Circulation mechanisms in the atmosphere of the Northern Hemisphere in the XX century. Materials of meteorological research published by IG USSR Academy of Sciences and International Studies. Geoffiz. Committee of the Presidium of the USSR Academy of Sciences, 240 p

Kholoptsev AV, Kurochkin LE (2018) Influence of hydrometeorological conditions on navigation safety in the Black and Azov Seas. In: Proceedings of the Crimean Academy of Sciences, pp 70–80

Kholoptsev AV, Podporin SA, Kurochkin LE (2018) Arctic invasions and trends in meteorological conditions in the oceanic regions of the Moderate Climate Belt. In: Proceedings of the international scientific conference "Science: Discoveries and Progress". III international scientific conference. Czech Republic, Karlovy Vary-Russia, Moscow, September 28–29, pp 450–460

Mankin M (2011) Atmospheric dynamics. Cambridge University Press, London, 512 p

Salby ML (1996) Fundamentals of atmospheric physics. Academic Press, New York, 560 p

Variations of Seismic Scaling Before Strong Earthquakes

I. R. Stakhovsky

Abstract The results of multifractal analysis of microseismicity preceded several strong earthquakes are presented. It is shown that singularity spectra of seismic fields before these earthquakes got widen during last 2–2.5 years before the main shock. The effect of seismic field $f(a)$—spectrum widening before the strong earthquakes is estimated quantitatively. Physical interpretation of this effect is discussed.

Keywords Earthquake · Multifractal measure · Singularity spectrum · $f(a)$—spectrum widening

1 Introduction

The process of preparation of a strong lithospheric earthquake from the physical point of view can be considered as a transition of lithosphere material in the focal region from a weakly non-equilibrium to a strongly non-equilibrium state, which in the conditions of the Earth's bowels ends in the formation of a main rupture (Stakhovsky 2017). Thus, in terms of the theory of dissipative structures of Prigozhin (1980) seismic structures can be defined as dissipative structures of the seismogenerating system. It is repeatedly shown that the theory of multifractal measures is an adequate mathematical apparatus for describing the spatial organization of dissipative structures of different physical nature (Mandelbrot 1989).

Multifractal, i.e. scale invariant structure of the seismic process was detected and confirmed in many works on analysis of seismic catalogues (Geilikman et al. 1990; Hooge et al. 1994). It is obvious that in the process of preparation of a strong earthquake, i.e. in the process of transition of the lithosphere material from a weakly non-equilibrium to a strongly non-equilibrium state, the multifractal structure of the seismic process can change. The search for these changes in seismic data is important for the issues of monitoring the state of the seismic generating environment.

The main problem of such a search today is the lack of representativeness of the seismic data themselves. The threshold for modern seismological catalogues is

I. R. Stakhovsky (✉)
Schmidt Institute of Physics of the Earth RAS, Moscow, Russia
e-mail: stakhov@ifz.ru

C. T. Olegovna (ed.), *Processes in GeoMedia—Volume I*,
Springer Geology, https://doi.org/10.1007/978-3-030-38177-6_29

usually quite high, which leads to a significant lack of data for scaling analysis of seismic structures. However, in some cases, changes in the scaling structure of spatial distributions of seismicity before strong earthquakes can be detected and quantified. The number of such cases is still small, but the general features of these changes are supported by the conclusions of the theory of dissipative structures, which makes it possible to attribute them to the physically justified prognostic signs of strong earthquakes.

This paper presents examples of changes in the structure of spatial distributions of seismicity before several strong earthquakes, obtained by multifractal analysis of data on small-scale seismicity. Such changes are the seismic field $f(a)$-spectra widening in recent 2–2.5 years before the main shocks of strong earthquakes. The quantitative characteristics of this effect are estimated and its physical interpretation is presented.

2 The Counting Technique

In all cases described below, changes in the structure of spatial distributions of seismicity before strong earthquakes are detected by comparing multifractal spectra calculated from two samples of data from the corresponding seismological catalogues. In the vicinity of the strong earthquake epicenter a rectangular polygon is chosen, and the center of the polygon coincided with the epicenter of the earthquake. The first sample contained seismic data from this test site for the period ending approximately two years before the main shock. The second sample contained data on seismic events that occurred between the last event of the first sample and the main shock of a strong earthquake. Both samples contained an equal number of events, which allowed the calculation results to avoid errors caused by different representativeness of the initial data.

For the purposes of multifractal analysis (see Stakhovsky (2017) for a detailed description of the theory), the territory of the polygon is covered by a renormalizable scale grid of square boxes (cells). Spatial distributions of epicenters of micro earthquakes are modeled by measures P, the substance of which in the i-th boxes of the scale grid is estimated by normalization:

$$p_i = N_i / N_0 \tag{1}$$

where N_i is the number of events in the i-th box, N_0—total number of events in the sample, i—order index of the boxes. In the future, the measures P will be called seismic fields. Catalogue data on event coordinates are used in the calculations of p_i.

By renormalizing the grid, the cumulant-generating function is determined $\tau(q)$:

$$\tau(q) = \lim_{r \to 0} \frac{\ln\left(\sum_{i=1}^{N} p_i^q(r)\right)}{\ln(1/r)} \tag{2}$$

where r—grid box size (scale), N—total number of non-empty grid boxes, q—measure moment order. The Legendre transform was used to move to the coordinates a and $f(a)$:

$$a = -\frac{d}{dq}\tau(q) \tag{3}$$

$$f(a) = aq + \tau(q) \tag{4}$$

where a is the singularity index, $f(a)$ is the spectrum of singularities. Using the multiplier method (Chhabra and Jensen 1989), the $f(a)$-spectrum can be constructed without the Legendre transformation:

$$a(q) = \lim_{r \to 0} (\ln r)^{-1} \sum_{i=1}^{N} \bar{p}_i(q, r) \ln p_i(r) \tag{5}$$

$$f(a(q)) = \lim_{r \to 0} (\ln r)^{-1} \sum_{i=1}^{N} \bar{p}_i(q, r) \ln \bar{p}_i(q, r) \tag{6}$$

where

$$\bar{p}_i(q, r) = p_i^q(r) / \sum_j p_j^q(r) \tag{7}$$

The choice of a specific algorithm for calculating the $f(a)$-spectrum is determined by the nature of the source data.

3 The Joshua Tree Earthquake

The earthquake Joshua Tree has happened in California, USA on April 23, 1992, coordinates of the epicenter are 33.96° N, 116.32° W, depth of hypocenter is 12 km, magnitude $M \approx 6.1$. The seismic data used for the calculations are taken from the Southern California Earthquake Data Center Catalogue (SCEDC). The size of the polygon in the vicinity of the epicenter of this earthquake is chosen 80 × 80 km. The average depth of micro-events preceding this earthquake was 8.7 km and their magnitudes were in the range $1.0 \leq M \leq 4.8$. The first data sample covered the

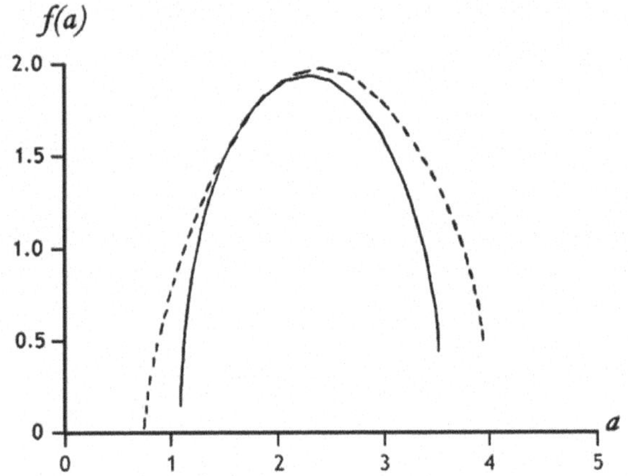

Fig. 1 Changes in the seismic field scaling before the 1992 Joshua Tree earthquake

time period 01.04.1988 to 31.03.1990, the second sample covered the time period 01.04.1990 to 23.04.1992. Both samples contained 1713 events.

Figure 1 clearly demonstrates the change of seismic field scaling in the vicinity of Joshua Tree earthquake epicenter. A solid line shows the $f(a)$-spectrum of the seismic field constructed from the first data sample, a dotted line shows the $f(a)$-spectrum of the seismic field constructed from the second sample. As we can see, in the last two years before the Joshua Tree earthquake, the seismic field $f(a)$-spectrum has undergone a significant widening.

Let's estimate the effect of $f(a)$-spectrum widening quantitatively. For this purpose we shall use the functional δS:

$$\delta S = S^{**} - S^{*} = \int_{a^{**}_{\min}}^{a^{**}_{\max}} f^{**}(a)da - \int_{a^{*}_{\min}}^{a^{*}_{\max}} f^{*}(a)da \tag{8}$$

where S—spectrum opening characteristic, the upper index in the form of two stars refers to the spectrum of the seismic field preceding the strong earthquake (i.e. to the second data sampling), the upper index in the form of one star refers to the spectrum of the earlier seismic field (i.e. to the first data sampling), and $a_{\max}$, $a_{\min}$- maximum and minimum values of the singularity indices respectively. Functions $f(a)$ in the area of integration are continuous, smooth and limited, i.e. the widening parameter δS can always be estimated numerically. In this case $\delta S = 0.94...$.

4 The Iceland 17.06.2000 Earthquake

This earthquake in Iceland has happened on 17.06.2000, the epicenter was located at the point with coordinates 63.97° N, 20.37° W, which is approximately in 80 km to

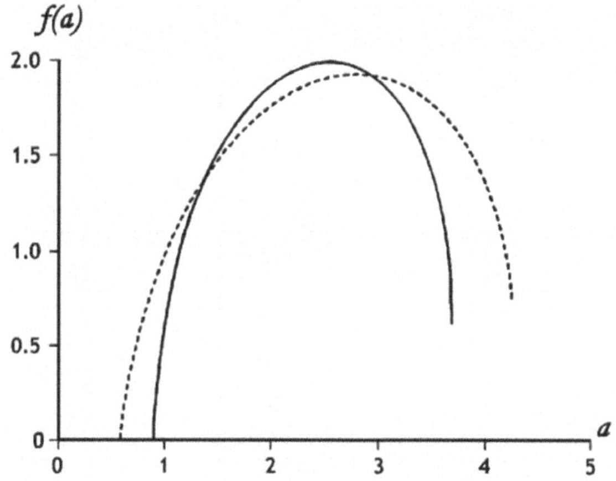

Fig. 2 Change of seismic field scaling before the earthquake of 17.06.2000 in Iceland

the east from Reykjavik. Magnitude of the earthquake was $M \approx 6.6$ and the depth of the hypocenter was 6.3 km. The seismic data used for calculations are taken from the Icelandic seismological catalogue (catalogue SIL, South Iceland Lowland). In the vicinity of the epicenter of this earthquake a polygon of the size of 60×60 km is chosen. Small-scale seismic events preceding this earthquake had magnitudes $M \leq 4.1$ and their average depth was 5.8 km. In this case, the first data sample covered the period 28.05.1995 to 16.06.1998, the second sample covered the period 17.06.1998 to 17.06.2000. Both samples contained 2230 events.

The solid line in Fig. 2 shows $f(a)$-spectrum of the seismic field, built from the first data sample, the dotted line shows the $f(a)$-spectrum of the seismic field, built from the second sample. As we can see, in this case the scaling of the seismic field has also changed significantly in the last two years before the main shock of the strong earthquake. The widening parameter of the seismic field $f(a)$-spectrum in this case was $\delta S = 1.08$....

5 The Northern Baja Earthquake

The earthquake Northern Baja (Spanish name—Sierra El Mayor Cucapah) has happened on 04.04.2010 in the territory of Mexico near the border with USA. Its magnitude was $M = 7.2$, coordinates of the epicenter—32.28° N, 115.29° W, the depth of the hypocenter is 10 km. Calculations are carried out within the framework of a polygon of 80×80 km, using data from the Southern California Catalogue. The first sample included events for the period 23.02.2002 to 31.12.2007, the second sample included events for the period 01.01.2008 to 04.04.2010. Both samples contained 1305 events in the magnitude range $2.0 \leq M \leq 5.4$ and the average depth of the hypocenters was 8.3 km. Calculated seismic field $f(a)$-spectra for both samples are presented in Fig. 3.

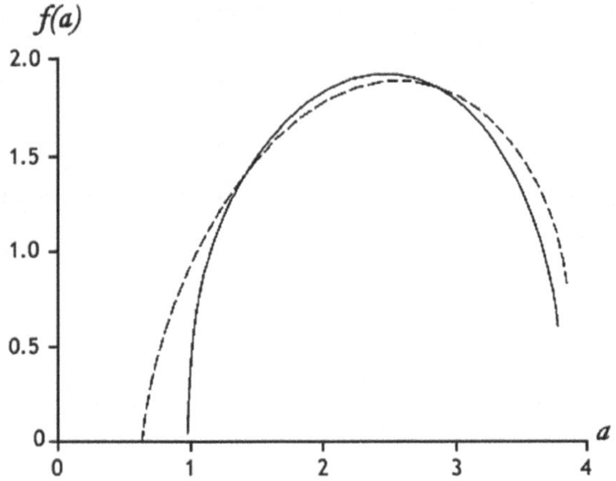

Fig. 3 Changes in seismic field scaling before the 2010 Nothern Baja earthquake

Just like in the previous plots, in Fig. 3 seismic field $f(a)$-spectrum determined from the first sample is shown as a solid line, seismic field spectrum determined from the second sample—dotted line. The spectrum of singularities of the seismic field immediately preceding the main shock of this earthquake in this case is much wider than the spectrum of singularities of the earlier field. The widening parameter is equal to $\delta S = 0.458...$.

The effect of seismic field $f(a)$-spectrum widening has characteristic repetitive features: it is connected with the expansion of the range of the singularity indices of the seismic field and the change of the shape of the $f(a)$-spectrum, while the values of the $f(a)$-spectra extrema practically do not change over time (the existing differences are insignificant and are explained only by the lack of data). Thus, the effect is determined by the behavior of the higher moments of multifractal measures that model the spatial distribution of seismic epicenters, while the monofractal dimension of seismic fields is actually insensitive to the process of preparation of a strong earthquake.

6 Discussion and Conclusions

The above examples show that changes in the structure of spatial distributions of seismicity before strong earthquakes have a systematic and repetitive character. Changes in the structure of the seismic field before strong earthquakes lead to seismic field $f(a)$-spectra widening in recent 2–2.5 years before the main shock. At the same time, examples of inverse evolution of the seismic field scaling before strong earthquakes in seismic catalogues have not been found.

The natural character of changes in seismic scaling before strong earthquakes can be justified from the perspective of the theory of dissipative structures. The theory (Prigogine 1980) states that at transition of physicochemical system from weakly non-equilibrium to strongly non-equilibrium state the properties of system begin to be defined by fluctuations (instead of average values of system parameters). In the case of seismic activity these fluctuations, i.e. short-term local bursts and local attenuation of activity cannot be identified in the field structure because of the high intermittency inherent in the spatial distributions of seismic events, but in the cumulative form they are manifested in the structure of the model fields and may be found by way of the effect of seismic field $f(a)$-spectra widening. In other words, the considered changes in seismic scaling are directly related to the transition of the seismogenerating system from a weakly non-equilibrium to a strongly non-equilibrium state, i.e. the process of preparation of a strong earthquake. In a certain sense, seismic scaling can be called a macro parameter of the seismic generating medium, which characterizes its state.

Thus, the effect of seismic field $f(a)$-spectra widening before strong earthquakes, apparently, should be attributed to the physical laws of the seismic process and the prognostic signs of strong earthquakes. It is important that this effect is detected by strictly formalized mathematical procedures in the processing of data from seismological catalogues. Thus, the results of calculations exclude the subjective factor, which is often found in the methods of searching and interpreting the precursors of earthquakes of other physical nature (emission of radon, electromagnetic bursts, animal behavior, etc.). Such precursors are rarely repeated in contrast to the effect of seismic field$f(a)$-spectra widening, which is repeatedly found in those cases, which are not yet frequent, when the representativeness of seismic data allows for statistically significant scaling analysis of the seismic catalogue.

References

Chhabra A, Jensen RV (1989) Direct determination of f(a) singularity spectrum. Phys Rev Lett 62(12):1327–1330

Geilikman MB, Golubeva TV, Pisarenko VF (1990) Multifractal patterns of seismicity. Earth Planet Sci Lett 99(1/2):127–132

Hooge C, Lovejoy S, Pecknold S, Malouin F, Schertzer D (1994) Universal multifractals in seismicity. Fractals 2(3):445–449

Mandelbrot B (1989) Multifractal measures, especially for geophysicist. PAGEOPH 131(1–2):5–42

Prigogine I (1980) From being to becoming. W.H. Freeman and Co., San Francisco, 200p

Stakhovsky IR (2017) Scale invariance of shallow seismicity and the prognostic signatures of earthquakes. Phys Uspekhi 60(5):472–489

Detecting and Analysis of Bubble Gas Emissions in Shallow Water by Method of Passive Acoustics

A. A. Budnikov, I. N. Ivanova, T. V. Malakhova, and V. V. Pryadun

Abstract The applicability of a passive acoustic method for monitoring bubble gas emissions in shallow water is being investigated. Experiments on the registration of audio signal generated by single gas bubbles with a diameter of 3–4 mm were conducted under laboratory and field conditions. Under laboratory conditions, one maximum in the frequency spectrum was observed, occurring at about 2500 Hz for bubbles of small diameter and 1570 Hz for larger bubbles. In situ experiments were carried out in Laspi Bay (the Black Sea). Several peaks in the frequency bubbles spectrum were shown: at 1550, 1100 and 600 Hz. The highest frequency recorded by hydrophone qualitatively coincides with the resonant frequency of a 4 mm diameter bubble calculated from the Minnaert equation.

Keywords Marine seep · Passive acoustic monitoring · Methane · Bubble plume · Black sea · Laspi bay

1 Introduction

Methane marine seeps were found along continental slopes around the world. Research related to the methane bubble seep emissions has important practical significance in many areas: assessing the contribution to the total amount of greenhouse gases, marine methane can be used as an alternative fuel source or it can be a marker of underground natural gas deposits. If the source of methane is located near the coastline, in the event of an earthquake there is a risk of catastrophic explosive emissions for the coastal area.

A. A. Budnikov (✉) · I. N. Ivanova · V. V. Pryadun
Faculty of Physics, M. V. Lomonosov Moscow State University, Moscow, Russia
e-mail: aa.budnikov@physics.msu.ru

I. N. Ivanova
e-mail: ivair@yandex.ru

T. V. Malakhova
The A. O. Kovalevsky Institute of Marine Biological Research of RAS, Sevastopol, Russia
e-mail: t.malakhova@imbr-ras.ru

C. T. Olegovna (ed.), *Processes in GeoMedia—Volume I*,
Springer Geology, https://doi.org/10.1007/978-3-030-38177-6_30

In the Black Sea region seeps were found along the coast of Crimea (Egorov et al. 2011), Bulgaria (Dimitrov 2002), the Caucasus (Tkeshelashvili et al. 1997) and in the Sea of Azov (Pasynkov et al. 2009). Over the past decade, many new areas of bubble seep emission have been discovered in the Crimea coastal waters (Egorov et al. 2012; Malakhova et al. 2015). In order to estimate the volume of gas emitted from such sources, various methods are used (Johnson et al. 2015), including passive acoustic (von Demling 2010; Dziak et al. 2018). However, there is no universal method that allows quantitative measurements of seafloor methane bubble streams with great accuracy.

Passive acoustic methods are the least energy consuming of the existing methods for detecting and monitoring natural seeps, which makes it possible to measure local fluxes over a long-term period. Long-term measurements, in turn, make it possible to more accurately analyze the temporal variability of volumes of gas emissions.

The shallow depth of coastal seeps allows to perform precision integrated research, with the simultaneous use of several duplicate methods. However, studies at shallow depths are complicated by the presence of a large number of heterogeneous noises, which are absent at large depths, at which similar studies are usually performed (Johnson et al. 2015; von Demling 2010; Dziak et al. 2018).

The purpose of this work was to study the applicability of the passive acoustic method for analyzing the characteristics of bubble gas emissions, as well as for long-term monitoring of gas seeps in the coastal shallow water zone.

At the first stage, the task was to experimentally determine the peak frequency of the audio signal produced by the evolved gas bubbles. At the second stage, it was necessary to determine the possibility of detecting this frequency in audiograms obtained in real conditions. For this, two experiments were carried out, the first in the laboratory at the Department of Physics of the Sea and Inland Waters, Faculty of Physics, Moscow State University, the second - in real conditions in the coastal shallow zone of the Black Sea (Laspi Bay, depth 2 m) in close proximity to the existing seafloor methane seep.

2 Object and Methods

In the laboratory experiment, a low flow bubble stream was modeled as follows: near the bottom of the salt water container (18‰) bubbles were squeezed through the needle with a low-power compressor (Fig. 1). The needles of 0.6 and 0.8 mm diameters were installed. The rate of bubble stream for a 0.6 mm needle was 105 bubbles per minute, for a 0.8 mm needle was 60 bubbles per minute. The size of the bubbles was estimated visually using a graduation bar located near the point of bubbles exit.

The diameter of the produced bubbles was 3–4 mm, which corresponds to the diameter of the bubbles observed in natural conditions in Laspi Bay. The depth of the liquid in the container was 0.5 m. The audio signal produced by the bubbles

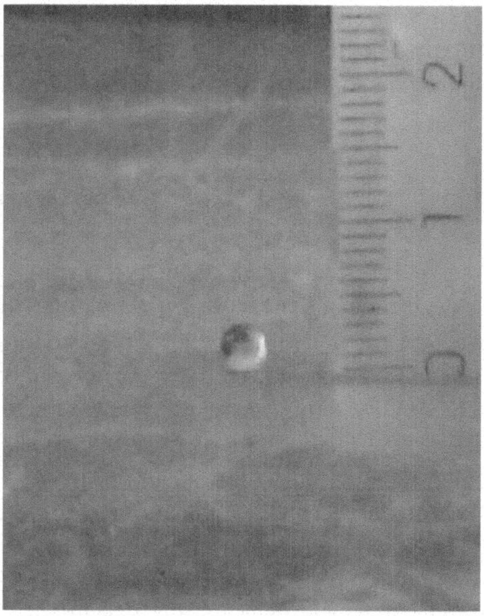

Fig. 1 The detachment of the gas bubble from the end of a needle, in a laboratory experiment

was recorded with a submerged broadband microphone. The records obtained were analyzed using Audacity (2.3.0).

For bubbles emanating from a 0.6 mm needle, an analysis of the spectrogram of the entire audio record shows that the bubble sound bandwidth to range from ~2 to 2.7 kHz with a peak frequency on 2.5 kHz (Fig. 2a).

For bubbles emanating from a 0.8 mm needle, the bubble sound bandwidth was in the range of ~1.5–2.1 kHz with a peak frequency of 1.6 kHz.

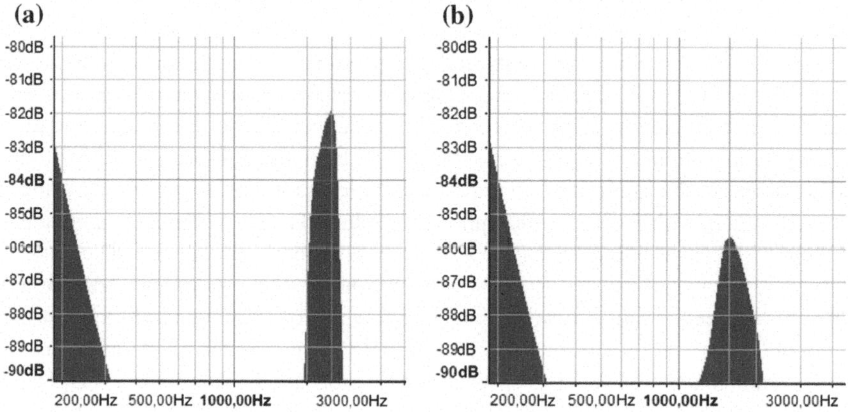

Fig. 2 Spectrograms of audio record from laboratory experiments: **a** the diameter of the needle is 0.6 mm. The rate of output −105 bubbles per minute, **b** the diameter of the needle is 0.8 mm. The rate of output −60 bubbles per minute. The diameter of the produced bubbles is 3–4 mm

In works (Johnson et al. 2015; von Demling 2010; Dziak et al. 2018), to estimate bubble radii the Minnaert equation is used, which relates the peak frequency of a bubble sound f with its equivalent spherical radius r_0:

$$f = \frac{1}{2\pi r_0}\sqrt{\frac{3\gamma p}{\rho}},$$

$\gamma \sim 1.32$ is the heat capacity ratio for methane, p is hydrostatic pressure, ρ is seawater density. In this case, the recorded frequencies of real bubbles do not coincide with the calculated frequency, although they are close to it in value.

The value of the frequency of the bubbles, given by the Minnaert equation for a laboratory experiment, was approximately 2.2 and 1.7 kHz for bubbles with a diameter of 3 mm and 4 mm, respectively, which coincides qualitatively with the values obtained from the spectrogram.

A model of a bubble emanating from a needle can be an oscillating bubble fixed on the substrate at one point. A numerical solution of the problem of natural oscillations of a bubble fixed at one point is given in (Gorskiy et al. 1988). It is noted that the frequency of the third mode of oscillation is closest to the resonant frequency of the volume pulsations of an oscillating free bubble in an ideal fluid. At the same time, there may be other modes determined by the configuration of the oscillating surface, with lower and higher frequencies. With bubble size increase, oscillations at additional frequencies may be more intense.

In accordance with the data obtained from the calculations, the frequencies of 2.5 and 1.7 kHz observed in laboratory experiments were chosen for the main frequencies generated during bubble formation and rise.

At the second stage of the work, an experiment was carried out on recording audio signals produced by seep bubble streams in real conditions. The experiment was conducted in Laspi Bay (southern coast of Crimea) at a depth of 2 m in close proximity to the proceeding methane seep. It was previously shown that bubble gas emissions in this area occur throughout the year and have a deep source (Malakhova et al. 2015). Thus, the nature of gas emissions is a point discharge of a series of one-dimensional bubbles (Fig. 3) and can be modelled by laboratory experiment.

The audiogram in Laspi Bay was recorded with an autonomous omnidirectional hydrophone. At the same time, an underwater video capture was performed using a GoPro4 camera with duplicate microphone recording. The environmental parameters at the measurement point during the experiment remained unchanged. The parameters were measured using a RCM 9 LW multiparameter profilometer (Aanderaa) with pressure, temperature, conductivity, turbidity and oxygen concentration sensors in continuous mode (Ivanova and Budnikov 2018) (Fig. 3). During the experiment, these parameters were: $T_w = 23.7\ ^\circ C$, $\sigma = 28.67$ mS m/sm, Tu $= 0.45$ NTU, $O_2 = 7.95$ g/l.

The size of the bubbles in Laspi Bay was determined by video processing and was approximately 3–4 mm. Methane bubbles were seeped out in clusters of 10–15 bubbles of approximately the same diameter. Before the appearance of a group of

Fig. 3 Underwater photography experiment in Laspi Bay **a** a general view of the measuring complex, **b** the point of exit of the bubbly gas (marked by a white arrow), **c** rising bubbles of methane (a measuring scale is applied in the graphical editor)

bubbles in the visibility zone of the camera, the characteristic sound of the bubbles leaving the emitting channel was recorded.

3 Received Data and Discussion

The analysis of audio recordings (primary and secondary) showed the presence of frequencies of 1550, 1100 and 600 Hz in the spectrum of the signal (Fig. 4). At the same time, the maximum peak value of the frequency in the audio signal from the bubbles, qualitatively coincides with the resonant frequency of the bubble with a diameter of about 4 mm, estimated using the Minnaert equation (Fig. 4).

In (Makarov 2016) acoustic signals of a bubble plume located at a depth of 40 m at the Selenga river delta (Lake Baikal) are shown. The main audio signal from the bubbles, obtained using a passive acoustic method, was in the frequency range

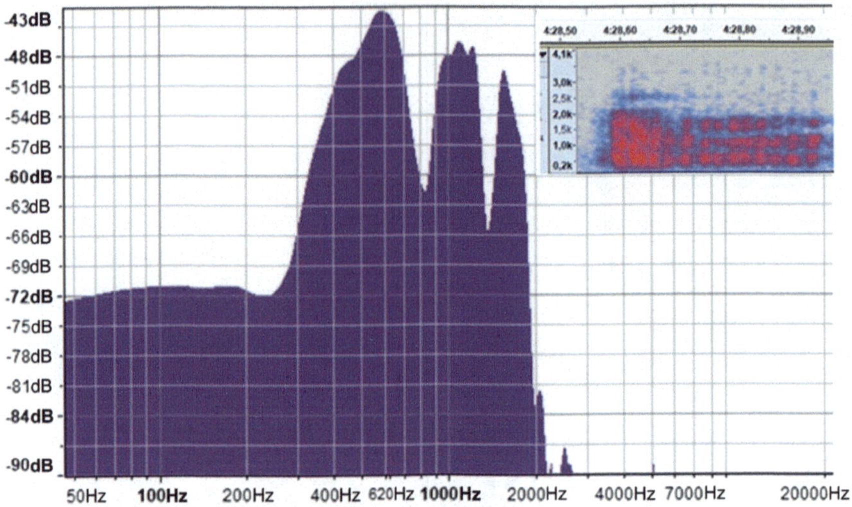

Fig. 4 Spectrograms of hydrophone record from Laspy Bay, Black Sea, depth 2 m. In the upper right corner 1.5 s of hydrophone data over the 4.1 kHz band

of ~0.5–4.5 kHz. The diameters of the observed bubbles varied greatly, but mostly ranged from 0.95 to 2.1 cm. The difference in the resonant frequency obtained in our work for bubbles in the Laspi Bay can be explained by a smaller depth compared with (Makarov 2016) and a significant variation in the diameters of the observed bubbles.

The measurements carried out in (Dziak et al. 2018) at a depth of 1228 meters show a more accurate coincidence of the observed diameters of the bubble with those calculated using the Minnaert equation, but the authors note the difficulty of observing millimeter-sized bubbles. In (Dziak et al. 2018), it was noted that the sound of bubble streams is a broadband series of short duration pulses that occur in clusters of dozens of pulses, which coincides with the records obtained in our work.

4 Conclusion

- The paper shows the principal possibility of detecting seep bubble streams in shallow water using a passive acoustic method, taking into account data on atmospheric pressure above the measurement site, as well as on the density of sea water.
- To estimate the size of evolved gas bubbles in natural conditions, it is possible to use the Minnaert equation.
- The technique, debugged in the most noisy shallow zone, may be further extended to deeper areas.

Acknowledgements The work was carried out as part of the student practice of the Physics Faculty of the M. V. Lomonosov Moscow State University together with O. Kovalevsky Institute of Marine Biological Research of RAS on the topic of the state task IMBR of RAS "Molismological and biogeochemical principles of the homeostasis of marine ecosystems", state number. registration AAAA-A18-118020890090-2.

References

Dimitrov L (2002) Contribution to atmospheric methane by natural gas seepages on the Bulgarian continental shelf. Cont Shelf Res 22:2429–2442

Dziak RP, Matsumoto H, Embley RW et al (2018) Passive acoustic records of seafloor methane bubble streams on the Oregon continental margin. Deep Sea Res Part II 210–217. https://doi.org/10.1016/j.dsr2.2018.04.001

Egorov VN, Artemov YG, Gulin SB (2011) Methane seeps in the Black Sea: Environmental and ecological role. In: Polykarpov GG, (ed) ECOSI-Hydrophysics, vol 405. NPC, Sevastopol, p (in Russian)

Egorov VN, Pimenov NV, Malakhova TV et al (2012) Biogeochemical characteristics of methane distribution in water and bottom sediments in areas of jet gas emissions in the waters of the Sevastopol bays. Sea Ecol J XI(3):41–52 (in Russian)

Gorskiy SM, Zinov'ev AY, Chichagov PK (1988) Own oscillations of a fixed gas bubble in a liquid. Acoust Mag XXXIV(6):1024–1027 (in Russian)

Ivanova IN, Budnikov AA (2018) Features of temperature stratification in the coastal zone of the Black Sea. Process GeoMedia 1(14):741–745 (in Russian)

Johnson HP, Miller UK, Salmi MS, Solomon EA (2015) Analysis of bubble plume distributions to evaluate methane hydrate decomposition on the continental slope. Geochem Geophys Geosyst 16:3825–3839. http://dx.doi.org/10.1002/

Makarov MM (2016) Bubble exits of methane from bottom methane sediments. Thesis for the degree of candidate of geographical sciences, Irkutsk, FGBUS LIN SO RAS (in Russian)

Malakhova TV, Kanapatskiy TA, Egorov VN et al (2015) Microbial processes and the genesis of jet methane gas emissions of coastal areas of the Crimean peninsula. Microbiology 84(6):743–752 (in Russian)

Pasynkov AA, Tihonenkov EP, Smagin YV (2009) Gas jet at the bottom of the Sea of Azov. GPIMO 1:77–79 (in Russian)

Tkeshelashvili GI, Egorov VN, Mestvirishvili SA et al (1997) Methane gas emissions from the bottom of the Black Sea in the mouth area of the river Supsa off the coast of Georgia. Geochemistry 3:331–335 (in Russian)

von Demling S (2010) Acoustic imaging of natural gas seepage in the North Sea: sensing bubbles controlled by variable currents. Limnol Oceanogr Methods 8:155–171. https://doi.org/10.1016/j.marpetgeo.2015.02.011

Paleogeodynamics of the Drake Passage in the Scotia Sea

A. A. Schreider, A. E. Sazhneva, M. S. Klyuev, A. L. Brekhovskikh, I. Ya. Rakitin, E. I. Evsenko, O. V. Grinberg, F. Bohoyo, H. Galindo-Zaldivar, P. Ruano, J. Martos, and F. Lobo

Abstract According to the results of geomagnetic, studies primarily made by Spanish research vessel "Hesperidas", in the region of the Drake Strait linear magnetic anomalies revealed. Modeling of linear magnetic anomalies within the frame of the lithospheric plates tectonics concept allowed to restore C3–C17 chrones. Analysis of the bottom geochronology and calculations of paleogeodynamic parameters showed that the destruction of the American-Antarctic continental bridge and the initialization of the Circum-Antarctic current in the area between South America and Antarctica occurred as a result of the bottom spreading in the Drake Strait region. At the initial stage, the bottom growth occurred in the sea of Scotia at the South-Eastern limit of the Strait in the range of chron C17r (36.969–37.753 million years ago or even earlier). On the next step during chron C11r (29.970–30.591 million years ago) began the formation of a new spreading center with simultaneous attenuation of spreading in the South-Eastern limit of the Strait, which stopped during chron C10n.2n (28.141–28.278 million years ago). The formation of a new centre for the spread of the bottom was accompanied by a heavy jump in spreading axis at the North-Western limit and the beginning of the formation of the Western Scotia paleorige. The maximum values of more than 6 cm/year growth of the bottom on the axis of the ridge reached in the range of C6–C6B (19.7–23.1 million years ago), after which the spreading began to fade and soon after the time of the C3An.1n (6.013–6.254 million years ago) occurred the death of the paleodrainage axis West Scotia ridge and the active phase of the destruction of the American-Antarctic continental bridge in the region of Drake Passage was over.

A. A. Schreider (✉) · A. E. Sazhneva · M. S. Klyuev · A. L. Brekhovskikh · I. Ya. Rakitin · E. I. Evsenko · O. V. Grinberg
Shirshov Oceanological Institute RAS, Moscow, Russia
e-mail: aschr@ocean.ru

F. Bohoyo
Spanish Geological and Mineralogical Institute, Granada, Spain

H. Galindo-Zaldivar
University of Granada, Granada, Spain

P. Ruano · J. Martos · F. Lobo
Andalusian Institute of Earth Sciences, Granada, Spain

C. T. Olegovna (ed.), *Processes in GeoMedia—Volume I*,
Springer Geology, https://doi.org/10.1007/978-3-030-38177-6_31

Keywords Scotia sea · Euler pole · Paleogeodynamics · Drake passage

1 Introduction

The formation of the Scotia Sea is inseparably connected with the destruction of the continental bridge between South America and Antarctica. As a result of geological and geophysical studies in the Western part of the Scotia Sea (Fig. 1) between 54° and 60° S 41° and 67° W has revealed dead axis of mid-ocean ridge spreading in Western Scotia (Atlas 2001; Schrader et al. 2012; Barker and Griffiths 1977; Bohoyo et al. 2007; www.bodc.ac.uk/products/bodc_products/gebco). The bottom ridge relief corresponds to a mountainous bottom elevation of more than 500 km wide with intensively dissected relief. The height of the ridge reaches 2 km above the adjacent abyssal plains of the Yagan basin from the North-West and Ona from the South-East. The depth of the pit is more than 3 km. From the West, the ridge is limited to the Shackleton disturbance zone and from the East to the South of the North Scotia ridge. Above the ridge axis, an intensive 200–500 nTl amplitude of magnetic anomaly up to 50 km wide is observed. From the South-East and North-West it is adjoined by linear magnetic anomalies with an amplitude of 100–300 t with wavelengths of 20–40 km, shifted by several transformer faults on the tens of kilometers. Linear magnetic anomalies and transformational faults in the Scotia Sea are not exactly defined until now and, being based on a limited amount of geological and geophysical data, differ in the drawing in various publications.

Thus, on the Western Scotia Ridge between the Transform Fractures Endurance and Quest in (Atlas 2001), the axes of the paleo-anomalies S5B and S5C are located in the interval between S5A-C5B and S5B-C5E, respectively, see (Eagles et al. 2005). The configurations of transform faults in the works (Atlas 2001; Eagles et al. 2005; Lodolo et al. 1998) do not coincide with each other. Between the Quest (Quest) and Shackleton transform faults in the Drake Strait, the most ancient confidently

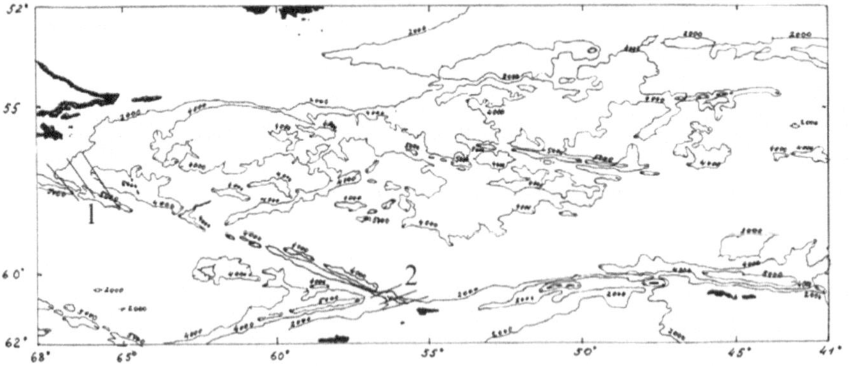

Fig. 1 Landscape map of the western part of the Scotia Sea by (www.bodc.ac.uk/products/bodc_products/gebco). Depths in m, study areas 1 and 2 are shaded

recognizable magnetic anomaly in the works (Lodolo et al. 1998) is considered to be the C10 paleo-anomaly, while in (Eagles et al. 2005) it is considered to be the C8 anomaly. For the Drake Strait region of the Scotia Sea, the literature does not contain paleogeodynamic reconstructions (including definitions of Euler poles of rotation and rotation angles) before the time of C8. The configuration itself of the border between the oceanic and continental lithosphere differs in the works (Galindo-Zaldivar et al. 2006; Lodolo et al. 1998) in individual details.

Thus, all of the above indicates the absence of a common understanding of the tectonic evolution of the bottom in the Drake Strait region. There is no quantitative basis for the reconstruction of the initial stages (before the C8 time) of the sea floor evolution, including the determination of the coordinates of the Euler poles and rotation angles. This paper is devoted to the first quantitative reconstruction of the initial stages (older than the C8 time) of the bottom opening in the Drake Strait region in connection with the subsequent evolution of the bottom of the water area and the initialization of the Circum-Antarctic current—the most important element of climate formation in the Southern hemisphere of the Earth and the planet as a whole.

2 Methodology and Results of Calculation of Geodynamic Parameters

The initial configuration and stages of destruction of the American-Antarctic bridge in time and space are ambiguous. The published studies do not have a single approach to determining the configuration and mutual position of the individual elements of the bridge, primarily at the initial stage of its development. In the absence of deep-sea drilling points in the Scotia Sea in general and in the Drake Passage region in particular, the main importance for the study of bottom geodynamics has acquired a comprehensive geological and geophysical interpretation of the results of the bottom relief study and the abnormal magnetic field (including the original data of the Spanish research vessel "Hesperidas"), from the point of view of the concept of lithospheric plate tectonics.

Computer technique was proposed in Bullard et al. (1965) for the first time for the best combination of conjugate isobaths on the slopes of the continents. The combination was done through trial and error, by minimizing the angular disagreement in measurements along the Euler latitudes. The technique illustrated the principle that the best combination can be made for any circuits that supposedly are or once were to be a single circuit. Implementing the principle of the best possible overlap, it is possible to achieve the restoration of the primary continuity of any contours, including isochronous, isobaths, isohypses, etc.

According to the data of the electronic bank on the bottom bathymetry (www.bodc.ac.uk/products/bodc_products/gebco), the profiles in the direction perpendicular to the slopes of the Fiery Land (Schreider and Sazhneva 2019), the Antarctic Peninsula and the Shackleton mountain structure separating them with an interprofile distance of 5 miles are constructed in the present paper. Analysis of the obtained slope profiles shows that almost all of them consist of three parts. The upper (shallower than 3.0 km) and lower (deeper than 4.1 km) parts of the slopes are characterised by a variable slope along the profile. The central part of the slopes, enclosed in the depth range of 3.3–4.0 km, is the steepest and has relative stability of slope along each individual profile.

The distance between the isobath 3.3 and 4.0 km of slope to the bottom surface is considered as a working basis for calculating the individual slope angle (steepness) along each of the profiles.

Sedimentation in different regions of the slopes led to their covering with sedimentary rocks, accompanied in time by the slopes' smoothing out (decrease of the slope angle) due to their covering with sediments. Fitting is uneven over time and space. Unevenness of fit is connected both with distribution and redistribution of drift areas and with sliding of a part of accumulated sediments down the slope due to instability. The instability of sediments is due to the accumulation of critical mass, which leads to the excess of the sliding force over the friction force (along the surface of the interface inside the sediments or along the foundation surface), which prevents sediments from sliding at a slope steepness.

The movement of specific portions of sediments have smooth or impulse in nature. The steepness of the foot of the moving mass also makes an important contribution to the mobility of the sedimentary mass on the slope. At low angles of inclination and all other things being constant, the movement of the sedimentary mass along the slope will be very slow. Estimates show that in order to overcome the adhesion force between the layers of sediments (in the Drake Strait it is mud, clay, sandstones) and avalanche-like failure of them down the slope with the development of significant rates of slippage, with all other conditions being equal, the slope of the sliding surface should exceed 3° (Zhmur et al. 2002). Among the reasons that trigger the slippage of sediments downhill is the impact on sedimentary masses of exogenous (e.g. regular currents) and endogenous (e.g. earthquakes) factors.

In the present paper the method of E. Bullard is applied for the first time for the case of combination of the isobaths of the Fiery Land slopes, the Antarctic Peninsula and the Shackleton disturbance zone separating them. Numerous tests of the dockability of different areas of different and eponymous isobaths have shown that the most suitable areas for paleo-geodynamic analysis are those in the range of 3.3–4.0 km. The slopes in this depth range are the steepest (the average slope surface slope angle exceeds 5°) and, according to the above information on the nature of sediment thickness slippage, which has the lowest sediment thickness (or even completely devoid of it).

Calculations of the Euler poles and rotation angles are carried out according to the original programs of the Laboratory of Geophysics and Tectonics of the World Ocean Floor of the IO RAS, incorporated into the Global Mapper software environment and the principles of calculation for which are described in Schrader (2011).

The deepest part of the Drake Passage is tens of kilometers wide and is characterized by depths of more than 3 km, which is stated in Schreider and Sazhneva

(2019). If the depression is formed according to modified circuit (Wernike 1985) is associated with the divergence of blocks of the continental crust, while numerous tests of the dockability of different sections of isobaths of the same name have shown that the sections of isobath are the most suitable for paleogeodynamic analysis in the range of 3.3–3.9 km.

According to calculations, at the position of Euler's final pole of rotation in the point with coordinates 67.81° S 75.80° W for about 100 km it is possible to obtain a very good combination of 3.5 km isobath in the areas 1 and 2 (Fig. 1) of the South-East of the Pacific Ocean tip of the Fiery Land and the West of the Antarctic Peninsula adjoining the zone of Shackleton disturbance (the areas of isobath 3.5 km between the AB points in Figs. 2 and 3). The rotation angle was 6.1° ± 0.7°. At the same time, the standard deviation at the calculated matching points was ±7 km (9 matching points). As a result, it is possible to restore the paleo-relief profile of the bottom before their splitting (Fig. 2b).

According to our calculations, at the position of Euler's final pole of rotation at the point with coordinates 69.76° S 79.53° W provides for about 90 km a very good combination of the 300 m isobath (Fig. 4) in the South of the Pacific Ocean tip of the Fiery Land (the area between the points of Fig. 4a, b) and in the area of the Antarctic Peninsula adjoining from the West to the zone of Shackleton disturbance (0.3 km isobaths between the D′E′ points in Figs. 4b and 5). The rotation angle was 6.5° ± 1.1°. The mean square deviation at the calculated matching points was ±8 km (7

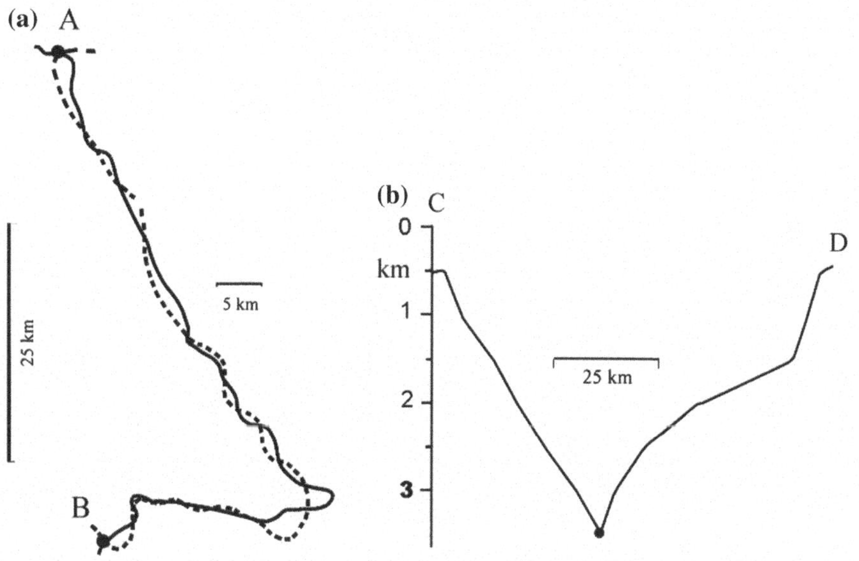

Fig. 2 The isobath junction (**a**) is located on the side of the Antarctic Peninsula (isobath 3.5 km is a solid line) and the fiery land (isobath 3.5 km is a dotted line); restoration on the basis of paleogeodynamic reconstruction of the bottom paleobathymetry (Fig. 3, line CD) profile (**b**) (dot indicates the location of the lithosphere rupture)

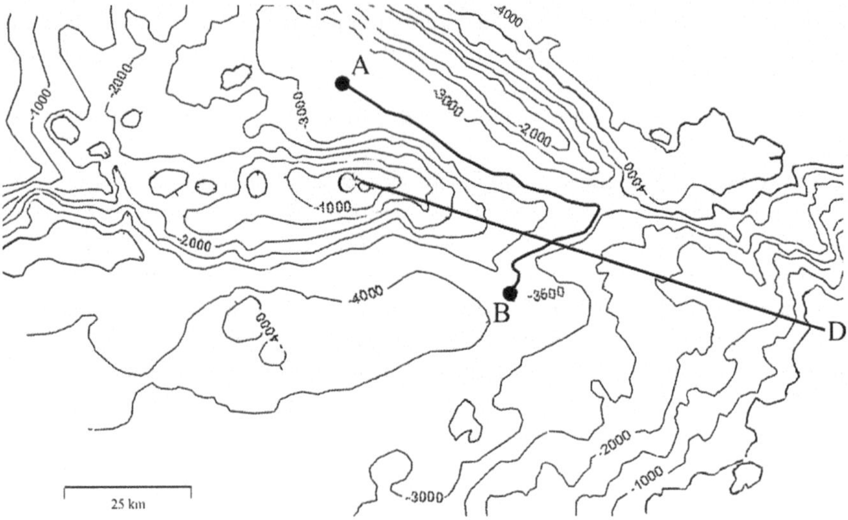

Fig. 3 The paleogeodynamic reconstruction of the drake strait based on the isobath junction of 3.5 km (Fig. 2a) of regions 1 and 2 (Fig. 1) shows the ends of the reconstructed lithosphere rupture zone axis and the position of the paleobathymetry profile

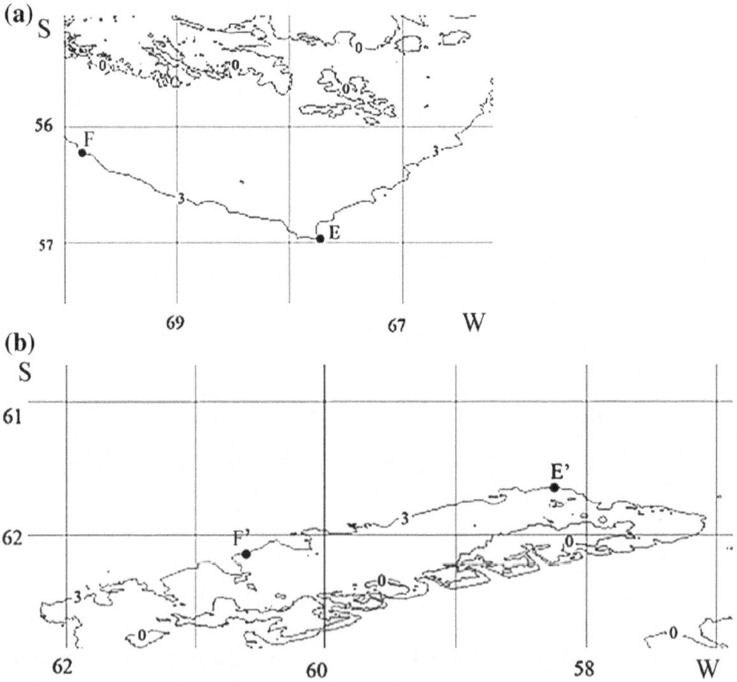

Fig. 4 The location of the sections being joined isobath 0.3 km from the Fiery Land (**a**) and from the Antarctic Peninsula D′E′ (**b**). Isobaths are hundreds of meters away

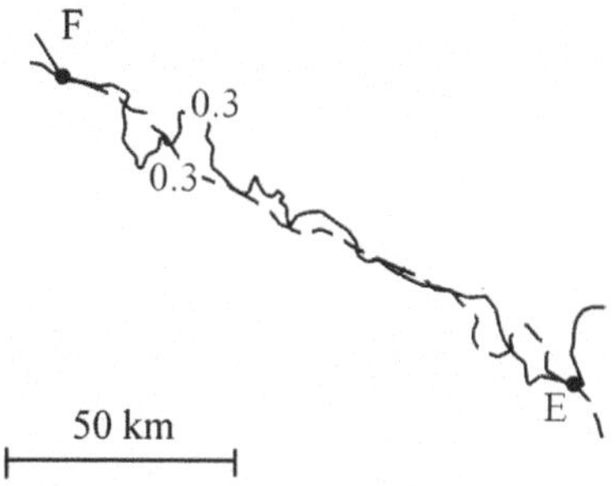

Fig. 5 The isobath joint from the side of the Antarctic Peninsula (isobath 0.3 km—solid line) and fiery land (isobath 0.3 km—dotted line)

matching points). As a result, it is possible to restore the paleo-relief profile of the bottom before splitting (Fig. 7). Calculations use the 300 m isobath and are estimates, as this isobath captures the upper edge of a continental slope, possibly covered with significant sedimentary cover. As a result of the calculations, the axis of the splitting zone of the peripheral fragments of the Fiery Land and the Antarctic Peninsula is reconstructed (bold line in Figs. 3 and 6).

Paleogeodynamic reconstructions allow to restore the mutual position of tectonic elements in the Drake Strait region, but do not provide information about the temporal dating of the analyzed events and remain a subject of discussion (Ivanov et al. 2008; Barker 2001; Bohoyo et al. 2007a, b, 2016; Eagles et al. 2005; Geletti et al. 2005; Lodolo et al. 1998). The solution of these issues is of fundamental importance for the restoration on a quantitative basis of the initial stages of the Scotia Sea opening and the related development of the Circum-Antarctic Current, which is the most important element of the Southern Hemisphere and Earth climate formation. It was mentioned above that the existence in the geological past of a continental bridge between the

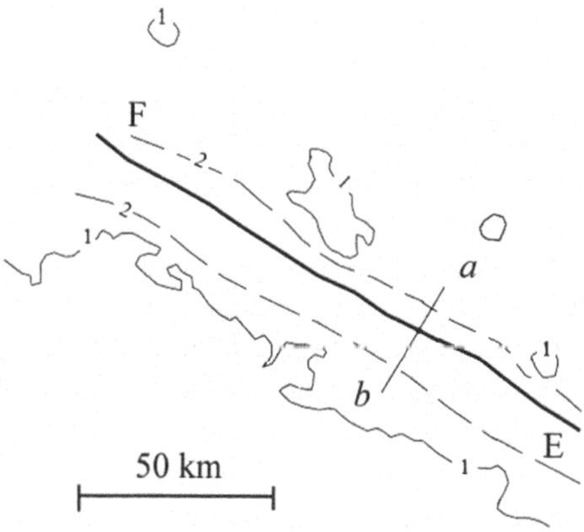

Fig. 6 Paleogeodynamic reconstruction of the Pacific periphery of the Drake Strait based on the isobath junction of 0.3 km of the regions of Fig. 4a, b shows the axis of the lithosphere rupture zone. The ends of the reconstructed axis of the lithosphere rupture zone (LRZ) and the position of the paleobathymetry profile (ab) are shown in hundreds of meters

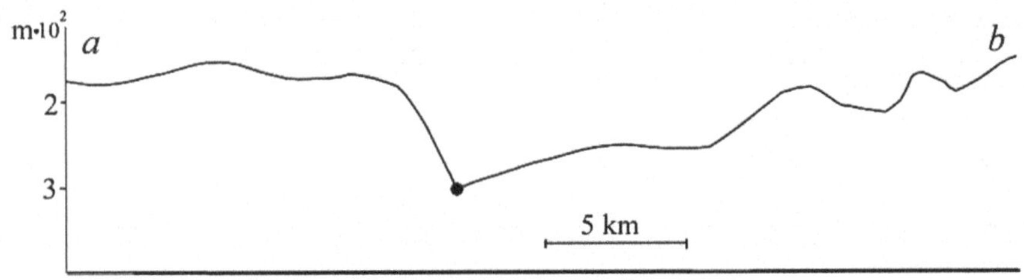

Fig. 7 Restoration of profile of the bottom paleobathymetry (Fig. 6, line ab) on the basis of paleogeodynamic reconstruction. The point denotes the location of the lithosphere rupture

South American and Antarctic continents in the world scientific literature is not in doubt. The beginning of the continental bridge split also differs in different works and is dated in the interval of 80–25 million years. Linear magnetic anomalies and transformational faults in the Yagan and Ona troughs, so necessary for the restoration of the region's paleogeodynamics, are still not precisely defined and differ in their drawing in various publications. It is mentioned above that deep-sea drilling in the Scotia Sea has not yet been carried out, so the main role in the study of the formation process is played by a comprehensive geological and geophysical interpretation of the results of the bottom topography study, as well as the anomalous magnetic field. In recent years, a certain amount of geological and geophysical data has been obtained in the West of the Sea of Scotia, especially on the expeditions of the Spanish research vessel (RV) "Hesperidas".

Complex geological-geophysical analysis of the seabed relief and anomalous magnetic field along the previously unpublished profiles of the Spanish expeditions to RV "Hesperidas" together with the materials of the world's geomagnetic databases, processed in the framework of the new global tectonics concept, allowed us to identify for the first time linear magnetic anomalies and make the first map-scheme of the bottom geochronology in the West of the Yagan and Ona trenches. It should be noted that all the data in the present work are carried out using the most modern version of the magnetic anomalies scale (Gradstein et al. 2012). As a result of modeling paleomagnetic anomalies (Fig. 8), local area of bottom expansion adjacent to the zone of Shackleton disturbance in the Drake Passage, in which the C10n.2n–C17n.1r (28.141–37.753 million years ago) chrons are developed, is identified here. The upper edge of the inversion magnetoactive layer coincides with the bottom topography (Fig. 8). The lower surface of the conformal layer is the bottom relief at the layer thickness of 0.5 km, and the magnetization is taken equal to 5 A/m. The angular parameters of the magnetization vector are taken in accordance with the parameters of the field of the axisymmetric terrestrial dipole, and the parameters of the modern magnetic field correspond to those in the DGRF field of the survey epoch.

According to calculations, bottom spreading occured at an average speed close to 0.8 cm/year and lasted about 10 million years. At the time of C11r (29.970–30.591 million years ago), the formation of a new spreading center on the axis of the West Scotia mid-ocean Paleo-Ridge (the oldest chrons of which are shown in Figs. 9

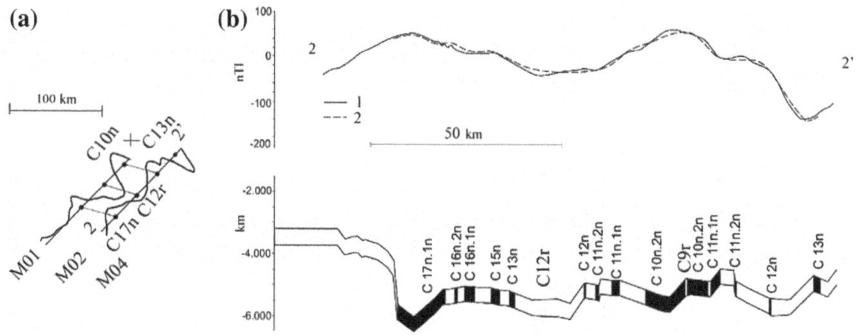

Fig. 8 Profiles of the anomalous magnetic field (a) along the transects M01, M02 RV "Hesperidas" near the SouthWestern limit of the Drake Passage (the cross corresponds to the point with coordinates of 60° S 56° W). Correlations of C17n–C12r chron are shown in accordance with the results of modeling the polarity chron. Observed (1) and theoretical (2) magnetic anomalies in the model of bottom expansion (b) along the line 2–2′ on the profile M02

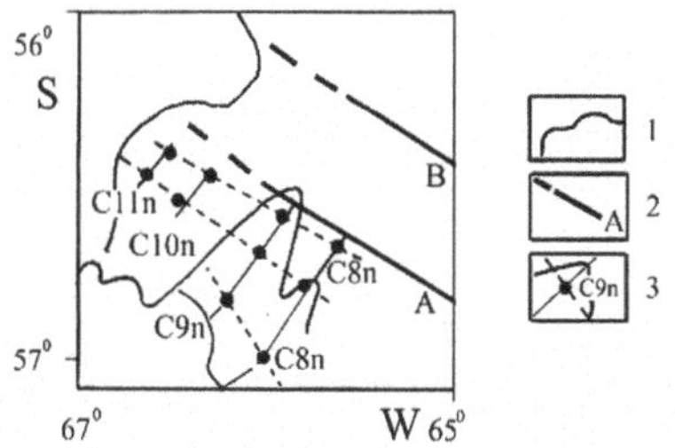

Fig. 9 Geochronology of the seabed as a result of the merger and partial reinterpretation of these niches. "Hesperidas", as well as the results of hydro-magnetic surveys from the works (Atlas 2001; Eagles et al. 2005). 1-isobath 2 km, 2-transform fracture, 3-position of the chron axis (dotted line) on the observation profile (dotted line) and its number according to the chronological scale (Gradstein et al. 2012)

and 10) began, with a simultaneous attenuation of the spreading at the South-Eastern limit of the strait, which stopped during the C10n.2n chron (28.141–28.278 million years ago). At the same time, the oceanic crust build-up to the South of the paleo-axis of the Western Scotia Ridge was approximately 20% more intensive than to the North of it. The maximum values of more than 6 cm/year of bottom expansion reached in the interval of C6–C6B chrons (18.748–22.564 million years ago), after which the spread began to fade (the youngest chrons of which are shown in Fig. 11). The axis of the mid-ocean paleo-ridge died during the C3n.1r chron (4.300–4.493 million years ago).

Thus, the destruction of the American-Antarctic Bridge in the area of the once unified massif connecting the South of the South American continent and the Antarctic Peninsula occurred at the beginning of the Drake Passage opening at the time when the C17r chron was older (36.969–37.753 million years ago).

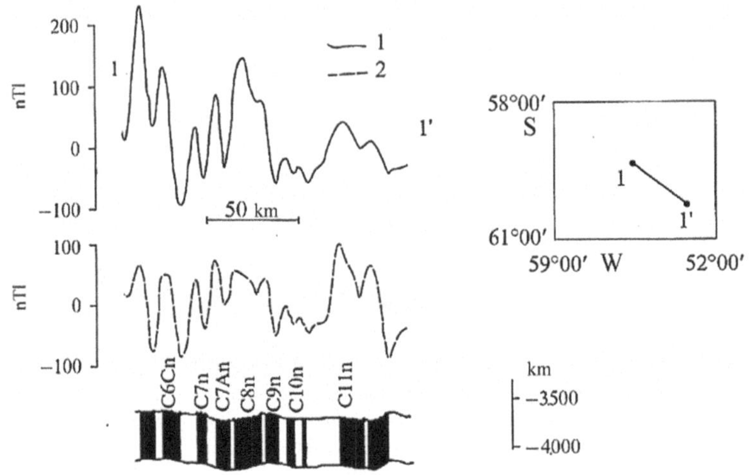

Fig. 10 Observed (1) and theoretical (2) magnetic anomalies along the observation profile in the SouthWest of the Drake Passage region (the profile location is given in the box). The modeling parameters are the same as in Fig. 8

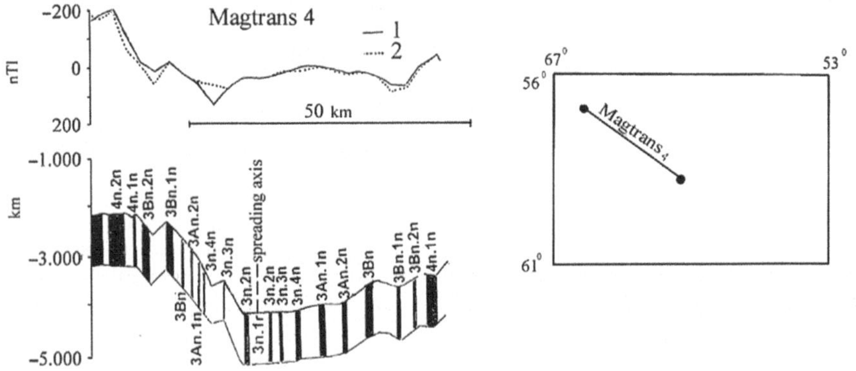

Fig. 11 Observed (1) and theoretical (2) magnetic anomalies along the observation profile in the central part of the Drake Passage region (the profile location is shown in the box). Parameters of modeling are the same as in Fig. 8

The formation of a deep-water passage with depths up to 3.5 km at the background of the destruction of the American-Antarctic bridge resulted in the initialization of the Circum-Antarctic Current in the space between South America and the Antarctic Peninsula.

3 Conclusion

Thus, as a result of the carried out researches the detailed paleogeodynamics of the fragmentation of the East American-Antarctic continental bridge was restored for the first time. In this regard, according to the results of geomagnetic research, primarily the Spanish research vessel "Hesperidas", linear magnetic anomalies have been identified in the Drake Strait region. Their modeling within the framework of the lithospheric plate tectonics concept made it possible to restore the C3–C17 chrons. The analysis of bottom geochronology and calculations of paleogeodynamic parameters showed that the destruction of the American-Antarctic continental bridge and the initialization of the Circum-Antarctic current in the area between South America and Antarctica occurred as a result of rifting and spreading of the bottom in the Drake Passage region. At the initial stage, the bottom was expanding at the South-Eastern limit of the spill in the interval of C17r (36.969–37.753 million years ago or even earlier). Subsequently, during the C11r chron (29.970–30.591 million years ago), a new spread center began to form, with a simultaneous attenuation of the spread at the South-Eastern limit of the strait, which stopped during the C10n.2n chron (28.141–28.278 million years ago).

The formation of a new floor-expansion center was accompanied by a multikilometer jump of the spreading axis to the North-West and the beginning of the formation of the Western Scotia Paleo-Ridge. The maximum values (more than 6 cm/year) of the bottom expansion on the ridge axis reached in the interval of C6–C6B chrons (19.7–23.1 million years ago), after which the spreading process began to fade and soon after the time of C3An.1 chron (6.013–6.254 million years ago) the axis of expansion of the Western Scotia Paleo-Middle Ridge dies and the active phase of destruction of the American-Antarctic continental bridge in the Drake Strait region is over.

Acknowledgements Part of the work related to the improvement of kinematic calculations methodology is carried out within the framework of the state assignment No. 0149-2019-0005. The main part of the research related to the restoration of paleo-geodynamics of the Drake Strait is carried out with the financial support of the Russian Foundation for Fundamental Research, Project-No. 17-05-00075.

References

Atlas (2001) Geological and Geophysical Atlas of the Pacific Ocean. GUGK 163p

Barker P (2001) Scotia sea regional tectonic evolution: implications for mantle flow and paleocirculation. Earth Sci Rev 55:1–39

Barker P, Griffiths D (1977) The evolution of the scotia ridge and scotia sea. Philos Trans Roy Soc Lond Ser 271:151–183

Bohoyo F, Galindo-Zaldivar J, Jabaloy A et al (2007a) Extentional deformation and development of deep basins associated with the transcurrent fault zone of the Scotia-Antarctic plate boundary. Geol Soc Lond Spec Publ 290:203–217

Bohoyo F, Galindo-Zaldivar J, Jabaloy A et al (2007b) Development of deep extensional basins with the transcurrent fault zone of the Scotia-Antarctica plate boundary. Revista de la Sociedad geologica de Espana 20(1–2):89–103

Bohoyo F, Galindo-Zaldivar J, Leat P et al (2016) Bathymetry and geological setting of the drake passage. British Antarctic Survey, 1sh

Bullard E, Everett J, Smith A (1965) The fit of continents around Atlantic. Symposium on continental drift. Phil Trans Roy Soc Lond 258A:41–51

Eagles G, Livermore R, Fairhead D, Morris P (2005) Tectonic evolution of the West Scotia sea. J Geophys Res 110(B02401):19

Galindo-Zaldivar J, Bohoyo F, Maldonado A et al (2006) Propagating rift during the opening of a small oceanic basin: The protector basin (Scotia Arc, Antarctica). Earth Planet Sci Lett 241(3–4):398–412

Geletti R, Lodolo E, Schreider A, Polonia A (2005) Seismic structure and tectonics of the Shackleton fracture zone (Drake Passage, Scotia Sea). Mar Geophys Res 26:17–28

Gradstein F, Ogg J, Schmitz M, Ogg G (2012) The geologic timescale 2012. Elsevier, Amsterdam, 1139pp

Ivanov VA, Pokazeev KV, Schrader AA (2008) Fundamentals of oceanology SP-B: From Lan, 576p

Lodolo E, Coren F, Schreider AA, Geccone G (1998) Geophysical evidence of a relict oceanic crust in the SouthWestern Scotia sea. Mar Geophys Res 19:439–450

Schrader AA (2011) Formation of a deep-water basin of the Black Sea. Sci World 216p

Schrader AA, Schrader AA, Galindo-Zaldivar H et al (2012) Initial stage of mid-ocean ridge spreading in Western Scotia. Oceanology 52(4):576–581

Schreider AA, Sazhneva AE (2019) Stage pole distribution for meso-cenozoic lithospheric plate rotation. In: Physical and mathematical modeling of earth and environment processes. Springer, Berlin

Wernike B (1985) Uniform sense normal simple shear of the continental lithosphere. Can J Earth Sci 22:108–125

www.bodc.ac.uk/products/bodc_products/gebco

Zhmur VV, Sapov DA, Nechaev ID et al (2002) Intensive gravitational currents in the near-bottom layer of the ocean. Izv NA Sir (Sighs) Phys 66(12):1721–1726 (in Russian)

Features Study of the Marks Movement on the Surface and in the Depth of Vortex Flow

T. O. Chaplina, E. V. Stepanova, and V. P. Pakhnenko

Abstract The paper is devoted to the measurement of the vortex flow characteristics and visualization of its structure by introducing marks (solid and soluble markers with different properties) in the flow. The technique of automatic processing of the displacements of markers registered in the experiments on the free surface of the vortex flow, based on the transformation of the raster image into a vector representation, allows to significantly quicken data processing. The taken experiments with a complex vortex flow with a free surface in a vertical cylindrical container show pronounced dependence of the parameters characterizing the movement of the marker on the vortex surface on the physical properties, shape and size of the marker. Solid-state markers revolve around the center of the surface cavern and simultaneously rotate around their own mass center. The angular velocities of rotation and revolution are related by a functional connection, the form of which is also determined by the flow parameters and marker properties (such as volume, mass, size and shape). The measured velocities of the flow central column filling with suspended particles introduced as a solution on the free surface are compared with earlier measurements of the soluble dye penetration velocities in a wide range of flow parameters, which showed a satisfactory agreement of the obtained data.

Keywords Complex vortex · Solid and soluble markers · Suspended particles · Vortex flow structure · Experimental data

T. O. Chaplina (✉) · E. V. Stepanova · V. P. Pakhnenko
Ishlinsky Institute for Problems in Mechanics RAS, Vernadskogo ave., 101-1, Moscow, Russia 119526
e-mail: chaplina_to@inbox.ru

E. V. Stepanova
e-mail: step@ipmnet.ru

T. O. Chaplina
Moscow, Russia

C. T. Olegovna (ed.), *Processes in GeoMedia—Volume I*,
Springer Geology, https://doi.org/10.1007/978-3-030-38177-6_32

1 Introduction

Spiral and helical structures are a common element in various natural phenomena. In liquid and gas flows, the spiral form often indicates of the presence of vortex flow, and vortices of various scales are elements of various hydrodynamic processes. Vortex flows can be found both in laboratory facilities and technical devices and in the form of whirlpools, vortex rings in the Ocean and in the Earth's Atmosphere—tornadoes and cyclones have vortices in origin. The wide spread of vortices of different scales makes the study of their structure and dynamics relevant.

Among the different types of vortex flows, the most studied are vortex rings observed in geophysical flows and technical devices (Alekseenko et al. 2003; Akhmetov 2007). In the study of vortices in natural full-scale conditions, the observation of the course of all processes is complicated, because the resulting motion is unsteady and there is no control over the conditions of the occurrence of vortex flows, which makes it difficult to obtain universal patterns. To observe and study vortex flows and transport of matter in them, laboratory modeling is often used, devoid of field observations disadvantages: in the laboratory it is easy to control the change in the characteristics of the arising vortex flows, monitor their dynamics and decay, and apply various methods to study the transport of matter in the studied flows. Experimental modeling of vortex structures is necessary at the present stage of development of science. A complete description of vortex flows has not yet been obtained analytically either on the basis of mathematical modeling. The study and formal description of the behavior of the soluble admixture in a stationary liquid was firstly started in (Thompson 1961), where the results of experiments with the introduction of an ink drop on the free surface of the water at rest are presented, the propagation of vortex rings arising from such a drop was described.

Observations of the transport of matter pattern in the vortex flow have shown that the soluble admixture with slightly higher density than the base liquid is not passive and is transported in a complex way both on the surface and in the depth of the liquid flow. On the free surface of a compact spot placed in an arbitrary point, two spiral arms are stretched out, directed along and against the flow on the surface (Stepanova and Chashechkin 2010). If the initial spot of the admixture is located exactly in the center of the rotating free surface of the vortex flow, one spiral arm is pulled out of it (Stepanova and Chashechkin 2010). In the depth of the liquid, the dye is transported along the helical lines wound on the cylindrical surface, the diameter of which depends on the position of the initial spot of the admixture and the age of the process (Shevtsov and Stepanova 2015).

In this paper, using an automated method for processing experimental data on the movement of solid markers on the surface of a complex vortex flow (based on the conversion of the experiment video into a sequence of individual raster images, followed by the removal of non-essential parts and the conversion of the final file into a vector format), as well as to track the velocity and structure features of the geometry of the penetration of the soluble dye and suspended particles into the liquid depth near the vertical axis of the complex vortex.

2 Experimental Setup and Method

Stationarity and reproducibility of flow parameters are necessary conditions for obtaining reliable results and a consistent study of the structure and dynamics of vortex flows. Observation of vortex structures showed their ability to self-displacement and prolonged attenuation, which complicate the implementation of both natural and laboratory experiments. At the same time, in facilities that create the rotation of the liquid (Escudier 1984), it is possible to generate stationary vortex flows, the parameters of which can be changed within a wide range and maintained with the necessary precision.

Experiments, the results of which are presented in the work was carried out on the stand “Vortex flow with torsion” (VFT), part of the complex UNU “Hydrophysical complex of the Institute for Problems in Mechanics, RAS” (HPC IPMech RAS). The scheme and detailed description of the facility is presented in (Chaplina and Stepanova 2018).

In the facility, a complex flow occurs, including vortex and wave components in the thickness and on the free surface of the liquid. A uniformly rotating disk spins the liquid around a vertical axis and simultaneously throws it along its surface to the side wall of the cylindrical container. The moving liquid rises along the walls of the container, moves to the center along the free surface and is immersed in the vicinity of the axis of rotation, forming a current flowing to the center of the disk, compensating for the constant transport along its surface. Directly above the surface of the disk, the particles rotate and simultaneously move from the center to the edge.

In the flow on the free surface, marks with different physical properties (solid-state markers of positive buoyancy, solution of suspended particles, soluble dye) are introduced. The characteristics of the flow occurring in a vertical cylindrical container under the action of the disc rotating at the bottom are determined by movement of all types of utilized marks.

3 Analytic Expression for the Free Surface Form

The form of the surface trough, calculated in the approximation of the ideal fluid taking into account the surface tension is in satisfactory consistent with the experimental data at low and moderate angular velocities of rotation of the disk, with an increase in the frequency of rotation of the disk surface is distorted by waves of different origin (inertial, gravitational, capillary).

The free surface of the flow created in the VFT stand is always not plane, even at low velocities of the inductor rotation there is a trough, the shape and depth of which are determined by the joint action of inertial, gravitational, centrifugal forces.

If the cavity does not touch the inductor, the surface deviation from the undisturbed level is described by the following expression (Kistovich et al. 2017; Chashechkin and Kistovich 2010).

$$\zeta = b(1 - \mu f(\tilde{r})), \quad \mu^{-1} = p^2\left(\frac{3}{2} - 2\ln p + \frac{2\tilde{r}_c^2}{\tilde{r}_*^4}\left(\frac{\tilde{r}_c I_0(\tilde{r}_c)}{I_1(\tilde{r}_c)} - \frac{\tilde{r}_c^2}{4} - 2\right)\right) \quad (1)$$

$$f(\tilde{r}) = \left(\frac{2\tilde{r}_c(I_0(\tilde{r}_c) - I_0(\tilde{r}))}{\tilde{r}_*^2 I_1(\tilde{r}_c)} + \frac{\tilde{r}^2 - \tilde{r}_c^2}{\tilde{r}_*^2}\right)\theta(\tilde{r}_c - \tilde{r}) + \left(2 - \frac{\tilde{r}^2}{\tilde{r}_*^2}\right)\theta(\tilde{r}_* - \tilde{r}) + \frac{\tilde{r}_*^2}{\tilde{r}^2}\theta(\tilde{r} - \tilde{r}_*)$$

where $\tilde{r} = kr$, $\tilde{r}_c = kr_c$, $\tilde{r}_* = kr_*$, $p = r_*/R_0$, $k^2 = g/\alpha$, $I_n(x)$—modified Bessel function of n-th order, $\theta(x)$—unit step (Heaviside) function.

The expression (1) includes three parameters: b, r_c, r_*, which are generally determined not only by the geometry of the problem (R_0—the radius of the cylindrical container, R—the radius of the inductor, H—initial depth of the liquid layer), the velocity of the disk Ω, the acceleration of gravity, viscosity and the surface tension coefficients, which are calculated from experimental data (in particular, the shape of the contour of the vertical section of the trough in the photos).

4 Buoyant Markers on the Free Surface of the Vortex Flow

To clarify the flow pattern and for indirect estimation of the velocity of rotation of the free surface of the liquid in the complex vortex, a series of experiments were carried out, where solid markers were used as indicators. The movement of the marker, which creeps along the surface of the liquid involved in the complex vortex flow, can be reduced to a combination of its rotation around the vertical axis of the flow and at the same time rotation around its own center of mass. For the description of the marker movement along free surface several coordinate frames are introduced. The origin of one XOY (rectangular Cartesian) coincides with the geometric center of the free surface, the axes are directed along the sides of the frame of the video record of the flow pattern, the plane of the coordinate axes of the system coincides with the level of the undisturbed free surface (until the inductor rotation is turned on) (Fig. 1). Another coordinate system—polar $(r, \ \psi)$—is located in the same plane and is most

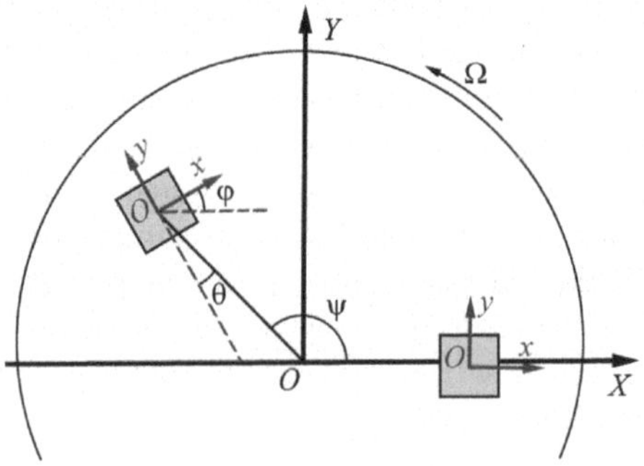

Fig. 1 The scheme of marker movement on a free surface with the introduced coordinate frames

convenient for registering marker movements around the vertical axis of the flow. This system is used to compute the coordinates of the marker center depending on time.

An attached (moving) Cartesian coordinate system xOy is associated with the center of the marker, used to set the angle of "twisting" φ, an additional angle θ describes the deviation between the instantaneous position of the x-axis of the xOy coordinate system and the radius vector of the marker r. It should be noticed that for all types of markers they have special surface labels, allowing to determine their angular position.

The radial and angular coordinates of the marker in the polar coordinate frame, as well as the values of the angles φ and θ, were independently determined from the images taken from the video-record of the experiment with a known time step.

The following markers were used in the experiments: square, round, ellipsoidal, ring-shaped, pentagonal and cross-shaped—the vertical size of all markers is 0.5 mm. Most of the experiments were carried out with the following parameters: the depth of the liquid is 40 cm, the rotation frequency of the inductor is 220 RPM and less, which corresponds to a free surface with a slight bend in the center for any size of the inductor.

According to the results of experimental data automatic processing, the dependence of the change in the radius of rotation of the markers of different shapes around the center of the rotating free surface is obtained and the dependence of the radial position of the marker on time is calculated (see Figs. 2 and 3).

A feature of all presented in Figs. 2 and 3 graphs are periodic changes in the coordinates (bouncing) of marker centers when approaching the center of a rotating

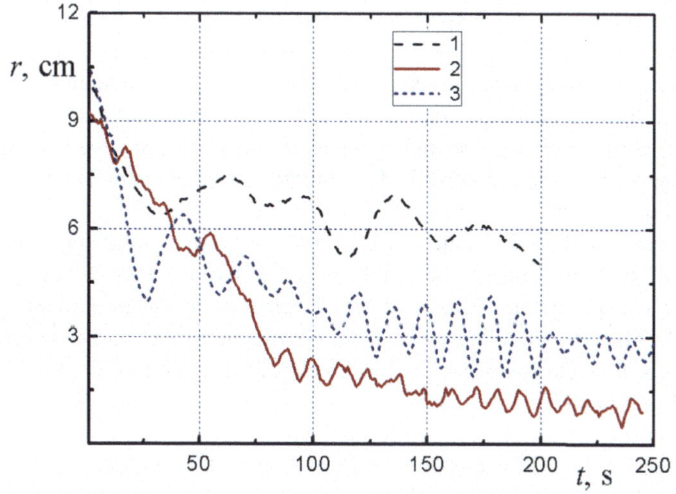

Fig. 2 The dependence of the change in the radius of rotation of the markers of different shapes around the center of the rotating free surface (H = 40 cm, R = 7.5 cm, Ω = 200 RPM): 1—pentagonal, 2—oval and 3—round markers with the largest size of 1 cm

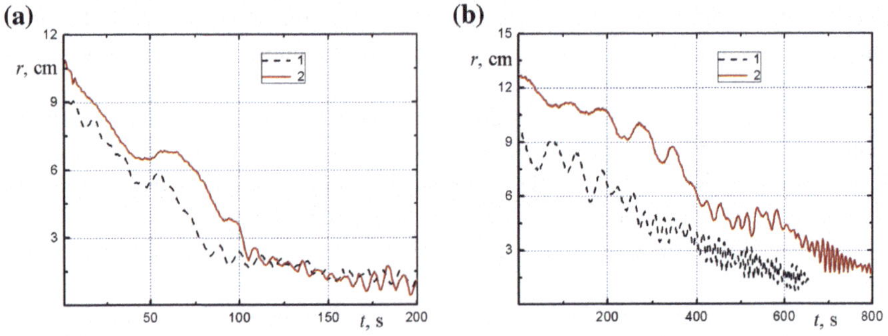

Fig. 3 The dependence of the change in the radius of circulation of markers with different forms around the center of the rotating free surface at different depths of the liquid ($R = 7.5$ cm, $\Omega =$ 200 RPM, liquid depth 1–20, 2–40 cm): **a** oval marker $L = 1.5$ cm long, **b** cross-shaped marker 1.5 cm

free surface. Small fluctuations in the coordinates of the marker center are associated with elliptical distortions of the free surface shape, as well as with small fluctuations in the position of the center of rotation of the free surface relative to the fixed walls of the container.

After being placed on a rotating free surface, the marker gradually gains velocity and moves to the center—the radial coordinate decreases, and the bouncing frequency increases. Processing of the experiments (Eisenberg 2014; Toe and Gonzales 1978; Kanasevich 1985) demonstrate the influence of the shape of the marker on the circular velocity of the marker on the character of the variation of the radius of the marker movement around the center. The shape of the marker is related to the rate of velocity gain at the beginning of the experiment—markers with smooth boundaries (round and oval) gain velocity more slowly and move to the center of the flow more slowly than square and cross-shaped markers (see Figs. 2 and 3). Predictably the gain of velocity and approach to the center of the flow the slower, the greater the mass of the marker, even if the shape has not changed. At the same time, the rate of velocity gain and approach to the center of the flow practically do not depend on the depth of the initial liquid layer involved in the vortex flow (Fig. 3).

The measurements of the radius of rotation of a large-scale solid-state marker from the center of the free surface show that the best match of the approximating function $R(t)$ with the experimental data is obtained using the logarithmic dependence as $R_m = A\ \ln(t) + B$, where R_m—the radial coordinate of the marker relative to the center, t—the time elapsed since the placement of the marker on the free surface of the flow (Figs. 2 and 3).

The smoothed interpolation of the dependence of the angular position of the marker on time allowed to calculate the local values of its rotation frequency. Over time, the marker gradually gains velocity and moves to the center of the free surface. It is also possible to calculate the threshold value of the frequency of the marker rotation when approaching the center. This value is directly related to the velocity of the activator, but does not coincide with it—for the velocity of the activator 160 RPM, the square marker rotates near the center of the free surface at an angular velocity of 1.4 RPM. The shape of the marker affects the rate of velocity gain, so the round marker gains its slower and slower approaches the center of the rotating free surface, compared to the square one.

When the marker is placed on the surface on its lower (submerged) surface, viscous shear stresses from the liquid begin to act, which transport it as a whole along some trajectory lying on the surface and rotate around its own center of mass.

As it moves to the center of the vortex trough, the marker enters the transition region, crosses the boundary between the solid and peripheral types of vortex rotation and gradually changes its own rotation relative to the center of mass to the opposite one, which remains after the complete transition of the marker to the region of "solid-state" vortex motion.

The nature of the "twisting" of the marker depends on its shape and the angular velocity of rotation of the inductor Ω. The round marker after a short interval of involvement ($\Delta t = 5$ s) rotates with almost constant angular velocity, the value of which increases monotonically with the increase of the disk rotation velocity: $\omega_1 = d\varphi_1/dt = 66.6\ \text{s}^{-1}$, $\omega_2 = d\varphi_2/dt = 99.3\ \text{s}^{-1}$, $\omega_3 = d\varphi_3/dt = 134.8\ \text{s}^{-1}$(Fig. 4).

The square marker also rotates at a nearly constant angular velocity at a low angular inductor velocity of $\omega_4 = d\varphi_4/dt = 30.1\ \text{s}^{-1}$ (at $\Omega = 3.3\ \text{s}^{-1}$) and at large $\omega_5 = d\varphi_5/dt = 128.2\ \text{s}^{-1}$ ($\Omega = 16.7\ \text{s}^{-1}$). At intermediate values of the angular velocity of rotation of the disk $9.2\ \text{s}^{-1}$, the square marker is rotated unevenly (curve 2, Fig. 7b), which reflects the complex nature of the interaction of the involved square marker with the main current. It is possible that the irregular "twisting" is due to the interaction of the marker with inertial and spiral waves appearing on the surface of the rotating liquid (Chashechkin and Kistovich 2010), which is significant in this flow regime. The dependence of the angle "twisting" for square marker from time are approximated by functions $\varphi_s = 40.8\ t^{1.4}$ (for $\Omega = 3.3\ \text{s}^{-1}$), $\varphi_s = 0.77\ t^{2.68}$ (for $\Omega = 9.2\ \text{s}^{-1}$), $\varphi_s = 1.4\ t^{2.4}$ (for $\Omega = 16.7\ \text{s}^{-1}$).

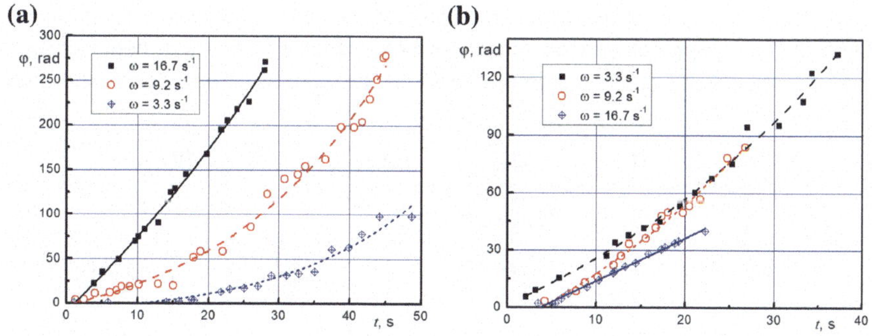

Fig. 4 The angle of rotation of the marker around its own axis from time ($H = 40$ cm, $R = 7.5$ cm, Re = 18,750, 51,750, 93,750): **a** circular cylinder with radius 0.5 cm and height 0.3 cm, **b** parallelepiped 1.0 × 1.0 × 0.3 cm

5 Experimental Study of the Soluble Dye Transport in the Depth of Vortex Flow

The complex structure of the vortex flow is clearly manifested in the transport of a soluble admixture both along the surface (Stepanova and Chashechkin 2010) and in the thickness of the complex vortex. At the free surface the steady-state vortex flow with selected parameters is placed in a fixed volume of soluble dye in such experiments continuous flow pattern registration is performed.

A characteristic form of surface perturbations (waves) in a complex vortex is a logarithmic spiral. Analytical expressions showing that the trajectories of liquid particles near the surface of the vortex are three-dimensional spirals along which the flow goes from the periphery to the center of the vortex are obtained in (Kistovich et al. 2019). It is shown that the calculated and visualized trajectories of liquid particles are in good agreement with each other and belong to the class of spatial logarithmic spirals (Fig. 5).

The transport of a soluble marker (or aqueous solution of suspended particles) occurs mainly (out of the size and color intensity of the colored region) in the cylindrical region surrounding the vertical axis of the flow. The size of the painted area is distinctly related to the size of the inductor, and the rate of penetration of the dye/suspended particles into the depth of the flow is an indicator of the vertical velocity of the flow as a whole. Such transport of the coloring substance is of interest, because despite the intensity of the entire flow (Re = 7200), the downwards movement of the dye in the central painted column is quite slow (Fig. 6).

The developed algorithm for determining the intensity of coloring of different parts of the flow by the brightness of their image obtained using different digital cameras allows seeing qualitative and quantitative state of the dye distribution in the flow. The position and geometric characteristics of the central painted column are most reliably determined, as well as its elongation as the dye penetrates into the depth of the complex vortex.

A characteristic form of the dependence of the depth of dye penetration into the depth of the complex vortex on time is shown in Fig. 7. The graph begins with a mark corresponding to the maximum depth of the trough on the surface, the lines

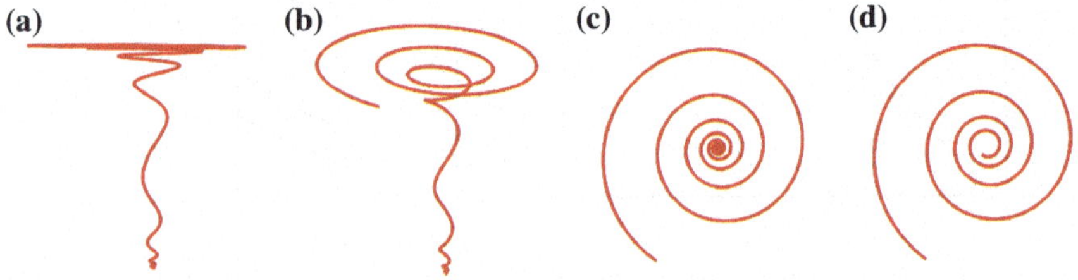

Fig. 5 Form of the trajectories of liquid particles near the surface of the complex vortex: **a** side view, **b** side view at an angle, **c** top view, **d** an enlarged image of the central part of the top view

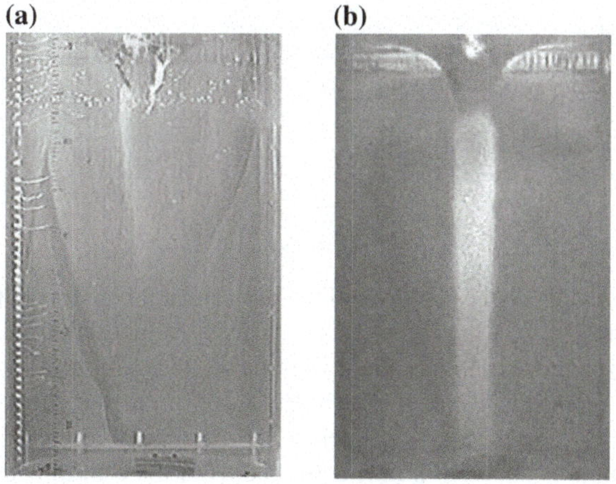

Fig. 6 Transport of admixtures in the liquid depth (H = 50 cm, Ω = 820 RPM, R = 7.5 cm): **a**, **b**—t = 28, 78 s

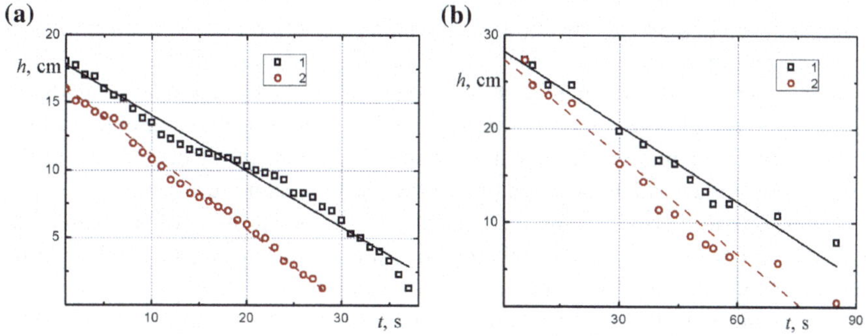

Fig. 7 Changing of the dye penetration depth over time: **a** H = 20 cm, Ω = 1600 RPM, R = 5.0 cm, **b** H = 30 cm, Ω = 250 RPM, R = 14.5 cm; 1—the lowest level of brightness changes compared to the background brightness, 2—the lowest level of the full width colored column

correspond to the depth of full filling (coloring) of the column for the entire width (symbols 1) and the penetration of the "admixture tip" (symbols 2) over time.

The linear interpolation of the data in Fig. 7a shows that the dye penetration rate (about 4.1 mm/s) is slightly higher than the column filling rate for the entire width (2.7 mm/s). The width of the painted column remains constant at all depths and at a radius of the activator disk of 7.5 cm is about 3.5 cm. In the case when the small admixture portion is introduced on the free surface, by the time when the lower edge of the central painted column reaches the rotating disk, the upper part of the painted column loses the intensity of the color, i.e., there is a decrease in the admixture concentration (Table 1).

Experiments have shown that the width of the column is not constant along the vertical coordinate, while the analysis of the movement of the left and right "front"

Table 1 The linear interpolations coefficients for velocity of dye penetration into the depth of liquid

		A (cm/s)	B (cm)
Fig. 7a	Line 1	−0.413 ± 0.010	18.181 ± 0.217
	Line 2	−0.550 ± 0.010	16.656 ± 0.169
Fig. 7b	Line 1	−0.271 ± 0.013	28.468 ± 0.581
	Line 2	−0.351 ± 0.020	27.698 ± 0.930

showed that the change in the positions of the right and left borders of the colored area occurs simultaneously, i.e. the colored area makes bending oscillations. The average thickness of the painted column is about 18 mm (i.e., the radius of the central painted column is 0.12 from the size of the activator $R = 7.5$ cm) for the experiment at a frequency of $\Omega = 225$ RPM, at a frequency of $\Omega = 435$ RPM the average depth of the painted column is about 25 mm (i.e. the radius of the central painted column is 0.09 of the activator size $R = 14.0$ cm). Summarizing the data of all experiments, it should be noted that the central column in the experiment with a higher frequency is wider than at lower frequencies, it is also worth noting that the colored central part of the flow is slightly wider near half of the total depth of the flow. The graph of dependence of average deviations of the painted column from vertical in time is presented in Fig. 8.

It is easy to see that deviations from the vertical of the right and left boundaries of the colored region occur almost in phase, which confirms the assumption of the bending, rather than the varicose nature of the oscillations of the central part of the vortex flow, within which the spreading of the marking admixture occurs.

Analysis of the periodic component of the change in the position of the boundaries of the colored area over time by analyzing the statistics of zero crossing and counting the corresponding lengths of the series shows that at a relatively low velocity of rotation of the activator (see Fig. 8) waves of greater length (25 mm) are manifested, and for the case with a higher velocity—short (15 mm). At high frequencies there are practically no oscillations with a wavelength exceeding one third of the height of the

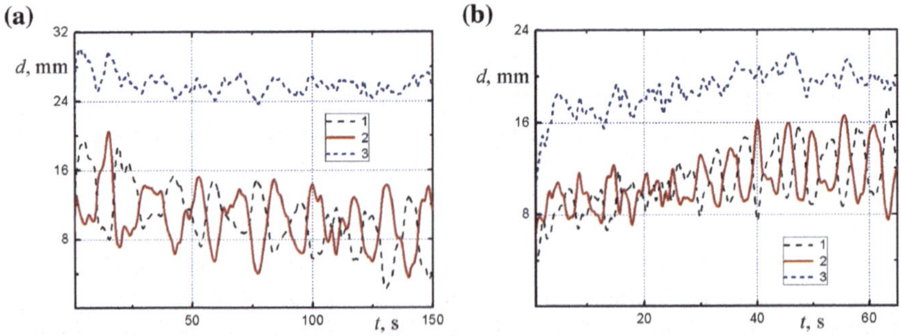

Fig. 8 The dependence of the average horizontal deviation of the painted area border on time ($H = 30$ cm, $\Omega = 225$ RPM): **a** $R = 7.5$ cm, **b** $R = 14.0$ cm (1—left and 2—right sides of the painted column, 3—the total width)

undisturbed liquid layer, at lower frequencies there are waves with a characteristic length of half the height of the liquid column.

6 Conclusions

The carried out experiments have confirmed that in addition to the rotation of the marker around the center of the free surface of the vortex flow, it slowly rotates around its own axis. The combination of the obtained data on the images in vector representation with the original and reverse video mounting allows improving the further experimental studies design.

Data on the displacements of solid-state markers on the free surface of the complex vortex are obtained. The transport of suspended particles of aluminum powder to the thickness of the flow along the central vertical axis is observed, the process rates are determined, and comparison with the data of previous experiments with other types of soluble dyes is made.

The software basis for tracking the velocity and structure features of the soluble dye and suspended particles penetration geometry into the liquid depth near the vertical axis of the complex vortex is developed and tested. The comparison with the previously obtained data confirms the coincidence of the vertical penetration rates of the admixture and the size of the painted area at constant flow parameters.

The work is support in part by the Russian Foundation for Basic Research (project 18-01-00116).

References

Akhmetov DG (2007) Vihrevye koltca. Akademicheskoe izd-vo « Geo », Novosibirsk (in Russian)

Alekseenko SV, Kujbin PA, Okulov VL (2003) Vvedenie v teoriju koncentrirovannyh vihrej. Institut teplofiziki S.S. Kutateladze, Novosibirsk [in Russian]

Chaplina TO, Stepanova EV (2018) Features of the different types of markers angular displacements in the complex vortex flow. Process V Geosredah 1:793–803 [in Russian]

Chashechkin YuD, Kistovich AV (2010) Deformation of the free surface of a fluid in a cylindrical container by the attached compound vortex. Dokl Phys 55(5):233–237

Eisenberg DJ (2014) SVG essentials, 2nd edn. Sebastopol, O'Relly Media

Escudier MP (1984) Observations of the flow produced in a cylindrical container by a rotating endwall. Experim Fluids 2:189–196

Kanasevich ER (1985) Analiz vremennyh posledovatelnostej v geofizike. Nedra, Moscow (in Russian)

Kistovich AV, Chaplina TO, Stepanova EV (2017) Vortex flow with free surface: comparison of analytical solutions with experimentally observed liquid particles trajectories. Int J Fluid Mech Res 44(3):215–227

Kistovich AV, Chaplina TO, Stepanova EV (2019) Spiral structure of the liquid particles trajectories near free surface of the vortex. Comput Tech 24(2):67–77. https://doi.org/10.25743/ict.2019.24.2.006 [in Russian]

Shevtsov NI, Stepanova EV (2015) Application of photometry method to some hydrodynamic objectives. Moscow Univ Phys Bull 70(3):208–212
Stepanova EV, Chashechkin YuD (2010) Marker transport in a composite vortex. Fluid Dyn 45(6):843–858
Thompson DW (1961) On growth and form. Cambrige University Press, Cambridge
Toe G, Gonzales R (1978) Principy raspoznavanija obrazov. Mir, Moscow (in Russian)

Influence of the Stresses on Flow Rate of Wells and Their Stability

V. I. Karev and T. O. Chaplina

Abstract The results of theoretical and experimental studies performed at the Institute for Problems in Mechanics of the RAS in the field of geomechanics of oil and gas wells are presented. Mathematical analysis of the stress state and size of fracture zones that arise near the well due to bottomhole pressure decrease is carried out. A new approach to the solution of geomechanical problems, based on tests of rock samples on the unique experimental facility Triaxial Independent Load Test System is proposed.

Keywords Stress state · Permeability · Rocks · Wellbore stability · Anisotropy

1 Introduction

Numerous data on exploratory drilling and operation of oil and gas fields, especially at great depths, indicate that stresses appearing in the well bore zone have a significant effect on permeability and filtration processes in formations and, therefore, on oil and gas inflow rates. But the role of the stresses in forming of filtration processes in rocks is poorly understood now. Meanwhile, a number of scientists in the mechanics of the last century, Khristianovich (1941) and others expressed the idea, that one of the decisive factors affecting the filtration in rocks is the process of deformation and destruction of rocks under the action of stresses arising in them.

V. I. Karev · T. O. Chaplina (✉)
Ishlinsky Institute for Problems in Mechanics RAS, Moscow, Russia
e-mail: chaplina_to@inbox.ru

V. I. Karev
e-mail: wikarev@ipmnet.ru

C. T. Olegovna (ed.), *Processes in GeoMedia—Volume I*,
Springer Geology, https://doi.org/10.1007/978-3-030-38177-6_33

2 Stress State Near the Well and Contribution of Filtration Forces

Early studies of the stress state, processes of deformation and fracture of rocks in the vicinity of mine workings were carried out in relation with the problems of development of solid mineral deposits, for example (Baklashov and Kartozia 1975). But the issues of the filtration impact, pressure gradient, compressibility, viscosity were not considered in these studies. Here are the results of analysis of the stress state and the size of fracture zones that occur in the vicinity of an oil well with a decrease in pressure on its face. The detailed analysis carried out in (Zhuravlev et al. 2014). The effect of additional mass forces due to the presence of pressure gradient in reservoir on the value fracture zones is studied for two conditions of the rock fracture. In the first case, the condition of the fracture is the achievement of a shear stress a certain limit—the criterion of Tresca. In the second case, the condition of local fracture can be presented in the form of Coulomb-Mohr criterion, according to which the achievement of the rock a limiting state on the site with the normal n is given by $\tau_n = [\tau]$, where $[\tau]$—tensile strength, i.e. $[\tau] = k + S_n \operatorname{tg} \rho$, where S_n, τ_n—normal and shear stresses at the site, k—coefficient of coupling, ρ—friction angle of rock. The contribution of the compressibility and viscosity of the fluid in the stress distribution and size of fracture zones is determined for each criteria of fracture.

The following main conclusions are made on the base of the analysis.

1. In case of an incompressible fluid the size of fracture zone increases with the growth of pressure gradient which causes flow into the well. Radius of fracture zone can be significantly increased for sufficiently weak rocks. This finding is of a great importance in terms of selection the optimum operating conditions of the wells, since the permeability of rocks, especially sandstones, essentially depends on the stresses acting in them—it can irreversibly increase or decrease.
2. The account of the compressibility shows that an increase effective compressibility of the fluid extends the size of the fracture zone.
3. Accounting for increasing the viscosity of the fluid with growth pressure has opposite effect, i.e. reducing the size of the fracture zone. However, since the effect of pressure on the viscosity is less then on the density of the fluid, the change in the compressibility has greater impact on the size of the fracture zone.

3 The Experimental Study of the Effect of Stress-Strain Behavior on the Filtration Properties of Rocks

It is impossible to answer the question of how the stress-strain behavior affects permeability of some specific rock types using calculations because this depends on the properties of the rock itself. Such dependence can be studied using a unique experimental apparatus developed at the Institute for Problems in Mechanics of the Russian Academy of Sciences—Triaxial Independent Load Test System (TILTS), Fig. 1. The facility allows independent loading of cubic rock samples with a 40 or 50 mm face in three axes, which is possible due to the original kinematic structure used in the design of the load assembly (Karev and Kovalenko 2011). Permeability measurements are performed in a continuous mode during the test of rock samples. This offers the possibility to recreate any stress conditions that appear in the bottom-hole zone when the well is being drilled, completed, or operated during the tests and to study their effects on the filtration properties of rocks.

The rocks that form the reservoir of oil and gas fields are characterized by strong dependence of the filtration properties on pressure drawdown, and, changes are irreversible starting from a certain level of pressure in the well and accordingly the value of shear stresses in the rock. The effect of irreversible permeability increase by forming of micro and macrocracks was found for many types of rocks such as fine and medium sandstone with a low clay content, siltstone and limestone (Karev 2015). As example, the results of test of a specimen from depth 6115 from reservoir of an oil field located at Orenburg region are presented at Figs. 2, 3 and 4. Figure 1 shows a loading program, Fig. 2—strain curves, Fig. 3—the dependence of rock permeability on stress. As it can be seen from the graphs, the specimen begin to deform plastically when maximum principal stress reaches to 150 MPa while permeability sharply significantly increases.

This effect got the name geoloosening and formed the basis of a new method to increase the production rate of oil and gas drilling—the geoloosening method (Khristianovich 2000). It was employed at several dozens of wells on fields in West Siberia and the Perm Region and showed high efficiency.

Fig. 1 Triaxial independent load test system of IPMech RAS

Fig. 2 Loading program of sample

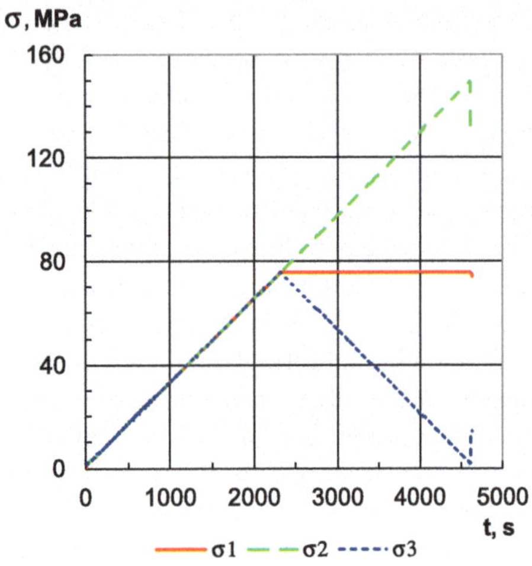

Fig. 3 Strain curves

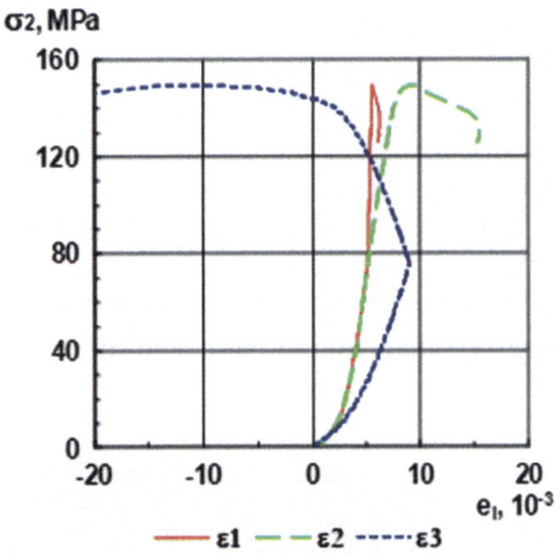

Fig. 4 Dependence of rock permeability on stress

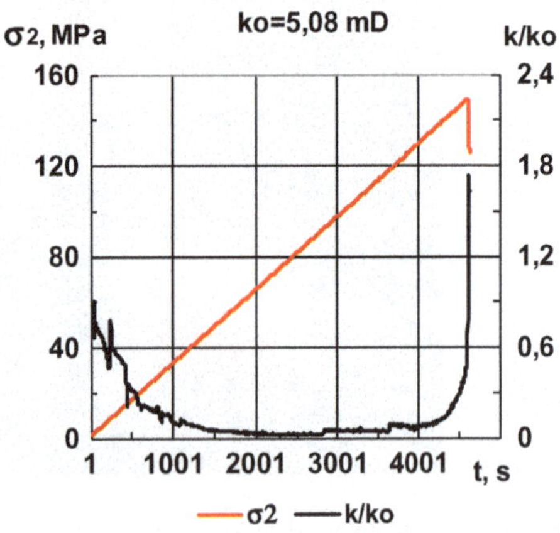

4 Directional and Horizontal Wellbore Stability

A method has been developed to identify parameters for safe directional and horizontal drilling and operation of horizontal wells. It's based on direct physical TILTS modeling of the conditions that appear in the vicinity of the wells during drilling and operation. The relationship between well stability and well geometry is due to anisotropy of elastic properties and especially strength properties of rocks. Destruction of the borehole wall rock is related to higher shear stresses in bedding planes, which are surfaces of weakness. At some hole angle depending on the properties of the rock, depth and bottomhole pressure, shear stresses in the bedding plane reach critical values and trigger destruction of borehole walls.

According to calculations, the points where the vertical plane passing through the borehole axis crosses a borehole wall are the most dangerous areas in terms of destruction (Klimov et al. 2013). For this reason, changes in principal stresses at such points triggered by changes in borehole pressure were selected as a sample load program for TILTS modeling the conditions in the vicinity of the well at various hole angles.

In order to study stability of directional wells, samples are cut out of core material above the occurring rocks. The samples should be cut at different angles to the vertical axis. Bottomhole pressure and, respectively, mud density at which the rock is destroyed is determined for each hole angle using the sample cut at this angle.

The studies at the Ulyanovskoye field in West Siberia are given here as an example. Figure 5 shows creep diagrams for several rock samples cut at different angels. During the tests of each of the samples, when radial stress reached the bottomhole pressure value corresponding to mud density of 1.2 g/cm^3, loading was stopped and changes in sample strain were measured over time. As one can see from the

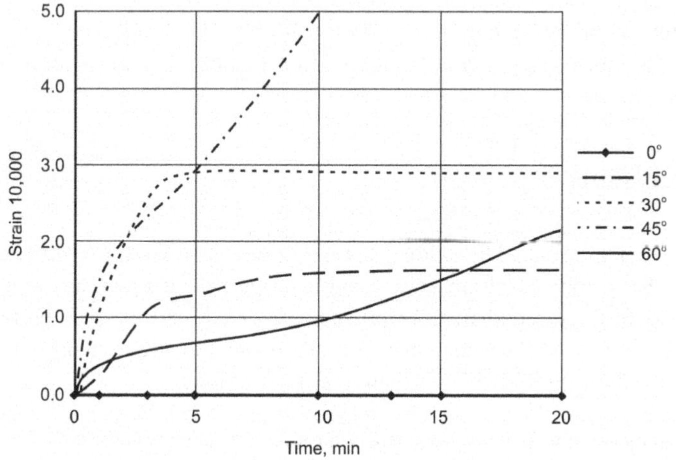

Fig. 5 Creep curves for various angels of borehole inclination. Mud density is 1.2 g/cm^3

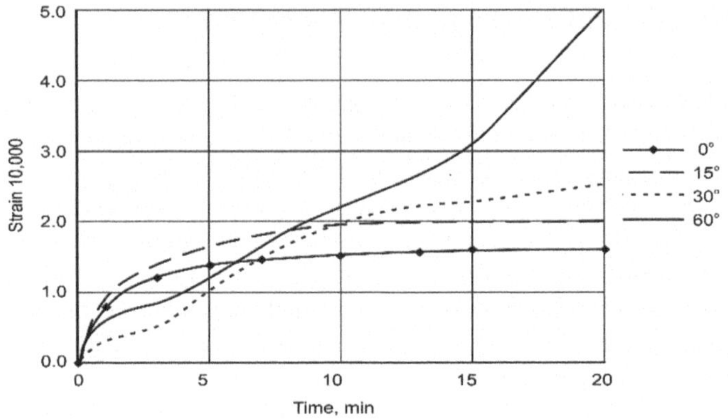

Fig. 6 Creep curves for various angels of borehole inclination. Mud density is 1.12 g/cm^3

diagrams, the sample cut at a 0° angle did not creep at all; the samples cut at 15° and 30° angles demonstrated limited creep, while the samples cut at 45° and 60° angles reached a preset load point, started creeping at an ever-increasing rate, and were finally destroyed. Figure 6 shows creep diagrams at mud density of 1.12 g/cm^3, which corresponds to lower bottomhole pressure and, therefore, higher stresses in the rock in the vicinity of the well. As one can see from the graph, unlimited creep and destruction are observed at smaller sample angles, starting from 30°. The findings are quite consistent with practical data.

Such studies have been conducted for a number of major Russian companies such as Gazprom, Lukoil, Surgutneftegas (Karev and Kovalenko 2013), and the findings of the studies have been confirmed by practical experience. In 2009, at the request of Stockman Development, an international company created by Gazprom, Total and StatoilHydro, there were conducted studies which offered the opportunity to validate engineering and process solutions for optimization of the development strategy and policy for the Stockman gas condensate field located offshore in the Barents Sea.

5 Conclusions

Stresses in rock matrix and filtration processes in reservoir have a significant mutual influence. Anisotropy of strain and strength properties of rocks has a significant impact on deformation and fracture processes as well, and it needs be taken into account. The study of the behaviour of rocks under the stress change is a key to solving geotechnical problems of oil and gas production.

Acknowledgements The study was supported by the Russian Science Foundation No. 16-11-10235 P.

References

Baklashov IV, Kartozia BA (1975) Rock mechanics. Nedra, Moscow

Karev VI et al (2015) Experimental study of the influence of a triaxial stress state with unequal components on rock permeability. Mech Solids 6:633–640

Karev VI, Kovalenko YF (2011) Triaxial loading system as a tool for solving geotechnical probPress/Balkema, pp. 301–310

Karev VI, Kovalenko YF (2013) Well stimulation on the basis of preliminary triaxial tests of reservoir rock. In: Rock mechanics for resources, energy and environment. Proceedings of EUROCK 2013. The ISRM international symposium. Wroclaw, Poland, 23–26 September Leiden, CRC Press/Balkema, pp 935–940

Khristianovich SA (1941) On flow gas-saturated liquid in porous rocks. J Appl Math Mech 2:277–282

Khristianovich SA et al (2000) Increase productivity of oil well by geoloosening method. Oil Gas Eurasia 2:90–94

Klimov DM et al (2013) Mechanical-mathematical and experimental modeling of well stability in anisotropic media. Mech Solids 4:357–363

Zhuravlev AB et al (2014) The effect of seepage on the stress-strain state of rock near a borehole. J Appl Math Mech 1:56–64

The manufacturer's authorised representative in the EU is Springer Nature Customer Service Centre GmbH, Europaplatz 3, 69115 Heidelberg, Germany. If you have any concerns regarding our products, please contact ProductSafety@springernature.com

Printed and bound by CPI Group (UK) Ltd, Croydon, CR0 4YY
05/10/2026
02228120-0010